Pharmacology for the Health Care Professions

Pharmacology for the Health Care Professions

Christine M. Thorp

University of Salford, UK

WILEY-BLACKWELL

A John Wiley & Sons, Ltd., Publication

This edition first published 2008© 2008 by John Wiley & Sons, Ltd.

Wiley-Blackwell is an imprint of John Wiley & Sons, formed by the merger of Wiley's global Scientific, Technical and Medical business with Blackwell Publishing.

Registered office: John Wiley & Sons Ltd, The Atrium, Southern Gate, Chichester, West Sussex, PO19 8SQ, UK

Other Editorial Offices:
9600 Garsington Road, Oxford, OX4 2DQ, UK
111 River Street, Hoboken, NJ 07030-5774, USA

For details of our global editorial offices, for customer services and for information about how to apply for permission to reuse the copyright material in this book please see our website at www.wiley.com/wiley-blackwell

Library of Congress Cataloging-in-Publication Data

Thorp, Christine.
  Pharmacology for the health care professions / Christine Thorp.
    p. ; cm.
  Includes bibliographical references and index.
  ISBN 978-0-470-51018-6 (hb : alk paper) – ISBN 978-0-470-51017-9 (pb : alk paper)
  1. Pharmacology. 2. Chemotherapy. I. Title.
  [DNLM: 1. Pharmaceutical Preparations. 2. Drug Therapy. 3. Pharmacology. QV 55 T517p 2008]
  RM300.T52 2008
  615′.1–dc22

                                                                                            2008021458

ISBN: 978-0-470-51018-6 (HB)
ISBN: 978-0-470-51017-9 (PB)

A catalogue record for this book is available from the British Library

Typeset in 10/12pt Times by Laserwords Private Limited, Chennai, India

First impression   2008

This book is dedicated to
the memory of my mother

Contents

Foreword   xiii

Preface   xv

Acknowledgements   xvii

1   Introduction   1

   1.1   Pharmacology and health care professionals   2
   1.2   Patient compliance   4
   1.3   Drug names   4

Part I   Principles of pharmacology   7

2   Drug disposition   9

   2.1   Chapter overview   9
   2.2   Administration of drugs   9
   2.3   Absorption of drugs   13
   2.4   Drug distribution   17
   2.5   Drug metabolism   20
   2.6   Excretion of drugs   23
   2.7   Summary   26

3   Effects of drugs on the body   29

   3.1   Chapter overview   29
   3.2   Adverse reactions to drugs   29
   3.3   Variation in response to drug therapy   36
   3.4   Targets for drug action   40
   3.5   Summary   45

Part II   Systemic pharmacology   49

4   Cardiovascular and blood disorders   51

   4.1   Cardiovascular disorders   51

4.2	Hypertension	58
4.3	Drugs used to treat cardiovascular disorders	60
4.4	Blood disorders	68
4.5	Anaemias	73
4.6	Lipid metabolism	75
4.7	Lipid-lowering drugs	77
4.8	Summary	79

5 Respiratory Disorders — **85**

5.1	Chapter overview	85
5.2	Asthma	88
5.3	Chronic bronchitis	88
5.4	Drugs used to treat respiratory disorders	89
5.5	Treatment of other respiratory conditions	92
5.6	Summary	94

6 Disorders of the endocrine system — **99**

6.1	Chapter overview	99
6.2	Pituitary gland	99
6.3	Thyroid gland	102
6.4	Parathyroid glands	104
6.5	Adrenal glands	105
6.6	Pancreas	106
6.7	Treatment of diabetes mellites	108
6.8	Summary	111

7 Disorders of the musculoskeletal system — **115**

7.1	Chapter overview	115
7.2	Rheumatic diseases	115
7.3	Drugs used to treat rheumatic diseases	116
7.4	Gout	123
7.5	Drugs used to treat gout	124
7.6	Osteoarthritis	125
7.7	Paget's disease	126
7.8	Treatment of Paget's disease	126
7.9	Osteoporosis	127
7.10	Drugs used to treat osteoporosis	127
7.11	Osteomalacia	128
7.12	Myasthenia gravis	128
7.13	Treatment of myasthenia gravis	129
7.14	Motor neuron disease and multiple sclerosis	130
7.15	Summary	132

8 Disorders of the skin   **137**

8.1 Chapter overview 137
8.2 Eczema 138
8.3 Treatment of eczema 139
8.4 Psoriasis 140
8.5 Treatment of psoriasis 141
8.6 Warts 144
8.7 Treatment of warts 145
8.8 Other viral infections of the skin 146
8.9 Fungal infections of the skin and nails 146
8.10 Drugs used to treat fungal infection of the skin and nails 147
8.11 Bacterial infection of the skin 150
8.12 Summary 151

9 Chemotherapy of infectious diseases   **155**

9.1 Chapter overview 155
9.2 Bacteria 155
9.3 Antibiotic drugs 157
9.4 Treatment of tuberculosis 160
9.5 Viruses 161
9.6 Antiviral drugs 162
9.7 Treatment of HIV infection 165
9.8 Fungi 165
9.9 Antifungal drugs 166
9.10 Protozoa 168
9.11 Antimalarial drugs 169
9.12 Drugs for toxoplasma and pneumocystis pneumonia 170
9.13 Parasitic worms 170
9.14 Anthelmintics 172
9.15 Summary 173

10 Cancer chemotherapy   **177**

10.1 Chapter overview 177
10.2 Biology of cancer 177
10.3 Principles of chemotherapy 180
10.4 Drugs used in cancer chemotherapy 182
10.5 Summary 188

11 Disorders of the central nervous system   **191**

11.1 Chapter overview 191
11.2 Affective disorders 194
11.3 Drugs used to treat depression 198

11.4 Drugs used to treat bipolar depression and mania 200
11.5 Psychoses 201
11.6 Drugs used to treat schizophrenia 203
11.7 Anxiety and insomnia 205
11.8 Anxiolytics and hypnotics 207
11.9 Treatment of anxiety 207
11.10 Treatment of insomnia 209
11.11 Attention deficit hyperactivity disorder (ADHD) 210
11.12 Treatment of ADHD 211
11.13 Parkinson's disease 211
11.14 Drugs used to treat Parkinson's disease 213
11.15 Epilepsy 216
11.16 Drugs used to treat epilepsy 217
11.17 Alzheimer's disease 221
11.18 Treatment of Alzheimer's disease 221
11.19 Summary 221

12 Anaesthesia and analgesia **229**

12.1 Chapter overview 229
12.2 General anaesthesia 229
12.3 Inhalation anaesthetics 231
12.4 Intravenous anaesthetics 232
12.5 Premedication and adjuncts to general anaesthesia 234
12.6 Local anaesthesia 236
12.7 Local anaesthetics 241
12.8 Analgesia 243
12.9 Peripherally acting analgesics 244
12.10 Centrally acting analgesics 247
12.11 Neuropathic pain 247
12.12 Summary 248

13 Contrast agents and adjuncts to radiography **253**

13.1 Chapter overview 253
13.2 Contrast agents 253
13.3 Cautions in use of contrast agents 257
13.4 Complications of intravenous administration of contrast agents 257
13.5 Adverse reactions to contrast agents 257
13.6 Management of acute adverse reactions to contrast agents 259
13.7 Adjuncts to radiography 260
13.8 Summary 264

Part III Prescribing and the law 269

14 Medicines, the law and health care professionals 271

14.1 Chapter overview 271
14.2 Legislation 271
14.3 Non-medical prescribing 279
14.4 Summary 284

15 Prescribing in practice 291

15.1 Chapter overview 291
15.2 Podiatry 291
15.3 Extension of access to prescription-only medicines
in podiatry and podiatic surgery 292
15.4 Podiatry in the community 299
15.5 Radiography 300
15.6 Medicines in radiography: prescription, supply and administration 300
15.7 Physiotherapy 315
15.8 Summary 319

Appendices 321

Appendix A: Drug Names 323

Appendix B: Glossary 331

Appendix C: Examples of Patient Group Directions 337

Bibliography 343

Useful websites 345

Index 349

Foreword

Students of pharmacology are well served by a number of academic textbooks on their subject but the majority are written from a traditional academic viewpoint. This new book is different in that it is written specifically for the audience of the Health Care Professions and the author Dr Christine Thorp is particularly well qualified in this respect.

Dr Thorp graduated from the School of Pharmacy and Pharmacology at the University of Bath, first with a BSc in 1975 and then with a PhD in 1979 providing her with a traditional academic view of pharmacology and experience of research. Since then she has undertaken a number of roles, most recently in the Faculty of Health and Social Care at the University of Salford, with responsibility for teaching pharmacology to students in a variety of Health Care Profession disciplines.

Dr Stephen Moss BPharm, MSc, PhD, FRPharmS
Department of Pharmacy and Pharmacology
University of Bath
July 2008

Preface

The need for a book such as this one has arisen as a result of recent changes in legislation and expansion in the numbers of health care professionals involved in administration and/or prescription of medicines.

The book is an introduction to pharmacology for health care professionals. Although anyone involved in the care of patients is a health care professional, this book has been specifically written for physiotherapists, podiatrists and radiographers (otherwise known as *allied health professionals*). However, the book may be of interest to other health care professionals.

The book aims to provide the knowledge of pharmacology necessary for undergraduates of all three professions and practitioners on post graduate programmes for accreditation of supplementary prescribing or access and supply of prescription-only medicines. It may also be of more general use to any health care professional involved in patient care, especially those who administer medicines under patient group directions.

The book is arranged into three parts. In the first part, Principles of Pharmacology, two chapters cover administration, absorption, distribution, metabolism and excretion of drugs (Chapter 2) and adverse drug reactions, drug–drug interactions, individual response to drugs and targets for drug action (Chapter 3).

The second part is Systemic Pharmacology, which covers common disorders of the major body systems and their treatment. The cardiovascular, respiratory, endocrine, musculoskeletal, skin and central nervous systems are considered. An outline of normal physiology of the systems is included where appropriate and relevant diseases described briefly. This is not intended to be a physiology book or a pathophysiology book. Should the reader need to consult such books, suggestions are given in the bibliography. Major groups of drugs are discussed, with emphasis on areas of relevance to the three professions for whom the book is intended.

In addition to drugs used to treat diseases of the major systems, the treatment of infections and parasites, the use of cancer chemotherapy, the use of anaesthetics and analgesics and the use of contrast agents and adjuncts to radiotherapy are included in Part 2.

The final part has two chapters. The first of the two (Chapter 14) is about legislation around the use of medicines with discussion of salient points from the Medicines Act 1968 and the Misuse of Drugs Act 1971. Specific exemptions for podiatrists, the use of patient group directions, supplementary prescribing and independent prescribing and a brief history of non-medical prescribing are considered.

The final chapter (Chapter 15) 'Prescribing in Practice' consists of contributions from podiatry, radiography and physiotherapy colleagues. They have described the use of various forms of access, supply, administration and prescription of medicines in their professions today and considered future developments in the light of the recent legislation allowing pharmacists and nurses to train as independent prescribers. Hopefully this will give the reader a realistic view of what is currently happening and what might happen in non-medical prescribing.

Useful web sites are listed at the end of each chapter, to encourage the reader to use the Internet for sources of reliable and respectable up-to-date information about disease, medicines and therapeutics. Although all websites were accessible at the time of writing, their existence cannot be guaranteed in the future.

Each chapter is followed by one or more case studies to illustrate the clinical use of drugs and problems that may arise from drug–drug interactions and adverse reactions. The situations are not based on any particular individuals; rather information has been gathered from many sources including my colleagues in physiotherapy and podiatry and used to construct the cases.

Finally, the chapters are finished off with review questions to test the reader's understanding of key concepts.

In the appendices, a list of drug names with their main therapeutic uses and a glossary of key terms used in the text are provided.

Drugs in current use are not all covered in this text; neither is this work intended as a recommendation for any drug use. Professionals should always consult the latest edition of the *British National Formulary* for definitive information about medicines.

Acknowledgements

I would like to thank friends and colleagues who encouraged and supported me in the writing of this book from its early inception through to final completion. I especially want to thank Leah Greene for her technical expertise and unfailing assistance with computer applications. I am grateful to Alison Barlow and Peter Bowden for their helpful ideas with matters relating to podiatry and Louise Stuart, MBE (Consultant Podiatrist) for an insight into supplementary prescribing; to Jan Dodgeon for help with topics relevant to radiography and Chris Frames and Chris O'Neal for their help with devising physiotherapy case studies.

Special thanks are due to those who contributed to Chapter 15, namely Professor Peter Hogg (Nuclear Medicine) and his co-authors, and Anthony Waddington (Podiatric Surgeon). Without their experience in practice this book would have had far less relevance to the health care professionals for whom it was written.

I have to thank students past and present for their inspiration, comments and suggestions over the years and I hope future students and practitioners will benefit from this.

Thanks to staff at Wiley (in particular Rachael Ballard, Fiona Woods and Jon Peacock), to Neil Manley for creating the index, and to Wendy Mould, who copyedited the book.

Finally, thanks to Alex for his understanding and patience.

1

Introduction

Pharmacology is the science of drugs and their effects on biological systems. A drug can be defined as a chemical that can cause a change in a biological system; the important biological system to be considered in this book is the human body. A drug is the active ingredient in a medicine; a medicine is the formulation of a drug into a tablet, capsule or other delivery system. The Medicines Act 1968 refers to drugs as medicinal products.

Drugs can be naturally occurring substances, for example hormones; everyday substances, for example caffeine and alcohol; synthetic chemicals marketed for therapeutic activity, for example aspirin; or substances used for recreation.

Pharmacology as a science encompasses the following:

- the action of natural chemicals in the body;

- the origins and sources of drugs;

- their chemical structure and physical characteristics;

- their mechanisms of action;

- their metabolism and excretion;

- studies of their action on whole animals, isolated organs, tissues and cells, enzymes, DNA and other components of cells;

- ultimately studies of their actions in humans and their therapeutic uses.

Pharmacology is also the study of the toxic effects of drugs and chemicals in the environment. All drugs are capable of being toxic and all drugs can produce unwanted effects at high doses, or if used incorrectly. The difference between a medicine and a poison is often merely a matter of concentration. In therapeutics, the treatment of disease is intended to have a beneficial effect with adverse effects kept to an acceptable minimum. The science of modern pharmacology is a relatively recent development. Prior to the 1930s, there were very few medicines available, and those that were available came from natural sources. Examples of drugs originally from natural sources and still in use today are quinine (from the bark of the cinchona tree and used to treat malaria), digitalis (from the foxglove and used for heart failure) and aspirin (extracted from the bark of willow tree and originally used to treat fever).

Development of new drugs can happen in many ways. Drugs have been developed following observation of side effects when being used for other purposes. It is now known

Pharmacology for the Health Care Professions Christine M. Thorp
© 2008 John Wiley & Sons, Ltd

that the site of action of many drugs is a cellular receptor. As knowledge of receptor structures has developed, this has allowed drugs to be designed to fit with receptors. The human genome project and mapping of genes has led to work on the development of drugs to alter genes.

1.1 Pharmacology and health care professionals

The importance of pharmacology to health care professionals cannot be overestimated. Members of the three professions, physiotherapy, podiatry and radiography, encounter patients on a daily basis, many of whom will be on drug therapy. Patients are increasingly likely to be receiving at least one drug; many older patients are likely to be on more than one drug, and prescription of eight or nine drugs at the same time is not uncommon. This is known as *polypharmacy* and it increases the chance of patients experiencing adverse effects or the effects of drug–drug interactions.

Depending on the nature of their work, health care professionals may spend some considerable time with individual patients who might have questions about their drug therapy. Some health care professionals may be treating mainly older patients, or younger patients or high-risk patients, and will become experienced and familiar with drugs in their areas of expertise.

Health care professionals can be ideally placed to spot adverse drug reactions and to play an important role in the long-term monitoring of commonly prescribed drugs. As professionals, they should be able to advise patients or know when to refer them to other experts in the health care team. Drug therapy of disease is ever expanding; new drugs exist for effective treatment or cure of more diseases than ever before. Correct use of drugs is paramount. It is therefore important for health care professionals to have an understanding of therapeutic uses of medicines, normal doses, adverse effects, interactions with other drugs, precautions and contraindications. It is equally important to be able to judge whether a change in a patient's condition is caused by drug therapy, or a change in the disease process. Medication can lead to symptoms such as dizziness, fatigue, dry mouth, constipation and patients may or may not associate new symptoms with drug use.

Health care professionals are increasingly involved in the administration of drugs to patients, either as an exemption to the Medicines Act 1968, under patient group directions, or as supplementary prescribers. The Medicines Act 1968, and additional secondary legislation since then, provides a legal framework for the manufacture, licensing, prescription, dispensing and administration of medicines. An exemption to the Medicines Act allows certain professionals, including podiatrists, access to specified prescription-only medicines, providing they are appropriately registered with the Health Professions Council. The use of patient group directions allows many health care professionals to administer prescription-only medicines to specific groups of patients without a normal prescription. Podiatrists, radiographers and physiotherapists are now included in the list of health care professionals who can train to prescribe medicines alongside doctors (and dentists) as supplementary prescribers.

Prior to 1994, only doctors, dentists and veterinary practitioners were allowed to prescribe medicinal products in the United Kingdom. That year the law was changed to enable district nurses, midwives and health visitors to prescribe from a limited formulary of dressings, appliances and some medicines. This formulary of medicines was extended in 2002.

A review of prescribing, supply and administration of medicines for the Department of Health (1999) (Crown Report 2) recommended two types of prescriber: independent and supplementary.

Over the next few years, supplementary prescribing by nurses and pharmacists was introduced and legislation to allow this was changed in April 2003.

A similar process occurred with podiatry, physiotherapy and radiography and led to extension of supplementary prescribing to these professions in April 2005. In a further development in 2006, nurses and pharmacists became eligible to train as independent prescribers.

Non-medical prescribing is now the term applied to prescribing by members of the health care professions who are not 'medically' qualified.

Prescribing can be described in the following three ways:

1. to order in writing the supply of prescription-only medicine for a named patient;

2. to authorize by means of an NHS (National Health Service) prescription the supply of any medicine (prescription-only, pharmacy or general sales list item) at public expense;

3. to advise a patient on suitable care or medication, including over-the-counter drugs, and therefore with no written prescription.

All health care professionals who are involved in prescribing, and/or administration of medicines have to abide by standards set out by their respective professional bodies. For podiatrists, radiographers and physiotherapists, this is the Health Professions Council. Health care professionals have a responsibility to consult documentation produced by the professional bodies and be accountable for prescribing and administering drugs. All members of health care professions have a responsibility to reduce the risk of errors in prescribing, must assess and appraise their own practice and show a commitment to continuing professional development. This is essential not least because information about drugs and associated legislation is constantly changing. New drugs come on the market, and others are withdrawn or reclassified. Reliable sources of information are the *British National Formulary* (*BNF*), the *Monthly Index of Medical Specialities* (*MIMS*), the *British Pharmacopoeia* (*BP*), patient information leaflets (PILs) and summaries of product characteristics (SPCs) supplied by medicines manufacturers. Official bodies concerned with the use, quality and safety of medicines are the Commission on Human Medicines (CHS, formerly the Committee on the Safety of Medicines), the Medicines and Healthcare Products Regulatory Agency (MHRA) and the National Institute for Health and Clinical Excellence (NICE).

1.2 Patient compliance

Patient compliance is important for successful drug therapy. Compliance in this context is defined as the extent to which the patient follows the clinical prescription. Non-compliance and reasons why patients do not always take drugs as prescribed should be appreciated. Some common reasons for non-compliance are that the patient has doubts about a drug's effectiveness, they believe they are cured, they misunderstand instructions, dosage regimes are too complicated, or they experience unacceptable side effects.

Health care professionals play an important role in improving compliance. This is particularly important if a drug is for serious conditions like epilepsy, glaucoma or hypertension, or is for infection because of the problem of drug resistance.

Well-informed patients are more likely to be compliant.

It is worth spending time explaining what the medication is, how it is taken and why, how long it is to be used for, what adverse effects to look out for and any alternatives if appropriate. The importance of the drug therapy can be explained and what might happen if the patient did not comply. Aids to help compliance can be suggested, for example packaging of daily doses can be arranged with pharmacists, special containers can be obtained, the help of relatives can be sought, suitable time of day for administration can be chosen and provision of written information can all help. Patient information leaflets must be included in packaging of medicines.

1.3 Drug names

All drugs have at least three names: the chemical name, the generic name and the proprietary name. Chemical names can be complicated and difficult to remember and are not used in this book. A generic name is a drug's official name and the majority of drugs in this book are referred to by their generic names. The proprietary name is the name given to a drug by the manufacturing company. As the same drug can be manufactured by several different companies, a drug can have multiple proprietary names and this can be confusing. Hence, proprietary names have been avoided in this book except where the proprietary name is in common usage. In the United Kingdom, the generic name is known as the *British approved name* (*BAN*). Following European Directive 92/27/EEC, European Law requires the use of the recommended International Non-proprietary Name (rINN). This ensures that all countries, in Europe at least, recognize the same drug. In most cases, the BAN and the rINN were the same, but some British names have been changed. For example, amphetamine is now spelt amfetamine and lignocaine is now lidocaine. Where this has happened, both names are listed in the BNF. Wherever possible, drugs should be prescribed by their generic name; this allows any suitable product to be dispensed and in many cases, it saves the health service money. The only exception to this rule is when a patient must always receive the same brand of a drug because different preparations can result in different blood levels of the drug. No details of dosages are given in this book (except in some of the case studies), because these are subject to change and often have to be varied to suit individual patients. In practice, the *BNF* or *MIMS* should be used as a guide to dosages.

Examples of individual drugs have been kept to a minimum in the text, with usually just one or two examples given in each section. It would be impractical to try to remember the names of all drugs available. In practice, health care professionals quickly become familiar with drugs commonly used in their area.

Nevertheless, the examples used in this book amount to over 300 drug names, which are listed for easy reference in Appendix I.

Part I
Principles of pharmacology

2
Drug disposition

2.1 Chapter overview

If a drug is to have a therapeutic effect on the body, it first has to reach its site of action. In order to do this a drug has to be administered in some way. Unless the route of administration is directly into the blood stream, the drug has to be absorbed, usually by diffusion. Once absorbed, distribution of the drug to different parts of the body follows. This necessarily includes passage through the liver. Most drugs are treated as potentially toxic substances and are metabolized by the liver. This detoxifies them and some drugs are almost totally inactivated on first pass through the liver. Eventually a drug will be excreted from the body. This usually occurs via the kidneys, although some drugs can be lost in faeces or exhaled air. This chapter discusses the processes of administration, absorption, distribution, metabolism and excretion of drugs together with factors affecting these processes. Collectively, these processes describe drug disposition, the way in which the body handles drugs. The study of the fate of drugs in the body is known as *pharmacokinetics*.

2.2 Administration of drugs

In order to get to their site of action in the body, drugs have to be administered in some way. There are two major routes of drug administration: enteral and parenteral. Enteral means to do with the gastrointestinal tract and includes oral and rectal administration. The parenteral route includes all other means of drug administration. There are many routes of parenteral administration, some of which are intended for a drug to have a systemic effect and others for a local effect. See Figure 2.1.

(In some definitions, parenteral is synonymous with injection (for example in the Medicines Act), but here the term is used to describe all routes of administration that are not enteral.)

Pharmacology for the Health Care Professions   Christine M. Thorp
© 2008 John Wiley & Sons, Ltd

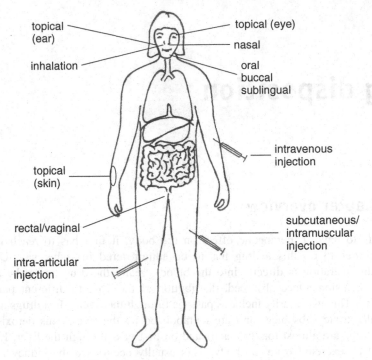

Figure 2.1   Sites of drug administration

2.2.1   Oral administration

The vast majority of drugs are administered by mouth as pills, capsules, tablets or liquids. Following oral administration, absorption of a drug is from the stomach or intestine directly via the hepatic portal system to the liver before reaching the general circulation. The liver is the main site of drug metabolism and inactivation (see page 20). Many factors affect drug absorption from the gastrointestinal tract, including lipid solubility of the drug; its molecular weight; the pH of the local environment; the surface area of the absorbing membrane; gastric emptying time; the rate of removal from the gastrointestinal tract by the blood and the degree of plasma protein binding of the drug once in the blood stream.

Because by mouth is a common route of drug administration, it is considered in more detail in Section 2.3.5.

There are advantages and disadvantages of administering drugs by the oral route. Advantages are that it is a safe and convenient route, generally acceptable to the patient and requires no particular skills. Disadvantages are that many drugs do not taste particularly nice; some can upset the stomach and cause nausea and vomiting or even ulcerate the stomach lining, while others may be destroyed by stomach acid or digestive enzymes or be extensively metabolized in the liver. The oral route requires a co-operative and conscious patient.

2.2.2 Rectal administration

Rectal administration avoids drug inactivation by stomach acid and digestive enzymes and about 50% of that absorbed bypasses the liver and goes directly into the systemic circulation. In some circumstances, it may be advantageous to administer a drug rectally, for example if a patient is unconscious or vomiting, or uncooperative in some way, but generally, it is considered an unpleasant method.

Many of the general factors considered above that affect absorption from the gastrointestinal tract apply equally to this route of administration.

2.2.3 Sublingual and buccal administration

Sublingual (under the tongue) and buccal (oral cavity) administration of a drug allows the drug to go directly into the systemic circulation without first passing through the liver. This route can provide a rapid means of absorption of a drug. It is a route commonly used, for example, to treat attacks of angina with glyceryl trinitrate.

Factors affecting absorption from the oral mucosal membranes include lipid solubility and molecular weight of the drug, pH of the saliva and the rate of removal by the blood as well as the application of the correct technique.

Some drugs administered this way are intended to have a local effect, for example to treat oral or throat infections.

The main disadvantage of this route of administration is that most drugs do not taste nice.

2.2.4 Nasal administration

Nasal administration is often intended to have a local effect, as in the use of nasal decongestants, although certain drugs are given this way to have a systemic effect. For example, antidiuretic hormone used to treat diabetes insipidus and hormones used for infertility treatment can be given by this route. This route can be another way of avoiding destruction of a drug by liver enzymes or stomach acid and digestive enzymes. Factors affecting absorption across mucosal membranes, similar to those considered under oral administration.

2.2.5 Topical administration

Topical administration to the epidermis of the skin is generally used for drugs intended to have a local effect. This route is of particular relevance to podiatrists and can be used to treat a local infection or other conditions. For example, the use of amorolfine cream in the treatment of fungal skin infections.

Most drugs are not easily absorbed through the skin but some are formulated into dermal patches for systemic absorption and others may penetrate damaged skin. For example, dermal patches can be used to administer nicotine replacement therapy.

Other routes of topical administration include application of drugs to the conjunctiva of the eye, the external ear, the vagina and the urethra, usually to treat local infection.

2.2.6 Inhalation administration

Many drugs are given by inhalation and may be intended to have a local or systemic effect. The lungs are adapted for absorption of oxygen having a large surface area for diffusion and a good blood supply. Particle size is a major determinant of absorption from this site of administration. Small particle size favours systemic absorption whereas large particle size discourages absorption into the systemic circulation.

A drug commonly given by inhalation for a local effect is salbutamol, used to treat asthma. Many general anaesthetics are given in gaseous form clearly intended to have a systemic effect.

2.2.7 Administration by injection

Drugs can also be given by injection. Methods of injection include subcutaneous, intra-muscular, intravenous, intra-arterial, intra-articular, intraspinal and epidural.

Drugs injected subcutaneously or intramuscularly have to diffuse between loose connective tissue or muscle fibres. The rate of absorption depends on the usual parameters for passage across cell membranes but also on the blood or lymphatic supply to a particular region. Increasing the blood supply by applying heat or massage will increase the rate of absorption. Conversely, for a local effect, addition of a vasoconstrictor to the injection decreases the rate of removal of the drug from the site of injection. Depot preparations are designed to give a slow sustained release of drug.

Drugs injected intravenously go directly into the blood stream and are rapidly distributed around the body. An advantage of intravenous injection is that it is possible to get high concentrations of a drug very quickly to its site of action, although this may also lead to toxic effects in other tissues. Disadvantages of intravenous injection are that it requires trained personnel using sterile techniques and once the drug has been given, mistakes cannot be rectified. Drugs can be given by continuous intravenous infusion, for example in cancer chemotherapy.

Intra-arterial injections are rarely used. Radio-opaque substances and cytotoxic drugs are sometimes injected into arteries in the diagnosis and treatment of cancer.

Intra-articular injections are sometimes used to administer a drug directly into a joint, for example with a corticosteroid in the treatment of arthritis or a contrast agent for imaging.

Intraspinal and epidural injections are given under certain circumstances to have a local effect, either as anaesthesia or to treat infection of the central nervous system. For details of injection techniques, see Chapter 12, page 236.

2.3 Absorption of drugs

Whatever the route of administration, a drug must reach its site of action. In order to do this, the drug will have to cross several cell membranes to reach the blood (unless it is injected intravenously).

The three ways by which substances, including drugs, can cross cell membranes are simple diffusion, facilitated diffusion and active transport.

2.3.1 Diffusion

Diffusion is the mechanism by which the vast majority of drugs pass across cell membranes. Both simple diffusion and facilitated diffusion are passive processes in that no energy is required other than the kinetic energy of the molecules themselves.

Several factors are known to influence the diffusion of substances across the cell membrane:

- the membrane must be permeable to the substance in question;
- there must be a concentration gradient;
- the molecular size/weight of the substance must be small enough;
- a large surface area is necessary for efficient diffusion;
- a short distance is necessary for efficient diffusion.

In practice, there is a concentration gradient because the drug is given in sufficient dose, most drug molecules are small enough to be absorbed (otherwise, they would be of no use) and the surface area and distance of the absorbing membrane are favourable.

2.3.2 Simple diffusion

Simple diffusion is depicted in Figure 2.2a, membrane transport mechanisms.

Simple diffusion of drug molecules depends mostly on lipid solubility.

The structure of the cell membrane can be a barrier to diffusion of drugs because it is essentially a lipid bilayer with proteins embedded in the inner and outer surfaces.

Lipid-soluble substances diffuse easily through the lipid bilayer and include oxygen, carbon dioxide, fatty acids, steroids and fat-soluble vitamins. The lipid solubility of a drug depends on its state of ionization. Certain small ions, for example sodium, potassium, calcium and chloride can pass through ion channels in the cell membrane. Such channels are highly specific and do not allow the passage of relatively large ionized drug molecules.

Drugs in the unionized form are generally lipid soluble whereas ionized drugs are not. The extent to which a drug is ionized depends on the pH of the local environment and

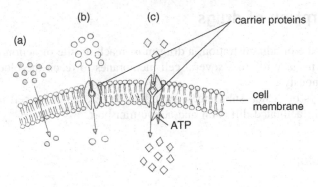

(a) simple diffusion
(b) facilitated diffusion
(c) active transport

Figure 2.2   Membrane transport mechanisms

the pKa of the drug. pH is a measure of hydrogen ion concentration – the lower the pH, the higher the hydrogen ion concentration and the greater the acidity of a solution. The pKa of a drug molecule is the pH at which the drug is 50% ionized and is different for different drugs. Chemically, most drugs are either weak acids or weak bases. In an acidic environment, as in the stomach, acidic drugs are unionized according to the following simple equation:

$$HA \rightleftharpoons H^+ + A^-$$

where A^- is an acidic drug and the excess hydrogen ions (H^+) drive the equation to the left.

In an alkaline environment, as in the small intestine and the majority of body fluids, basic drugs are unionized according to the following simple equation:

$$BH^+ \rightleftharpoons B + H^+$$

where B is a basic drug and the deficit of hydrogen ions drives the equation to the right.

Thus, acidic drugs are preferentially absorbed in the stomach and basic drugs are preferentially absorbed in the intestine. In practice however, because of the large surface area of the small intestine, the majority of drug absorption takes place there. Nevertheless, alteration of stomach pH can alter the absorption characteristics of acidic drugs.

A few drug molecules are small enough to diffuse through aqueous pores in the cell membrane with water, for example alcohol. However, the majority of drugs are too large to diffuse in this way.

2.3.3   Facilitated diffusion

Facilitated diffusion is depicted in Figure 2.2b.

Many nutrients and a few drugs can pass across the cell membrane by facilitated diffusion. In this case, in addition to the concentration gradient, a membrane protein acts as a carrier to transport a substance from one side of the membrane to the other. Carrier proteins are specific and only transport molecules that they 'recognize'. Glucose enters many body cells by facilitated diffusion and the process appears to be more efficient than simple diffusion.

Carrier systems exist for the transport of some amino acids and vitamins and the same carrier can transport drugs that are structurally similar to them.

2.3.4 Active transport

Active transport is depicted in Figure 2.2c.

Active transport involves a carrier protein but differs from diffusion in two important ways. Cellular energy in the form of ATP (adenosine triphosphate) is required to drive the process and transport goes against the concentration gradient. By such a mechanism, substances can be concentrated in certain parts of the body. Active transport mechanisms are particularly important in the transport of ions, nutrients and neurotransmitters and may be involved in the transport of some drugs. Many drugs have been developed that interfere with the active transport of neurotransmitters (see Chapter 3, page 44).

2.3.5 Absorption from the gastrointestinal tract

Since the vast majority of drugs are administered by mouth, it is important to consider factors that affect absorption of drugs from the gastrointestinal tract. See Figure 2.3 for a diagram of the digestive system.

The function of the digestive system is to provide nutrients for the body through the processes of mechanical degradation and liquefaction and the action of enzymes on the food we eat. Drugs taken orally are also subjected to these processes.

The digestive system consists of the mouth, oesophagus, stomach, small intestine, large intestine, rectum and anus together with the liver and pancreas.

The mouth is where food is chewed and mixed with saliva before being swallowed. The oesophagus conveys food to the stomach.

In the stomach food is stored while further digestion takes place. The stomach produces acid and enzymes to begin protein digestion. It can take 2–4 hours before food is passed onto the small intestine.

Stomach acid and enzymes can destroy some drugs and they have to be protected by an enteric coat so they pass unharmed into the small intestine.

Semi-digested and liquidized food passes gradually from the stomach into the duodenum, the first part of the small intestine.

The small intestine is the most important part of the digestive system for digestion because it produces many enzymes and is highly adapted for absorption of nutrients. It has a large overall surface area because of its length and because its inner lining is folded into villi.

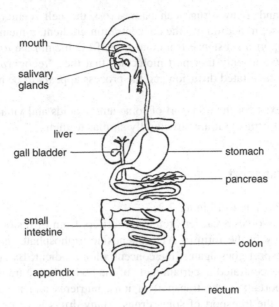

mouth

salivary
glands

liver

gall bladder

stomach

pancreas

small
intestine

colon

appendix

rectum

Figure 2.3   The digestive system

As food enters the small intestine secretions are added from the gall bladder and pancreas. Pancreatic secretions contain many digestive enzymes and sodium bicarbonate, which neutralizes stomach acid. Bile contains bile acids, which are essential for the emulsification of fats prior to their digestion.

The pH in the small intestine is slightly alkaline. While this favours the absorption of basic drugs because they will be unionized, most drug absorption takes place in the small intestine anyway because of the large surface area.

Nutrients (and drugs) absorbed pass directly to the liver in the hepatic portal system before going to other parts of the body.

The liver excretes some drugs into the intestine via bile. Once back in the small intestine the original drug can be reabsorbed.

The large intestine is where water is reabsorbed from the remains of digested food. Here some drug metabolites that have been excreted in bile can be regenerated by the action of bacteria in the large intestine.

The absorbing membrane for nutrients and drugs is the mucous membrane of the epithelial cells lining the gastrointestinal tract. General factors affecting diffusion across cell membranes, considered above, apply in addition to the pH of gastrointestinal contents: surface area of the gastrointestinal tract; gastric emptying and intestinal transit time; blood flow from the gastrointestinal tract; plasma protein binding; active transport mechanisms and drug formulation. Table 2.1 gives some of the effects these factors have on drug absorption.

Once absorbed from the gastrointestinal tract a drug passes directly to the liver in the portal circulation and may be subjected to metabolism before further distribution round

Table 2.1 Factors affecting absorption from the gastrointestinal tract

Factor	Effect
pH of gastrointestinal contents	Acidic drugs are unionized in acidic conditions and preferentially absorbed in the stomach; basic drugs are unionized in alkaline conditions and preferentially absorbed in the small intestine
Surface area	The gastrointestinal tract and the small intestine in particular has a huge surface area, adapted for absorption; because of this and despite the effects of pH most drugs are predominantly absorbed in the small intestine
Gastric emptying and intestinal transit time	Drugs given with a meal take longer to be absorbed; this may be necessary for drugs principally absorbed in the stomach or for those that irritate the stomach lining; some very lipid-soluble drugs are better absorbed with a fatty meal; other drugs can delay gastric emptying or increase/decrease intestinal transit time
Blood flow	The intestine has a good blood flow that creates a concentration gradient as the drug is constantly being removed from its site of absorption
Plasma protein binding	Many drugs are bound to plasma proteins and this helps maintain the concentration gradient because the bound drug is effectively removed
Active transport mechanisms	Of minor importance but drugs related to nutrients can be absorbed more rapidly by transport mechanisms
Formulation	Some drugs can be made to be rapidly dissolving for quick effect; others may have an enteric coat to protect the stomach lining/protect the drug from stomach acid

the body. This is known as *first pass metabolism*, which can result in considerable loss of activity for some drugs on first pass through the liver.

Some drugs are recycled by enterohepatic shunting (or cycling). Enterohepatic shunting describes the process whereby a drug is first metabolized and then excreted into the intestine via bile. Once in the intestine gut bacteria or intestinal enzymes convert the drug back to its original form, which is then reabsorbed. This effect, which can be repeated many times, prolongs the duration of action of the drug until it is eventually excreted by the kidneys.

2.4 Drug distribution

Only a free drug at its site of action can have a pharmacological effect, therefore it is important that a drug is distributed around the body effectively.

When a drug is administered, it does not achieve an equal concentration throughout the body. Unless a drug is injected directly into the blood stream it will be absorbed from its site of administration, then enter the systemic circulation and be transported to the tissues in plasma. The body can be considered to be made up of aqueous and lipid compartments. Lipid compartments include all cell membranes and adipose tissue. Aqueous compartments include tissue fluid, cellular fluid, blood plasma and fluid in places like the central nervous system, the lymphatic system, joints and the gastrointestinal tract. The distribution of a drug into these different compartments depends on many factors.

2.4.1   Aqueous solubility

Aqueous solubility affects distribution because water-soluble drugs have difficulty crossing cell membranes and therefore tend to remain in the circulation. Consequently, water-soluble drugs are not well distributed throughout the body. They exist in large amounts in the plasma or tissue fluid and are rapidly cleared by the liver or kidney. In practice, such drugs have little therapeutic use.

2.4.2   Blood flow

At equilibrium, drugs are partitioned between plasma, plasma proteins and the different tissues. The rate of distribution to different tissues depends largely on the rate of blood flow through them. Some areas of the body have a relatively good blood supply, for example, the major organs; muscles and skin have a moderate supply; and bone and adipose tissue have a poor supply. Thus, major organs receive a relatively high concentration of a drug whereas it can be difficult to get drugs into less well-perfused areas.

Although the brain has a very good blood supply, distribution of drugs into the central nervous system is restricted. This is because of the so-called 'blood-brain barrier'. This is not an anatomical barrier as such, rather a combination of the tight junctions between endothelial cells of brain capillaries and the close association of glial cells with the outside of the capillaries. This arrangement makes diffusion of lipid-soluble drugs into the brain difficult and diffusion of water-soluble drugs almost impossible.

2.4.3   Plasma protein binding

A large number of drugs have a high affinity for albumin and other plasma proteins. Binding to plasma protein inhibits distribution outside the blood since only unbound drug will be further distributed. Plasma protein binding therefore reduces active drug concentration and ultimate response to the drug.

Drugs can compete for the same protein binding sites and this is a form of drug interaction. A well-known and important example is that of warfarin and aspirin. Warfarin is an anticoagulant, which binds extensively to plasma proteins, and this is taken into account when dosages are worked out. Aspirin taken with warfarin competes for the same protein binding sites, which means that they each displace the other and the amount of free drug in the plasma is increased for both drugs. Patients stabilized on warfarin should never take aspirin because the effect of increased free plasma concentration of warfarin can be severe haemorrhaging. Coincidental increased activity of aspirin is not as serious.

2.4.4 Lipid solubility

Lipid-soluble drugs enter cells readily. Distribution of such drugs is widespread unless plasma protein binding is extensive. Elimination of lipid-soluble drugs is usually slow because clearance from plasma via the kidneys removes only a small proportion of the drug in any given time.

2.4.5 Tissue sequestration

Considerable amounts of drug may be stored in certain tissues, particularly fat and muscle. Sequestration in this way gives an apparent large volume of distribution (see below) but also means that only a small proportion of total drug concentration will reach its site of action. This can create difficulties with the usage of certain drugs. For example, general anaesthetics are highly lipid-soluble drugs. Sequestration into adipose tissue can make anaesthetizing obese people hazardous because it is difficult to control the amount of free drug in the circulation. Similarly, benzodiazepines (antianxiety drugs) can be difficult to clear from the body because they are stored in large amounts in adipose tissue. This can complicate withdrawal from their use.

Apart from storage in lipid tissue, certain drugs can be preferentially taken up or sequestered into other tissues.

For example, griseofulvin has an affinity for keratin. Since this drug can be used to treat fungal infections of the skin and nails its sequestration into keratin is something of an advantage.

The antibiotic tetracycline has an affinity for bones and teeth. It should never be used in children as its accumulation can damage teeth and stunt growth.

2.4.6 Metabolism and excretion

The rate at which a drug is metabolized will affect its distribution. Similarly the rate of elimination or excretion also affects distribution and vice versa.

2.4.7 Volume of distribution

Volume of distribution is a concept that describes the body compartments into which a drug could be distributed. If a drug is water soluble it is likely to remain in the blood stream and its volume of distribution will be relatively small and equal blood volume. Similarly, acidic drugs tend to bind to plasma albumin and therefore also remain in the blood stream and have a small volume of distribution. If a drug is highly lipid soluble, then it will be distributed to many parts of the body and have a large volume of distribution. In addition, basic drugs tend to bind to tissue proteins and as a result have a large volume of distribution.

2.5 Drug metabolism

On entering the body, drugs are treated as if they are toxic substances, which need to be detoxified, if a mechanism exists, and eliminated as soon as possible.

This means that most drugs are subjected to some kind of metabolism and then excreted.

Metabolism involves changes to the molecular structure of a substance and these changes are produced by the action of enzymes. This has two important effects on drug molecules:

1. The drug is made more water soluble and therefore more easily excreted by the kidneys;

2. The metabolites are usually less pharmacologically active than the parent drug. This is not always the case. A drug metabolite may have a new and completely different pharmacological activity, or it may be as active as or more active than the original drug.

Some drugs have to be metabolized in order to become active.

Prodrugs are drugs that have been designed to remain inactive until they have been metabolized by the body.

They may be formulated in this way because the active drug is difficult to administer, whereas the inactive prodrug does not cause the same complications, or because the active drug is not absorbed or distributed to its intended site of action easily.

Most tissues in the body have the enzymes capable of metabolizing a variety of substances. However, as one of the main functions of the liver is the metabolism of toxic substances produced during normal metabolic processes, it is not surprising that the majority of drug metabolism takes place in the liver.

Some drugs are almost completely inactivated by first pass metabolism in the liver. The extent of first pass metabolism varies from individual to individual and can lead to unpredictable effects for some drugs administered orally.

Other tissues where significant metabolism of drugs can occur include the intestinal mucosa, the lungs and plasma.

2.5.1 Metabolic reactions

There are two general types of metabolic reactions, which are known as *Phase 1* and *Phase 2 reactions*. Some drugs undergo both Phase 1 and Phase 2 reactions, but, depending on its chemical nature, it is possible for a drug to be metabolized by either type of reaction only. See Figure 2.4 for possible routes of metabolism of drugs.

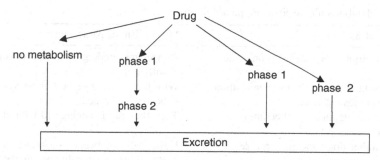

Figure 2.4 Possible routes for metabolism of drugs

2.5.1.1 *Phase 1 reactions*

Phase 1 reactions involve the biotransformation of a drug by one or more of the following reactions to a more water-soluble metabolite, which is more likely to be excreted by the kidney or go on to Phase 2. This may increase the toxicity of some drugs.

2.5.1.1.1 Oxidation Oxidation is the most important and commonest type of metabolic reaction, which involves the addition of oxygen to the drug molecule.

In the liver, oxidation reactions are catalysed by a group of enzymes known as the *microsomal mixed function oxidase system* or the cytochrome P450 enzyme family.

Oxidative reactions take place in many other tissues as well.

2.5.1.1.2 Reduction Reduction reactions involve the removal of oxygen or the addition of hydrogen to the drug molecule.

Enzymes capable of catalysing reduction reactions are found in many body tissues, including the liver and in the intestinal bacteria.

2.5.1.1.3 Hydrolysis Hydrolysis involves the splitting of a drug molecule by the addition of water.

Enzymes capable of catalysing hydrolysis are found in many body tissues but particularly in the small intestine.

2.5.1.2 *Phase 2 reactions*

Phase 2 reactions make drugs or Phase 1 metabolites into more hydrophilic, less toxic substances by conjugation with endogenous compounds in the liver.

This too encourages renal excretion.

The most important conjugation reaction is with glucuronic acid to form a glucuronide. Other conjugations occur with sulfate, acetyl, methyl and glycine groups.

Many drugs are metabolized by a combination of routes and this can vary from individual to individual and depends on the dose of drug, the presence of interacting drugs and the state of the liver. Metabolism of aspirin and paracetamol are given as examples in Table 2.2 to illustrate this.

Table 2.2 Metabolism of aspirin and paracetamol

Metabolism of aspirin	Metabolism of paracetamol
The main metabolite is a glycine conjugate	Paracetamol undergoes glucuronidation and sulfation
With increasing dose the glycine conjugation system becomes saturated	With increasing dose both these systems become saturated
Glucuronide conjugation then becomes important	Then the drug is conjugated with glutathione
With even higher doses the glucuronide system becomes saturated	If this pathway becomes saturated a hepatotoxic metabolite accumulates (hence the danger of overdose)
A greater proportion of the drug appears in the urine as salicylic acid	
The rate of elimination then depends on urinary pH	

2.5.2 Factors affecting metabolism of drugs

There are many factors that can affect the metabolism of drugs.

2.5.2.1 Enzyme induction

Some drugs are capable of increasing the activity of drug-metabolizing enzymes. This is known as *enzyme induction* and occurs when a drug is administered over a period of time. Its significance is that other drugs metabolized by the same enzymes are metabolized faster and therefore circulate in a lower concentration than expected for a given dose. This could mean that they become ineffective. Alternatively, a drug could be metabolized faster than expected into a toxic compound. The mechanisms involved in enzyme induction are unclear but somehow DNA sequences are affected so that production of the relevant enzyme is increased.

Examples of drugs that are well known to cause enzyme induction are carbamazepine, phenytoin (both antiepileptics) and alcohol.

2.5.2.2 Enzyme inhibition

Inhibition of enzyme systems may cause adverse reactions due to higher than expected drug concentrations. The effects of inhibition can appear as soon as the inhibiting drug reaches a high enough concentration to compete with another drug for the same enzymes.

Cimetidine, a drug used to treat stomach ulcers, can inhibit the metabolism of several potentially toxic drugs such as phenytoin (antiepileptic), warfarin (anticoagulant) and theophylline (bronchodilator).

Erythromycin, an antibiotic, similarly increases the activity of theophylline, warfarin and digoxin (used in cardiac failure).

2.5.2.3   *Individual variation*

Response to drugs varies between individuals and although the reasons are often multi-factorial there are cases where gene polymorphism is responsible (see Chapter 3).

There are some well-known examples of individual variation in response to drugs. Propranolol is a drug commonly used to treat high blood pressure. About 8% of the general population are poor hydroxylators of propranolol. Because propranolol is extensively metabolized affected individuals show exaggerated and prolonged responses to this drug.

Another example is the metabolism of isoniazid, a drug used to treat tuberculosis. This drug is metabolized by acetylation and 50% of the population acetylate it slowly, which can lead to toxic effects.

Suxamethonium is a neuromuscular blocking drug frequently used as an adjunct to anaesthesia; it causes muscle paralysis (see Chapter 12). Normally it is metabolized by an enzyme, plasma pseudocholinesterase. One in 2500 of the population has a less active form of the enzyme. The effect of this is that the duration of action of suxamethonium is extended from about 6 minutes to 2 hours.

2.5.2.4   *Age*

Age can have an effect on the way the body metabolizes drugs. Both liver function and the number of hepatic enzymes are reduced at birth (especially in pre-term infants). However, these develop rapidly during the first 4 weeks of life. In the elderly the metabolism of drugs declines because of a reduction in hepatic enzymes.

2.6   Excretion of drugs

Drugs are potentially toxic substances and must be eliminated by the body as quickly as possible. This is often at odds with their intended therapeutic use, but nevertheless is what generally happens.

The main route of excretion of drugs and drug metabolites is via the kidneys. Elimination varies amongst individuals and depends on the rate of metabolism of a particular drug, the rate of production of urine and the pH of urine.

The nephron (or kidney tubule) is the basic structural and functional unit of which there are about one million in each kidney. It is a blind ending tube, which begins with the Bowman's capsule. The Bowman's capsule is the dilated closed end that surrounds the knot of capillaries known as the *glomerulus*. Blood is supplied to the glomerulus by an afferent arteriole and removed by an efferent arteriole. From the Bowman's capsule the nephron becomes the proximal convoluted tubule, followed by the loop of Henlé, the distal convoluted tubule and finally the collecting duct. See Figure 2.5 for diagram of a nephron.

There are three processes to the production of urine. Drugs, like waste products from normal metabolism, can be subjected to all of them.

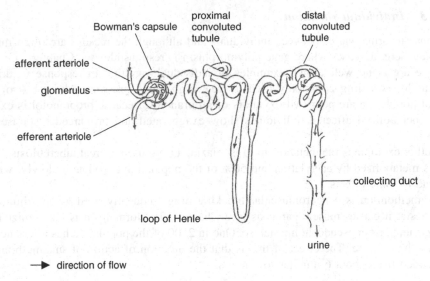

Figure 2.5   Structure of a nephron

2.6.1   Glomerular filtration

The glomerulus is adapted for the filtration of water and substances dissolved in it from plasma. Pressure in the capillaries of the glomerulus forces fluid and dissolved substances into the Bowman's capsule. Most drugs are small enough to be filtered, although if they are lipid soluble they are readily reabsorbed from other parts of the kidney tubule. Proteins are generally too large to be filtered at the glomerulus and do not appear in the urine. Therefore, drugs that are extensively bound to plasma proteins pass through the glomerulus slowly.

2.6.2   Tubular secretion

Tubular secretion of drugs occurs in the proximal convoluted tubule by active transport mechanisms normally used for the removal of waste products. Transport systems exist for the secretion of acidic substances, for example uric acid and basic substances, for example creatinine. Many drugs can be actively secreted by this mechanism, which increases their rate of removal from the body.

Active transport does not depend on concentration gradients and can overcome plasma protein binding. As the free drug is transported, more of the drug dissociates from the protein binding site and can be eliminated.

Substances transported by the same transport system compete with each other and the more slowly a drug is transported the more effective it is at inhibiting the transport of another.

A well-documented historical example of this is the competition between penicillin and probenecid. Penicillin is a drug that is approximately 80% protein bound, but is rapidly eliminated by a kidney transport system. This was a disadvantage in the days when penicillin was expensive and difficult to produce, because its action was prematurely terminated unless high doses were given. Penicillin used to be administered together with probenecid. Probenecid is slowly transported by the same system and therefore considerably reduces the elimination of penicillin.

(Probenecid is used to treat gout because it also inhibits the reabsorption of uric acid by a similar mechanism. See Chapter 7.)

2.6.3 Tubular reabsorption

Useful substances are reabsorbed from the kidney tubules back into plasma by diffusion or active transport. Much of this occurs early on in the nephron at the proximal convoluted tubule but also at the distal convoluted tubule where tubular filtrate concentration is high. Lipid-soluble drugs may be reabsorbed in this way and as a result, they are eliminated slowly from the body. Metabolism of drugs tends to make them less lipid soluble, more water soluble and therefore more likely to ionize. Tubular filtrate is normally slightly acid and this favours the excretion of basic drugs because they ionize more readily in acidic conditions and therefore do not diffuse easily back into plasma.

Change in pH of the tubular filtrate can affect drug ionization. Manipulation of pH to increase the rate of elimination of a drug has a practical use in cases of overdose as shown by the following examples.

Administration of sodium bicarbonate increases the alkalinity of the tubular filtrate therefore acidic drugs, for example aspirin and barbiturates become ionized and cannot diffuse back from the tubules to the plasma. Hence their rate of elimination is increased. This may be useful in the event of an overdose.

Conversely, the elimination of basic drugs such as amfetamine, antihistamines, morphine and tricyclic antidepressants can be increased by making tubular filtrate more acidic by administration of ammonium chloride.

2.6.4 Other routes of excretion

Although generally not as important as renal excretion, other routes of excretion account for elimination of some drugs.

It is possible for drugs to move back across the intestinal wall by diffusion or active transport and be lost in faeces. Some drugs, particularly hormones, can also be excreted into the intestine via bile. Such drugs are conjugated in the liver and passed into bile before they reach the systemic circulation. Often bacteria in the large intestine digest the conjugate releasing free drug, which can then be reabsorbed and recycled by the process of enterohepatic shunting (see page 17). Prolonged use of antibiotics can sterilize the large intestine and interrupt this cycle and increase drug elimination in faeces. Unexpected loss of a drug in this way can reduce its therapeutic action. Diarrhoea can have a similar effect. The effectiveness of oral contraceptives can be reduced in this way.

Certain drugs are excreted via expiration; the rate of excretion depends on plasma concentration, alveolar air concentration and blood-gas partition. Anaesthetic gases are the only group of drugs that are significantly excreted in expired air. A small amount of alcohol is excreted in this way but this accounts for only a small proportion of the overall elimination.

Loss of drugs in sweat and breast milk occurs, but is of minor importance although the appearance of drugs in breast milk can have serious consequences for the nursing baby (see Chapter 3).

2.7 Summary

Drugs can be administered to patients by a variety of routes and can be intended to have a local or systemic effect. The oral route is the most commonly used for systemic effect because it is convenient and generally acceptable to patients and requires no special skills. Absorption of a drug from the gastrointestinal tract is normally by diffusion and depends on many factors including lipid solubility and pH of the gastrointestinal contents. Once absorbed into the blood stream, a drug is distributed around the body. From the gastrointestinal tract, a drug goes to the liver in the hepatic portal system where it is likely to be metabolized. Metabolism usually makes a drug more water soluble and less pharmacologically active. Distribution to the rest of the body depends on how well a drug can pass across cell membranes. Some drugs can be sequestered in certain tissues, particularly fat and muscle. Many drugs are bound to plasma proteins and this can limit their distribution and action because only the free drug can have an effect. Eventually, whether metabolized or not, a drug is eliminated from the body. The usual route of excretion is via the kidneys and rate of elimination depends on rate of urine production, pH of urine and transport mechanisms. Although drugs do appear in faeces, expired air, sweat and breast milk, these are normally minor routes of excretion.

Useful website

www.merck.com./mmhe/sec02.html The Merck Manuals Online Medical Library; Section on
   drugs.

Case studies

The following two case studies could be of relevance to any health care professional.

Case study 1

Mr Robinson is a 54-year-old patient who you see on a regular basis about once every 3 months. Since you last saw him, Mr Robinson has had a mild stroke, from which he seems to be recovering well. He tells you that his doctor at the hospital has put him on a new drug 'to thin his blood' and told him that he must not drink alcohol.

What can you tell Mr Robinson about the drug he is taking and how can you explain to him the importance of not drinking with this medication? Is there anything else you could advise him about while taking the drug?

Case study 2

You are seeing a patient who is due to have minor surgery. Ms Clarke is 32 and apart from the minor surgical problem, she is fit, active and apparently healthy. However, she tells you she is really worried about the operation because she had a bad experience with an operation when she was a child. Ms Clarke cannot quite remember the details, but she says that when she came round from the previous operation she couldn't move and can remember being scared.

What do you think might have happened to Ms. Clarke after the previous operation? How can you reassure her that this will not happen again?

Chapter review questions

You should be able to answer these review questions from the material in this chapter.

1. Review the advantages and disadvantages of oral administration and intravenous injection of drugs.

2. Discuss the similarities and differences of simple diffusion, facilitated diffusion and active transport of drug molecules across cell membranes.

3. Explain the significance of pH in diffusion of drug molecules.

4. Explain the importance of plasma protein binding to absorption of a drug from the gastrointestinal tract, its distribution around the body and its excretion by the kidneys.

5. Review the factors that affect distribution of drugs around the body.

6. Review the factors that can affect metabolism of drugs.

7. Discuss the influence of glomerular filtration, tubular secretion and tubular reabsorption on drug excretion.

8. Explain why making the tubular filtrate alkaline would speed up the elimination of barbiturates.

9. Explain the implication of travellers' diarrhoea for someone using oral contraceptives. Describe the mechanism by which the problem arises.

3

Effects of drugs on the body

3.1 Chapter overview

In medicinal use, drugs are usually administered with the best of intentions, which is to have a beneficial therapeutic effect. However, all drugs have the potential to cause unwanted effects too, and these range from being minor or trivial to being life threatening. Unwanted effects are known as *adverse reactions* or side effects and can occur by a variety of mechanisms. It is important for health care professionals to be aware of potential adverse effects of drugs and to know how to report new ones when they occur.

Apart from the possibility of adverse drug reactions occurring, individuals show variation in response to drugs and this is particularly likely in the very young and the very old.

In order to have an effect on the body, therapeutic, adverse or otherwise, a drug must interact with its site of action. Sites of action are known as *targets* for drug action and include receptors, ion channels, enzymes and carrier molecules.

The study of the effects that drugs can have on the body and how they produce these effects is known as *pharmacodynamics*.

3.2 Adverse reactions to drugs

An adverse reaction is any reaction to a drug that is harmful to the patient. Side effects, unwanted effects and adverse effects are all adverse reactions; some are considered to be less serious than others, although the terms are used synonymously.

Adverse reactions can occur with any drug use and can be either related to or unrelated to the expected pharmacological effect. Variation in effect of a drug can occur in different individuals and indeed in the same individuals on different occasions. This can result from either differing concentrations of a drug at its site of action, or differing physiological responses to the same concentration of a drug.

Adverse drug reactions present a major clinical problem. It has been estimated that up to 5% of all admissions to hospital result from adverse reactions to drugs and in hospital up to 20% of patients experience an adverse reaction.

Pharmacology for the Health Care Professions   Christine M. Thorp
© 2008 John Wiley & Sons, Ltd

Many adverse reactions are well known. Information about them can be found in patient information leaflets supplied with medicines, in the *Monthly Index of Medical Specialties* (*MIMS*) or in the *British National Formulary* (*BNF*). However, not all patients will suffer all possible adverse reactions. Knowledge of family history can help predict who will suffer from adverse reactions. It is therefore good practice when taking a patient's history to ask about incidents of adverse reactions to drugs or allergies in the family.

There is a 'yellow card scheme' to notify the Medicines and Healthcare Products Regulatory Agency (MHRA) or the Commission on Human Medicines (CHM; formerly the Committee on Safety of Medicines) of unexpected adverse reactions. This is especially important for new drugs as adverse reactions can be missed during clinical trials. Patients as well as health care professionals can use the yellow card scheme. A new system of reporting adverse reactions, the National Reporting and Learning System (NRLS) has recently been set up by the National Patient Safety Agency. Any health care professional who becomes a supplementary prescriber or an independent prescriber would be expected to contribute to these schemes.

Adverse reactions to drugs can be divided into type A (augmented) and type B (bizarre).

3.2.1   Type A adverse drug reactions

These adverse drug reactions are generally dose-related and most can be predicted. They can be caused by an exaggeration of a drug's intended pharmacological effect or by an unwanted action or side effect. Type A adverse reactions are most likely to occur with drugs that have a steep dose–response curve (see Figure 3.1) and/or small therapeutic ratio (TR).

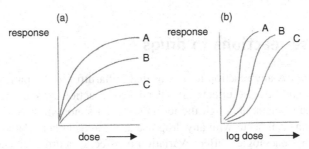

(a) relationship between drug dose (concentration) plotted against response to the drug for three drugs, drug A, drug B and drug C

(b) relationship between log dose plotted against response to drug
log dose curves are plotted because they are easier to interpret: at the steepest point, small changes in drug concentration result in large changes in response; drug A is more likely to produce an adverse reaction than drug B or C

Figure 3.1   Dose–response curve

The TR of a drug is the ratio of the maximum non-toxic dose to the minimum dose that has a therapeutic effect and is essentially a measure of a drug's safety margin.

$$TR = \text{maximum non-toxic dose/minimum effective dose}$$

The most dangerous drugs are those with a small TR where the maximum non-toxic dose is very close to the effective dose. Their use cannot be justified for trivial illnesses but they may be used for life-threatening conditions when the benefits outweigh the risks. Drugs with a TR of one or less cannot be used therapeutically.

3.2.1.1 Overdose

A type A adverse drug reaction is a possibility if a patient is given a higher than recommended dose for any particular route of administration due to practitioner error. In order to avoid this it is good clinical practice for drug dosages to be checked by another member of staff prior to administration. Where this is not possible, or impracticable, extreme care must be taken to ensure that a patient receives the right dose of the correct medicine by the route intended.

Overdose is also possible if the correct dosage of a drug is administered via the wrong route. For example, a local anaesthetic injected into a blood vessel rather than into the tissues produces a rapid rise in blood level and this increases the risk of unwanted effects of the drug.

3.2.1.2 Age

Age can affect the response to drugs, because in the old and the very young metabolism and excretion are not as efficient in comparison to the young healthy adult. Distribution can also be affected due to differences in body composition and the availability of plasma proteins for binding. At these extremes of life, drugs tend to produce greater and more prolonged effects (see page 36 onwards).

3.2.1.3 Disease

Many disease states can cause individual variation in response to drugs. Any disease that results in alteration in the pharmacokinetics of a drug will create these variations. Diseases of the liver and kidney, any disease that affects intestinal motility, mal-absorption syndromes and any condition that reduces plasma protein concentration are all implicated. Some diseases can alter the physiological sensitivity to a drug at its site of action.

3.2.1.4 Genetics

The way in which the body handles drugs can show genetic variation. Quite often, this is due to genetic differences in enzyme activity. This can lead to differences in the rate at which a drug is metabolized and therefore after a given period of time, plasma levels will be different in different individuals.

3.2.1.5 Drug–drug interactions

The administration of more than one drug simultaneously can potentially alter the actions of any of them. These are known as *drug–drug interactions* and there are many possible mechanisms. Some of these are summarized in Table 3.1.

Clinically, drug interaction only becomes important when a drug has a small TR.

If warfarin and aspirin are used simultaneously, both drugs appear in the plasma in higher than expected concentrations due to competition for plasma protein binding. This will increase the pharmacological activity of both drugs, but the action of warfarin is of most importance.

Drug interaction is not always an adverse reaction. In hypertension, for example, the additive effects of multiple drug therapy is often necessary to achieve a reduction in blood pressure.

A summary of type A adverse reactions is given in Table 3.2.

3.2.2 Type B adverse drug reactions

Type B adverse drug reactions are much rarer than type A, but they are unpredictable and not dose-related and they are potentially more serious. Many are due to drug allergy, but there are other causes.

Table 3.1 Mechanisms of drug–drug interactions

Mechanism	Effect	Example
Potentiation	Actions of drugs with similar actions are additive	Central nervous system depression with benzodiazepines and alcohol
Opposite effects	Drugs with opposite effects cancel out each other's activity.	β receptor stimulants (bronchodilators) and β receptor blockers (antihypertensives)
Altered absorption	Absorption can be increased or decreased by the actions of another drug	Opiates slow intestinal transit time and increase absorption of other drugs
		Tetracycline can reduce absorption of iron salts
Competition for protein binding	Drugs compete for the same plasma protein binding sites	Warfarin and aspirin displace each other from binding sites
Enzyme induction	Drug induces increased activity of enzymes [a]	Alcohol induces enzymes that metabolize warfarin
Enzyme inhibition	Drug inhibits activity of enzymes[a]	Cimetidine (for stomach ulcers) inhibits enzymes that metabolize warfarin
Altered excretion	Competition for transport systems in kidney	Elimination of methotrexate (cancer chemotherapy) inhibited by probenecid (for gout)

[a] See Chapter 2 for further explanation and examples.

Table 3.2   Type A adverse drug reactions

Cause	Effect
Overdose	Incorrect dosage/correct dose, wrong route; causes increased plasma levels
Age – old and very young	Metabolism and excretion less efficient Distribution varies with different body composition
Disease	Pharmacokinetics altered in disease states, especially of liver and kidney
Genetic variation	Different enzyme activity
Drug-drug interaction	Increased/reduced effect of one or more drugs

3.2.2.1   Idiosyncrasy

Idiosyncratic reactions occur when the patient experiences an effect unrelated to the expected action of the drug. The effect is usually harmful and occurs in a very small proportion of individuals. It is due to factors peculiar to the individual that may be genetic in origin, but the mechanisms are usually poorly understood.

3.2.2.2   Insensitivity and intolerance

The magnitude of the response to a given dose of a drug in a population of patients shows a normal distribution. The vast majority, approximately 99% of responses, lie within acceptable limits. A small percentage of patients (less than 0.5%) do not respond to the drug. These individuals are insensitive or unresponsive to the drug. An equally small percentage responds in an exaggerated manner. These patients are described as intolerant. They exhibit the expected response to the drug but in a greater magnitude than would be acceptable. See Figure 3.2.

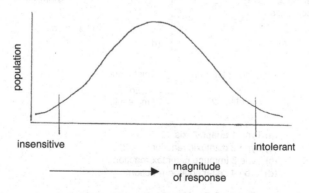

insensitive subjects show little response to a drug at any given dose
intolerant subjects show an exaggerated response to a drug at any given dose

Figure 3.2   Distribution of response to drug action

3.2.2.3 *Hypersensitivity*

Hypersensitivity reactions to drugs involve immunological reactions. Large molecules such as vaccines, insulin and dextrans can provoke immune reactions themselves but most drugs are too small to be antigenic on their own. In some patients, something that cannot be predicted is that a drug or drug metabolite can combine with tissue proteins to form an antigenic conjugate. In this situation, the patient shows a response, which can be described under one of the classical definitions of allergic response (that is, type 1–type 4, see below). Such reactions require initial exposure to the drug to cause sensitization, after which, subsequent exposure to the same drug triggers an immunological reaction. Although drug allergy is unpredictable, it is more likely to occur in individuals with a history of atopic disease, for example asthma.

3.2.2.3.1 *Type 1 anaphylaxis* This is the most common allergic reaction. See Figure 3.3a.

The allergen, whether it is a drug, pollen or dust mite, causes production of immuno-globulin E (IgE) antibodies on first exposure. These antibodies attach themselves to mast cells. On subsequent exposure to the same allergen the combination of IgE and allergen causes the mast cells to release a variety of chemicals including histamine. The effects of this can be local, for example hay fever, asthma and urticaria, or systemic causing whole body oedema and anaphylactic shock.

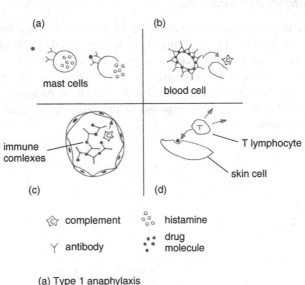

(a) Type 1 anaphylaxis
(b) Type 2 cytotoxic reaction
(c) Type 3 immune complex reaction
(d) Type 4 cell-mediated reaction

Figure 3.3 Hypersensitivity reactions

An example of a drug that is known to cause type 1 allergic reactions is penicillin. Radiological contrast agents can also provoke this type of allergic reaction.

3.2.2.3.2 Type 2 cytotoxic reaction   Some drugs bind to blood cell membranes. See Figure 3.3b.

This makes the blood cells antigenic and results in the production of IgG antibodies against them. The antibodies so produced activate the complement system. Complement destroys the blood cells. This can lead to haemolytic anaemia if red blood cells are involved and can occur in response to penicillin. If white blood cells are the target of the reaction then the result is agranulocytosis; this can occur for example with carbimazole (used to treat an overactive thyroid gland, see Chapter 6). Thrombocytopenia can be the result if the cells involved are platelets; this can happen in response to heparin (used in thrombotic disorders, see Chapter 4).

3.2.2.3.3 Type 3 immune complex reaction   Other drugs can form immune complexes with antibodies, IgG or IgM, which circulate in the blood and can be deposited in particular areas of the body, for example in the joints, skin or kidneys. See Figure 3.3c. Here the immune complexes can cause a local inflammation by activation of the complement system.

This type of allergic reaction can occur with penicillin.

3.2.2.3.4 Type 4 cell-mediated reaction   Some drugs can combine with proteins in the skin to form antigens. See Figure 3.3d. When this occurs T lymphocytes are activated and these cause damage to skin cells resulting in rashes, lumps and itchy weeping skin. This type of allergic reaction is possible in response to local anaesthetics.

3.2.3   Other adverse drug reactions

Teratogenesis and carcinogenesis can be considered as adverse reactions to drugs.

Teratogenesis is the occurrence of foetal developmental abnormalities caused by drugs being taken during pregnancy, usually in the first trimester. Most drugs cross the placenta to some extent and should be avoided during pregnancy (see page 39). Known teratogens include alcohol, anticancer drugs, warfarin, anticonvulsants and tetracyclines.

Carcinogenesis is the occurrence of drug-induced tumours. This can happen through damage to DNA causing mutation of genes, activation of proto-oncogenes or damage to oncogene suppressors (see Chapter 10). Due to strict testing of new drugs during the developmental stages this should be a rare phenomenon. However, some drugs through immunosuppression, are known to increase the risk of tumour formation. For example, drugs used to treat cancers and drugs used to prevent transplant rejection are known to do this.

3.3   Variation in response to drug therapy

In Chapter 2, we have seen that responses to drug therapy depend on the four processes of absorption, distribution, metabolism and excretion and there are many factors that affect these four processes.

It is important to realize that these are principles that have been derived from standard studies of drug action on 'average' or 'normal' individuals during the development process for new drugs.

Drug development involves investigation of drug action in order to determine suitable and effective dosage regimes, to allow likely adverse effects to be determined and to ensure safety in use as far as possible.

Clinical trials are usually done on young, healthy male volunteers. Obviously, in therapeutic use drugs are taken by many different sectors of the general population – least likely to be young, healthy men. Drug trials therefore continue after a drug is on the market through monitoring by the MHRA of the first patients using new drugs and through the 'yellow card' scheme of reporting adverse reactions thereafter.

It is possible to predict, and to some extent avoid, likely adverse effects in particular populations using knowledge of factors that affect the four processes above in different patient groups.

3.3.1   Drugs and older people

The elderly are known to suffer a higher frequency of adverse effects to medication and the reasons are many and varied. Knowledge of the possible reasons, together with consideration of the problems and changes caused by the ageing process, means that many of the adverse affects could be avoided or at least reduced in intensity. The lower the TR of a drug, the more important these considerations become.

3.3.1.1   Absorption

Drug absorption can be altered in older people for a number of reasons. The ability of parietal cells in the stomach to produce hydrochloric acid reduces with age. This may result in increased gastric pH, which could cause a slight delay in the absorption of orally administered acidic drugs.

The passage of food from the stomach to the duodenum becomes less efficient with age, increasing gastric emptying time. This may cause a slight delay in the absorption of drugs in the intestine.

With ageing, there is also reduced gastrointestinal blood flow, which can further delay absorption of drugs from the gastrointestinal tract. Alteration in absorption can also make older people more vulnerable to the ulcerogenic effects of some drugs, for example non-steroidal anti-inflammatory drugs.

3.3.1.2 Distribution

Distribution of drugs around the body is likely to be different in older people because of changes in body composition, plasma protein concentration and blood flow to major organs. With ageing, body composition changes such that there is more body fat with respect to total body weight, and the proportion of body water falls. The importance of this is that highly lipid-soluble drugs, for example benzodiazepines (antianxiety drugs), tend to have prolonged duration of action.

Plasma protein concentration also falls with age. This means that for extensively protein bound drugs there will be a greater concentration of free drug than expected. This could lead to increased incidence of side effects. This is the case with pethidine (an analgesic).

After the age of 30, cardiac output, renal blood flow and hepatic blood flow all begin to decline. Blood flow in other organs tends to remain constant. This could alter expected drug distribution.

3.3.1.3 Metabolism

Age-dependent changes in the liver have the most impact on drug metabolism. Over the age of 70 liver blood flow may only be 40–45% of peak levels. This decrease may be further reduced by concomitant pathologies such as left ventricular dysfunction or congestive heart failure, both of which are more commonly found in older patients.

The metabolism of many drugs is directly proportional to the rate of blood flow through the liver; that is they exhibit flow-dependent hepatic clearance.

The size of the liver also reduces with age in both absolute and relative terms. Thus, the number of functioning hepatocytes is reduced, and therefore so is the capacity to metabolize drugs.

The process known as *first pass metabolism* is drastically reduced with ageing. Drugs that would normally undergo this process, achieve far higher blood levels in older patients than would otherwise be expected. This is of particular significance in drugs administered orally and the oral dosage of such drugs must be reduced in older people.

Older patients appear to be more at risk of developing drug-induced hepatitis particularly with the use of non-steroidal anti-inflammatory drugs.

3.3.1.4 Excretion

The principal organ of excretion for the majority of drugs is the kidney. The physiological functioning of this organ declines with age. At 80 years of age the glomerular filtration rate is approximately half that at age 25. This means that drugs that are mainly excreted via the kidney are liable to accumulate in older patients. If these are drugs with a low TR, failure to excrete them at the expected rate may result in toxicity.

In addition, patients with reduced renal function may also show increased sensitivity to some drugs, decreased sensitivity to others and tolerate side effects less well.

3.3.1.5   Drug use

The number of drug prescriptions issued increases with increasing age of patients. Taking older people to be over 65, around 70% of them are taking one or more drugs on a regular basis. In the United Kingdom, older people account for approximately 16% of the population yet they consume 50% of the National Health Service drug budget. This is partly because of the increasing range of disorders that can be medically treated. In addition to this, many older patients use self-medication with non-prescribed remedies. It has been estimated that on average, for each prescription item older people receive, they use two additional non-prescription items. This increased drug usage in the elderly can give rise to a number of problems.

3.3.1.5.1   Drug–drug interaction   This occurs when two or more drugs, being taken concurrently, produce a pharmacological response not attributable to a single drug. (See Table 3.1 and page 32).

The practice of multiple drug prescribing is known as *polypharmacy*. The more drugs being taken together the greater the risk of drug–drug interactions becomes. For patients taking 2–5 drugs the estimated incidence of drug interactions is between 3 and 5% whereas with 10–20 drugs the estimated incidence rises to 20%.

Not all patients show clinical symptoms of drug–drug interactions. Only about 7% of potential interactions clinically manifest themselves. Factors that determine those individuals who will show symptoms of drug interaction are unknown.

The principal groups of drugs that are most likely to cause drug–drug interactions are anticoagulants, oral hypoglycaemics (for type 2 diabetes), monoamine oxidase inhibitors (for depression), cytotoxics (chemotherapy) and digitalis-like drugs (for heart disease).

3.3.1.5.2   Drug–patient interaction (adverse drug reactions)   These may be due to overdose, genetics, idiosyncrasy, insensitivity, intolerance or hypersensitivity (see page 31). Adverse reactions are observed more frequently in older people due to increased drug use in this sector of the population.

3.3.1.5.3   Compliance   A definition of compliance in this context is the extent to which the behaviour of the patient coincides with the clinical prescription. Although miscompliance is an avoidable error, as many as 50% of older people may not be taking their medicines as intended.

Factors affecting compliance in older people are many and include the complexity of multiple therapy, possible difficulty in understanding instructions, ambiguous or unclear patient information or labelling, preconceived ideas about medicines and their benefits, previous bad experience, difficult dosing regimes, unpalatable formulations and real or imagined side effects.

3.3.1.6   Altered pharmacodynamics

As we get older, there are physiological changes that can alter the effect a drug has on the body.

This is difficult to prove as it involves the increase in the sensitivity of drug receptors with age. It seems to be of importance with drugs such as the benzodiazepines and warfarin. In addition existing organ pathology such as renal or hepatic disease can affect drug response (for example with diuretics or probenecid).

3.3.2 Drugs and infants and children

Most drugs are not developed or evaluated in children. Children cannot be regarded as simply smaller versions of adults with respect to drug use for a number of reasons.

Absorption may be slower because gastrointestinal function is different in children. The percentage of body water is higher in children than in adults and children have a lower percentage of body fat. In very young children, the concentration of plasma proteins is low compared with that in adults. These factors will affect drug distribution. An infant's liver is unable to metabolize certain drugs; for example, inability to metabolize chloramphenicol resulted in 'grey baby' syndrome. Up until at least the age of 7 months, the kidneys are not as efficient (size for size) as an adult's, so drug elimination may be reduced.

Due to the existence of these factors, it is unsatisfactory simply to adjust doses for children because of weight alone. For the majority of drugs that have a low TR dose adjustment is best made in relation to metabolic rate, which can be estimated from body surface area. However, specialist knowledge is needed in this area. A special edition of *BNF for Children* is now produced.

Unexpected reactions to drugs can occur in children. For example, antihistamines normally cause sedation as a side effect but can cause hyperactivity in some children.

3.3.3 Drugs and pregnancy

The developing foetus is very susceptible to adverse effects of drugs; often much more so than the mother. All drugs, given in sufficient doses, are capable of passing from the mother via the placenta to the foetus. This gives rise to a number of potential effects including death of the foetus and/or spontaneous abortion, malformation of the foetus, possible modification of post-natal behaviour or intelligence or even development of malignancy in later life of the foetus.

There is a major risk of malformation up to eight weeks post conception; therefore, drugs are best avoided for the first trimester of pregnancy. However, pregnancy may not be confirmed until towards the end of this period. Before drugs are given or other potentially harmful procedures undertaken, it should always be determined whether a woman might be pregnant.

Particular groups of drugs known to be potentially harmful to the foetus in the first trimester include male and female hormones, iron preparations, aspirin-like drugs, alcohol and barbiturates.

These are only examples – many others exist. Drugs may also affect the foetus at later stages of pregnancy so the best advice is to avoid drug use during pregnancy unless absolutely necessary.

In cases of chronic maternal disease, the risks of stopping drug therapy have to be balanced against the benefits to the developing foetus. For example, it would be dangerous to cease treatment for chronic conditions such as asthma or epilepsy. In these situations, drugs least likely to harm the foetus should be used.

3.3.4 Drugs and breast feeding

Usually drug therapy does not interfere with breast feeding but it is possible for effective amounts of drug to pass into the milk, particularly if the drug behaves as a weak base. This is because breast milk generally has a slightly acid pH and in this environment, basic drugs would be ionized and therefore likely to be trapped in breast milk.

Many drugs are known to appear in breast milk and are listed in the *BNF* with advice about their use during breast feeding. Examples of some drugs that can have an effect on a nursing infant are given in Table 3.3.

3.4 Targets for drug action

Targets for drug action are the sites where a drug interacts with part of a body cell or other body component. These include receptors in cell membranes or within the cell itself, ion channels and carrier proteins in cell membranes and enzymes in body fluids.

3.4.1 Receptors

Receptors are either an integral protein in the cell membrane (type 1, 2 and 3 below) or a protein in the cytoplasm or the nucleus (type 4). They form the site where the chemical messengers of the body, hormones and neurotransmitters, interact with cells. Many drugs

Table 3.3 Drugs that appear in breast milk and effect on nursing infant

Drug	Used for	Effect on nursing infant
Phenobarbital	Epilepsy	Drowsiness, depression of suckling reflex
Chlorpromazine	Psychosis	Drowsiness
Lithium	Mania	Hypotonia, hypothermia, cyanosis
Theophylline	Asthma	Irritability
Laxatives	Constipation	Diarrhoea
Warfarin	Coagulation disorders	Haemorrhage

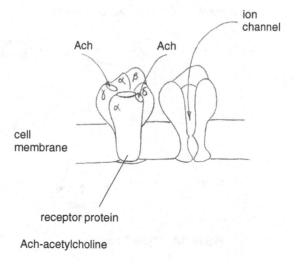

Ach-acetylcholine

Figure 3.4 Type 1 receptor

act either as agonists or antagonists of these endogenous receptors. Agonists are drugs that stimulate receptors and antagonists block the action of the natural chemical at the receptors. There are four main receptor types based on the mechanism by which they exert their effects.

3.4.1.1 Type 1 receptors

Type 1 receptors are coupled directly to an ion channel.

The nicotinic acetylcholine receptor is the most studied of this type and is considered to be typical. This receptor is found at the neuromuscular junction. It is a protein consisting of five subunits, two α and one each of β, γ and δ, which are arranged in the membrane around the channel. (See Figure 3.4.)

Acetylcholine binds to the α subunits. Thus, each receptor has two binding sites and both must be occupied for the receptor to be activated. Receptors for some other neurotransmitters, for example gamma-aminobutyric acid (GABA) and glutamate, are believed to be similar. This type of receptor controls the transient increase in postsynaptic membrane permeability to a particular ion at the synapse. This is a rapid event, occurring in milliseconds, which indicates that the receptor is directly connected to the channel.

3.4.1.2 Type 2 receptors

These are receptors for many hormones and neurotransmitters and are coupled to effector systems by a G-protein.

The β-adrenoreceptor was the first of this type to be characterized and others seem to be similar. The structure is a single polypeptide chain spanning the membrane, folded into seven transmembrane domains and coupled to a G-protein. (See Figure 3.5.)

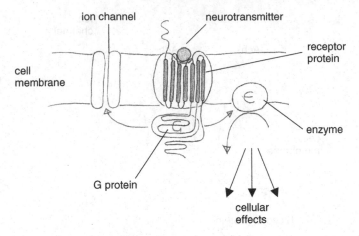

Figure 3.5 Type 2 receptor

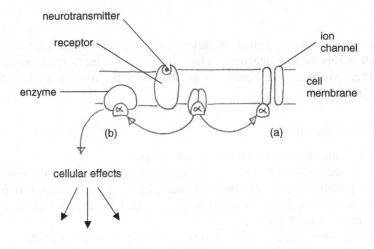

(a) regulation of an ion channel

(b) cellular effect through activation of second messenger system

α - a subunit combined with GTP (see text)

Figure 3.6 G-Protein

G-proteins are found on the inside of the membrane and act as transducers between receptors and cellular effectors. They are made up of three subunits (α, β, γ) one of which (α) has GTPase (guanine triphosphatase, an enzyme) activity.

When a neurotransmitter or hormone binds with its receptor, the α subunit of the G-protein combined with GTP (guanine triphosphate) detaches and activates an effector in the cell (see Figure 3.6).

The process stops when the GTP has been hydrolysed to GDP (guanosine diphosphate) and the subunit recombines with the others. A single neurotransmitter or hormone molecule can activate several G-protein molecules so amplification of effect is achieved. Activation of effector mechanisms results in a product being produced. The product is often (i) regulation of an ion channel or (ii) a second messenger. Second messengers are intracellular molecules that bring about the effect of the original neurotransmitter or hormone through activation of cellular enzymes.

3.4.1.3 Type 3 receptors

These are receptors for insulin and other growth factors, which are directly linked to tyrosine kinase (an enzyme).

Type 3 receptors have large extracellular and intracellular domains. See Figure 3.7.

The extracellular domain is the binding site for hormone and the intracellular domain includes tyrosine kinase. The binding of hormone results in activation of tyrosine kinase. Tyrosine kinase then activates cellular enzymes, which in turn stimulate transcription of particular genes. This brings about the cellular response to the original hormone, for example growth.

3.4.1.4 Type 4 receptors

These are receptors for steroid hormones and thyroid hormone.

Type 4 receptors are quite different to the other three types, being located in the nucleus or the cytoplasm of the cell. (See Figure 3.8.)

The hormone first has to enter the cell. Steroid hormones and thyroid hormones pass easily across the cell membrane, as they are lipid soluble. Once the hormone binds to its

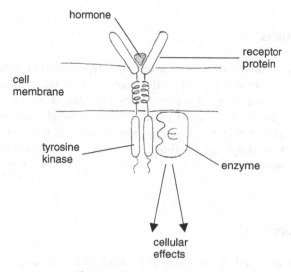

Figure 3.7 Type 3 receptor

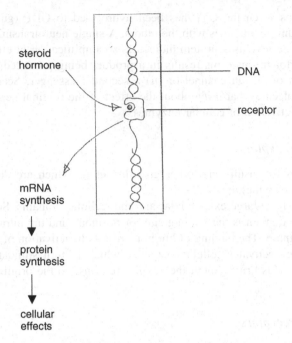

steroid
hormone

DNA

receptor

mRNA
synthesis

protein
synthesis

cellular
effects

Figure 3.8 Type 4 receptor

receptor, the receptor is thought to unfold exposing a DNA-binding domain. The receptor molecule then binds to a particular region of DNA and activates certain genes. The result is increased protein synthesis that mediates the cellular response.

3.4.2   Ion channels

Some ion channels are linked directly to a receptor (type 1 above), others are activated by G-proteins (type 2 above) and drugs that interact with these receptors will affect the functioning of the channels. However, drugs can also affect channel function by binding directly to parts of the channel protein. Some drugs physically block channels. For example, local anaesthetics block sodium ion channels in neuronal cell membranes (see Chapter 12). Other drugs can modulate ion channel function either by inhibiting or facilitating the opening of the channel, for example calcium channel blockers (see Chapter 4).

3.4.3   Carrier proteins

Carrier proteins are involved in the transport of many molecules across cell membranes. Neurotransmitter carrier mechanisms are of particular interest since many drugs have been developed that interfere with this process. Carrier proteins can be inhibited by drugs or persuaded to pick up false transmitters. For example, selective serotonin reuptake inhibitors (SSRIs, antidepressants, Chapter 11) inhibit membrane carriers for serotonin, and methyldopa (antihypertensive, Chapter 4) is taken up into neurons by the carrier for dopa and subsequently converted to the false transmitter methylnoradrenaline.

3.4.4 Enzymes

Enzymes are the targets for many drugs. Most commonly, drugs inhibit the enzyme competitively, which may be either reversible or irreversible. Aspirin, for example, inhibits an enzyme called *cyclo-oxygenase*. This prevents the formation of certain prostaglandins that would otherwise cause inflammation and pain. Other drugs act as false substrates for the enzyme so that an abnormal product is made instead of the normal one. The abnormal product then brings about the action of the original drug. Methyldopa can be used again as an example here. It is a false substrate for the enzyme that normally converts dopa into dopamine.

Some drugs need conversion by an enzyme from an inactive pro-drug into an active form before they can be of therapeutic use.

Methyldopa, above, could be said to be a prodrug and methylnoradrenaline the active drug.

3.5 Summary

Drugs are capable of producing adverse reactions as well as beneficial therapeutic effects.

Adverse reactions can occur by a variety of mechanisms and some of them are well known. Type A adverse reactions are dose-related and can be predicted to occur through accidental overdose, individual differences due to age, disease or genetics or drug–drug interactions.

Type B adverse reactions are rarer but potentially more serious. Many type B adverse reactions are due to allergy.

Variation in response to drug therapy can occur due to many factors. At extremes of life absorption, distribution, metabolism and excretion can all be different to normal so that drug action becomes unpredictable. In particular, plasma protein binding, liver and kidney function and the number of drugs being taken concurrently become important in older patients.

Special care is needed when prescribing for infants and children and drug use should be avoided if possible in pregnancy and breast feeding.

In order to have a therapeutic effect, a drug has to interact with a receptor or other site of action in the body. Receptors are divided into four types according to their location, structure and effects when activated. Types 1, 2 and 3 are found in cell membranes. Type 4 receptors are found in the cell cytoplasm or nucleus. Ion channels, carrier proteins and enzymes can also be targets for drug action.

Useful websites

www.askaboutmedicines.org General information about medicines and useful links to other sources of information.

www.emc.medicines.org.uk Electronic Medicines Compendium. Patient information leaflets and summaries of product characteristics.

www.emea.europa.eu/ European Medicines Agency. European Agency for the Evaluation of Medicines.

http://merck.com/mmhe/sec02.html The Merck Manuals Online Medical Library; Section on drugs.

www.mhra.gov.uk Medicines and Healthcare Products Regulatory Agency.

www.npsa.nhs.uk National Patient Safety Agency.

www.yellowcard.gov.uk Yellow card scheme for reporting adverse drug reactions for health care professionals and patients.

Case studies

The following two case studies are hypothetical, but any health care professional should be able to provide professional advice to patients in such situations.

Case study 1

You are seeing a patient, Annie Brown, aged 25 who is pregnant. Annie tells you that her doctor has said that she must not take any medication during her pregnancy without checking with the doctor first. Annie is a little puzzled by this but says she did not have time to talk to the doctor about it. She asks you to explain.

Discuss how you would address Annie's concerns.

Case study 2

Mrs Howe, aged 85, is a patient that you visit at home. She has been back at home for two days after being in hospital with a broken hip sustained after a fall. For the time being Mrs Howe is mostly bedridden while she recovers further. Before leaving the hospital, Mrs Howe's consultant prescribed several drugs, including analgesics and laxatives.

Despite her age and current frailty, Mrs Howe is mentally very alert but she is worried about how she will remember to take all her drugs at the right time. What advice can you offer her about how to make sure she remembers when to take the various drugs she has been prescribed? Considering she is taking so many drugs, what else do you think it might be helpful to discuss with Mrs Howe?

Mrs Howe also tells you that she has noticed an itchy skin rash that was not there before she left hospital. She wonders whether it could be due to the complex mixture of drugs she is taking and asks you what she should do about it. What should you tell her?

Chapter review questions

You should be able to answer the following review questions from the topics covered in the preceding chapter.

1. Where can you find information about adverse reactions to drugs?

2. Why should health care workers look out for potential allergic reactions in patients and how might you predict if a patient is likely to have an allergic reaction to a particular drug?

3. Discuss the reasons why absorption, distribution, metabolism and excretion of drugs are likely to be altered in older people.

4. Explain why adverse reactions to drugs occur more frequently in older people than in the general population.

5. Children have a different body composition to adults. How does this affect the way they handle drugs?

6. Briefly describe the four main types of receptors that can be targets for drug action.

Part II
Systemic pharmacology

4

Cardiovascular and blood disorders

4.1 Cardiovascular disorders

It is likely that many older patients will have one or more cardiac disorders and possibly hypertension as well. This is a huge area of therapeutics and there are many drugs available to treat these conditions. Common heart disorders are cardiac failure, ischaemic heart disease and disorders of heart rhythm (arrhythmias). Hypertension is included here because many of the drugs used to treat heart disorders are also used in hypertension.

A basic knowledge of the physiology of the heart, the mechanism of contraction, electrical activity and control of rate and rhythm, maintenance of blood pressure and what can go wrong is needed in order to understand how the drugs work.

4.1.1 Cardiac physiology

The heart consists of four chambers: two atria and two ventricles and is divided into right and left sides. (See Figure 4.1.)

The walls of the atria and ventricles consist of cardiac muscle, with that of the left ventricle being the thickest. Collectively cardiac muscle is known as the *myocardium*.

The heart functions as a double pump to circulate blood through the lungs and round the rest of the body.

The circuit of blood through the heart is as follows.

Blood enters the right atrium from the superior and inferior vena cavae. The atrium contracts and from here the blood passes through the tricuspid valve to the right ventricle. The ventricle contracts and blood flows through the pulmonary valve and pulmonary arteries to the lungs. Blood comes back from the lungs in the pulmonary veins into the left atrium. The atrium contracts and blood passes through the bicuspid valve into the left ventricle. As the ventricle contracts blood leaves via the aorta passing through the aortic valve. Blood flow through the heart and round the body is maintained by contraction of the atria in unison followed by contraction of the ventricles in unison. The pulmonary and aortic valves close while the atria contract; the tricuspid and bicuspid valves close while the ventricles contract. This prevents backflow of blood. The contraction phase is known as *systole* and the relaxation phase of the heart is known as *diastole*.

Pharmacology for the Health Care Professions   Christine M. Thorp
© 2008 John Wiley & Sons, Ltd

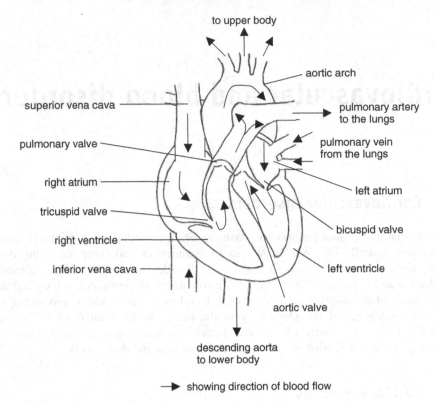

Figure 4.1   Structure of the heart

Cardiac output is the volume of blood pumped out of the left ventricle every minute and is a function of the stroke volume and heart rate. Stroke volume is the amount of blood ejected by the left ventricle during one contraction and the heart rate is the number of beats per minute. For an average person at rest, stroke volume is around 70 ml and heart rate is about 75 beats per minute. An average cardiac output at rest is therefore (70 × 75) or about 5.25 l per minute. Cardiac output is not constant and has to vary to meet changes in the body's demand for oxygen, for example during exercise. Normally, the force of contraction of cardiac muscle depends on its resting length. End diastolic volume (EDV) is the volume of blood in the ventricles just before they contract. If EDV increases, cardiac muscle is stretched and this produces an increase in the force of contraction. This will increase cardiac output. EDV depends on preload, which is determined by venous return and filling of the ventricles. During exercise preload increases and therefore cardiac output increases. Cardiac output also depends on end systolic volume (ESV), which in turn depends on afterload. Afterload is a function of total peripheral vascular resistance and blood pressure. If afterload becomes greater, ESV increases and cardiac output is reduced. As this happens, the EDV also increases and it follows that there is an increase in the force of contraction to restore cardiac output. In either case, the increase in force of contraction results in an increase in cardiac work and oxygen consumption.

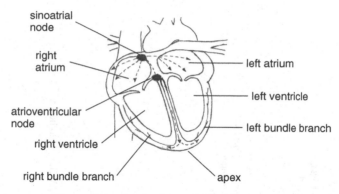

Figure 4.2 Conduction pathways in the heart

The basic rhythm of heart rate is maintained by a part of the heart known as the *pacemaker* or sinoatrial (SA) node. This is a group of cardiac muscle cells in the right atrium that depolarize spontaneously. Depolarization of cardiac muscle cells brings about contraction. The pacemaker forms part of a conduction system that transmits electrical activity through the heart by means of action potentials so that contractions are coordinated and the heart can function as an efficient pump. The conduction pathway of electrical activity goes from the SA node to the atrioventricular (AV) node between atria and ventricles, then down the septum between the two ventricles to the apex of the heart and finally round the ventricles themselves. The conduction pathway in the heart is shown in Figure 4.2.

Electrical activity in the heart can be picked up by electrodes placed on the skin and recorded as the familiar electrocardiogram (ECG). The ECG is a record of the sum of all action potentials in the heart as it contracts. Action potentials are generated by depolarization followed by repolarization of the cardiac muscle cell membrane. Depolarization is initiated by an influx of sodium ions into the cardiac muscle cells, followed by an influx of calcium ions. Repolarization is brought about by efflux of potassium ions. The phases of a cardiac action potential are shown in Figure 4.3 where the depolarization is the change in resting membrane potential of cardiac muscle cells from $-90\,mV$ to $+20\,mV$. This is due to influx of sodium ions followed by influx of calcium ions. Contraction of the myocardium follows depolarization. The refractory period is the time interval when a second contraction cannot occur and repolarization is the recovery of the resting potential due to efflux of potassium ions. After this the cycle repeats itself.

A typical ECG is shown in Figure 4.4, where the P wave corresponds to atrial depolarization and contraction; the QRST complex corresponds to atrial repolarization and relaxation and the onset of ventricular depolarization and contraction; and the T wave represents ventricular repolarization and relaxation. Analysis of the size and shape of the waves can indicate abnormalities of the heart.

Heart rate can be altered by the autonomic nervous system to meet changes in demand for blood supply. Sympathetic stimulation increases the rate and force of contraction in response to exercise, threats and emotions. Parasympathetic stimulation reduces heart rate and force of contraction under resting conditions.

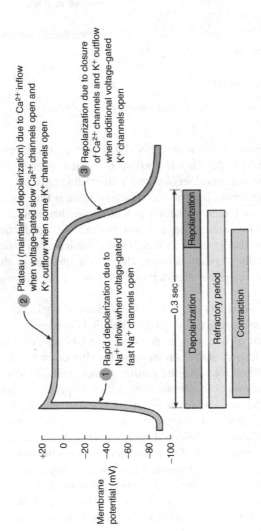

Figure 4.3 Cardiac action potential.
Taken from Jenkins, G.W., Kemnitz, C.P. and Tortora, G.J. (2007) *Anatomy and Physiology from Science to Life*, Wiley, Chichester

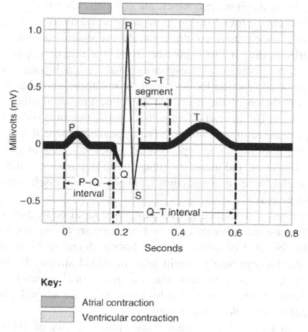

Figure 4.4 Electrocardiogram.
Taken from Jenkins, G.W., Kemnitz, C.P. and Tortora, G.J. (2007) *Anatomy and Physiology from Science to Life*, Wiley, Chichester

4.1.2 Blood pressure

Blood flows through the heart and circulation because it is under pressure. This is commonly known as *blood pressure* and can be measured in mmHg (millimeters of mercury). A normal blood pressure measurement would be 120/80 mmHg, where the higher figure is the (systolic) blood pressure when the ventricles are contracting and the lower figure is the (diastolic) blood pressure when the ventricles are relaxed. Blood pressure is highest in the aorta and reduces in blood vessels further away from the heart until it reaches 0 mmHg as blood returns to the right side of the heart.

As the heart and circulation is a closed system, the blood pressure is maintained with each contraction of the ventricles.

Blood pressure depends on the total volume of blood in the system and on the resistance provided by all the blood vessels. Peripheral vascular resistance depends on the diameter of blood vessels, the viscosity of blood and the total blood vessel length. Under normal circumstances, blood viscosity and total blood vessel length remain the same and do not affect blood pressure. Blood vessel diameter, however, can and does change from moment to moment by vasoconstriction (narrowing) or vasodilation (widening). For example, if body temperature rises above normal, one of the changes that happen is dilation of blood vessels in the skin to allow heat loss. Normally, blood pressure is maintained by feedback mechanisms, which are complex. Essentially, blood pressure is monitored by baroreceptors found in the aorta and carotid arteries and regulated by the cardiovascular centre in the medulla of the brain. Baroreceptors continually send impulses to the cardiovascular centre. If blood pressure rises, the rate of impulses being sent to the cardiovascular centre increases. In response to this, the cardiovascular centre reduces heart rate and force of contraction by decreasing sympathetic stimulation and increasing parasympathetic stimulation of the heart. At the same time sympathetic stimulation of blood vessels decreases. The overall effect is that cardiac output is reduced, blood vessels dilate and blood pressure returns to normal. The opposite happens if blood pressure falls. The rate of impulses from the baroreceptors decreases, the cardiovascular centre inhibits parasympathetic stimulation and increases sympathetic stimulation to the heart and blood vessels. Heart rate and force of contraction increase, blood vessels constrict and blood pressure returns to normal.

The baroreceptor reflex described above allows for rapid adjustments of blood pressure. There are longer-term controls on blood pressure involving several hormones. If blood volume falls (because of dehydration or haemorrhage) or blood flow to the kidneys is reduced, the kidneys secrete renin into the blood stream. Renin together with angiotensin-converting enzyme (ACE) produces angiotensin II, which is a powerful vasoconstrictor and stimulates the release of aldosterone from the adrenal glands. Figure 4.5 shows the renin-angiotensin-aldosterone pathway.

Vasoconstriction and the salt and water reabsorption activity of aldosterone result in restoration of blood volume and therefore blood pressure.

There are times when blood pressure has to rise, for example during exercise and in response to threats and emotions. In these circumstances, blood pressure control reflexes are temporarily overridden until the exercise or other response is over when blood

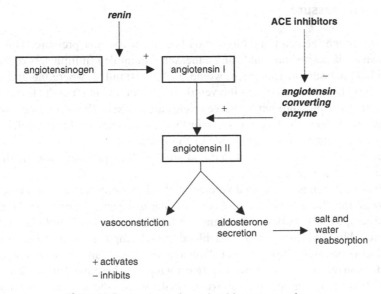

Figure 4.5   Renin-angiotensin-aldosterone pathway

pressure returns to normal. Persistently raised blood pressure is known as *hypertension* and needs treating because it can lead to heart attack, stroke or other serious illness.

4.1.3   Cardiac failure

In cardiac failure, cardiac output is reduced. This means initially that with exercise the organs and tissues do not have an adequate blood supply. As the condition progresses the supply may not be sufficient even at rest.

Cardiac failure can result if the myocardium is damaged by ischaemia, if the heart is inefficient due to valve disease or because of excessive peripheral vascular resistance and hypertension. Depending on the precipitating factors, cardiac failure can affect either side of the heart or both.

Left ventricular failure can occur due to hypertension, aortic valve disease or coronary artery disease. The symptoms are pulmonary oedema and breathlessness.

Right ventricular failure can occur due to chronic lung disease, pulmonary valve disease and congenital defects. The symptoms are systemic oedema, particularly of the legs and ankles.

Biventricular failure often follows left ventricular failure and consequently oedema develops in the peripheral tissues, causing swelling of the legs, and in the lungs, causing breathlessness.

There are no drugs to cure cardiac failure but the condition can be controlled by drug therapy.

Drug groups used to control cardiac failure are cardiac glycosides, diuretics, ACE inhibitors and angiotensin II receptor antagonists.

4.1.4   Ischaemic heart disease

Ischaemic heart disease arises if one or more of the coronary arteries become partially occluded (by atheroma or arteriosclerosis). This leads to reduction in blood flow to the myocardium.

Angina pectoris strictly translated is pain in the pectoral region of the chest. Usually this is referred to simply as *angina* and the typical presentation is that of a crushing pain in the upper left side of the chest which may radiate into the neck or down the left arm. The pain is usually severe and often produces significant distress in the sufferer. This is referred pain that is brought about because of insufficient blood flow to the myocardium. In severe cases, it may be a symptom of acute coronary thrombosis (heart attack).

Angina may be categorized as stable or unstable. Stable angina is a condition that is predictable and results from exertion or stress. Symptoms recede when the stress or exertion ceases. This type of angina is usually due to severe chronic coronary artery disease. Unstable angina is often unpredictable and occurs at rest or because of minor exertion. The underlying pathology is thrombus formation or extension in the coronary arteries. A variant of unstable angina at rest is due to unpredictable spasm of the coronary arteries.

Drugs that either improve the blood supply to the myocardium, or reduce its metabolic demand for oxygen or both can be used to treat angina.

These drugs include nitrates, potassium channel activators, β-blockers and calcium channel blockers.

Myocardial infarction occurs when there is a prolonged reduction in oxygen supply to a region of myocardium, due to narrowing of the arteries combined with sudden formation of a thrombus and blockage of a coronary artery. Death of a part of the myocardium occurs due a disruption in its blood supply. General treatment includes relief of pain, treatment of arrhythmia, treatment of shock and correction of heart failure. Anticoagulant therapy is also recommended as prophylaxis against further infarctions.

Death may result from ventricular fibrillation, heart block or asystole (see below under arrhythmias).

4.1.5 Arrhythmias

Disorders of the conduction system through the heart are known as *arrhythmias* (or *dysrhythmias*), because the inherent rhythm of the heart is disturbed. There are different kinds of arrhythmia depending on which part of the heart is affected.

Atrial flutter describes the condition where there is rapid but regular, beating of the atria of between 240 and 360 beats per minute.

Atrial fibrillation is said to occur when there is rapid but disordered contraction of the atria of between 400 and 600 beats per minute.

Paroxysmal supraventricular tachycardia is an intermittent increase in the rate of atrial contraction.

Ectopic beats are extra beats caused by the depolarization of a focus of cells other than the SA node.

Ventricular fibrillation is the asynchronous contraction of the ventricles, which rapidly leads to circulatory failure if not treated as an emergency.

Heart block occurs if there is AV node damage; in this situation, the atria and ventricles beat independently reducing the efficiency of the heart as a pump.

Asystole is when the heart stops beating.

Table 4.1 shows some causes of arrhythmia.

There are many drugs available to treat arrhythmias and the choice of drug often depends on the type of arrhythmia. Anti-arrhythmic drugs have been classified according to their mode of action into the so-called Vaughan–Williams classification. (See Table 4.2.)

Drugs used to treat arrhythmias are cardiac glycosides, β-blockers, calcium channel blockers, lidocaine and amiodarone.

4.2 Hypertension

Hypertension is defined as persistently high systemic arterial blood pressure above 140/90 mmHg, although this does depend on age. For older people, a more realistic

Table 4.1 Causes of arrhythmia

Cause	Effect
Ectopic pacemaker	A group of cardiac muscle cells starts to depolarize at a rate faster than the SA node; this can be caused by ischaemia or noradrenaline
Re-entry of delayed action potentials	Ischaemic damage of cardiac muscle cells delays action potentials, which can then depolarize other cells no longer refractory, setting up 'circus' movement (a loop of depolarization)
Delayed after depolarization	Increase in calcium levels inside certain cardiac muscle cells causes spontaneous depolarization, which can then speed up the part of the heart affected
Heart block	Because of damage to the AV node the depolarization is not passed onto ventricles; the ventricles beat at a slower, and often irregular, rate than the atria

Table 4.2 Vaughan–Williams classification of anti-arrhythmic drugs

Class	Mode of Action	Example
Class Ia	Membrane stabilizing	Quinidine
Ib		Lidocaine
Ic		Flecainide
Class II	Beta blockers	Celiprolol
Class III	Prolongs refractory period	Amiodarone
Class IV	Calcium channel blockers	Verapamil

The Vaughan–Williams classification is of academic interest. Clinically, the type of arrhythmia against which the drugs are effective is of more importance than their mode of action.

figure would be 160/90 mmHg. Hypertension can be divided into two types known as *secondary* and *essential hypertension.*

Secondary hypertension has a definable cause, which can be secondary to another disorder or can be drug induced.

Examples of secondary causes are Cushing's syndrome, phaeochromocytoma, hyperaldosteronism, renal disease or use of oral contraceptives and corticosteroids. (Phaeochromocytoma is a benign tumour of the adrenal glands resulting in excessive secretion of adrenaline and noradrenaline. See Chapter 6.)

Once the secondary cause has been identified, it should be treated rather than the hypertension itself.

Essential hypertension describes raised blood pressure of unknown cause. Because it is common for blood pressure to rise with age, it used to be thought that an increasing blood pressure was essential for survival, hence the name essential hypertension. It is now known that persistently raised blood pressure over a period of many years will

Table 4.3 Guidelines for the management of hypertension

Blood pressure mmHg	Grade	Management
120/80	Optimal	–
130/80	Normal	–
135/85	High normal	5 year monitoring
Hypertension		
140/90	Mild	Lifestyle changes
		Drug therapy if diabetic/[a] 10 year risk of CVD >20%
160/100	Moderate	Drug therapy and lifestyle changes
180/110	Severe	Drug therapy and lifestyle changes

[a] 10-year-risk for cardiovascular disease

Adapted from Williams, B., Poulter, N.R., Brown, M.J. et~al. (2004) The BHS Working Party British Hypertension Society guidelines for hypertension management, 2004-BHS IV: summary. *British Medical Journal*, 328, 634–40.

cause damage to blood vessels and hypertrophy of the left ventricle and can lead to renal failure, stroke, coronary artery disease and cardiac failure.

Blood pressure varies naturally from individual to individual. Optimal blood pressure is considered to be 120/80 mmHg; normal blood pressure 130/80 mmHg and high normal blood pressure 135/85 mmHg. Hypertension is conventionally diagnosed as blood pressure of 140/90 mmHg or above whatever the age of the patient. With advancing age, there is a decrease in the elasticity of the arteries (arteriosclerosis), which, through increased peripheral vascular resistance, produces a steady rise in systolic blood pressure.

Treatment of hypertension is symptomatic, aiming to maintain blood pressure below 140/85 mmHg, rather than curative and must be continued for the rest of the patient's life. Before treatment begins, it is important to weigh the benefits of blood pressure reduction against the possible consequences of drug treatment. Non-drug interventions such as weight reduction, increased exercise, salt restriction, stopping smoking, reduction of alcohol intake should always be included in the management of hypertension. British Hypertension Society guidelines for the management of hypertension have recently been changed to be in line with European guidelines. (See Table 4.3.)

Drugs that are used to treat hypertension include diuretics, ACE inhibitors, angiotensin II receptor antagonists, β-blockers, calcium channel blockers, α_1-antagonists and centrally acting antihypertensives.

4.3 Drugs used to treat cardiovascular disorders

Groups of drugs are described together with their relevant therapeutic indications because often the same drugs are used for different conditions.

4.3.1 Cardiac glycosides

Cardiac glycosides are used to treat cardiac failure and atrial flutter and atrial fibrillation.

Digoxin and digitoxin are cardiac glycosides that improve cardiac contractility, which in turn increases cardiac output. This is thought to occur through inhibition of the Na^+/K^+ ATPase pump in the cardiac cell membrane, which results in more calcium ions being available for contraction. This is because normally calcium ions are exchanged for sodium ions across the cardiac muscle cell membrane at the end of a contraction and thereby removed from the cell. If the Na^+/K^+ ATPase pump is inhibited, the concentration of sodium ions in the cell increases reducing the concentration gradient for exchange with calcium ions. Thus, the concentration of calcium ions within the cell rises (see Figure 4.6).

The effect of increased intracellular calcium ion levels is a more controlled forceful contraction of the myocardium. This is known as a *positive inotropic effect*.

In addition, through an effect in the central nervous system cardiac glycosides cause an increase in parasympathetic activity and therefore slow conduction through the AV node, hence their usefulness in atrial flutter and atrial fibrillation.

However, cardiac glycosides have a very low therapeutic ratio and can be very toxic. Slowing of AV conduction can lead to AV block and arrhythmias. These signs could be confused with clinical deterioration.

Withdrawal of the drug usually results in recovery, but severe digoxin toxicity can be treated in an emergency by intravenous injection of digoxin specific antibody fragment, which neutralizes toxic effects.

Other signs of toxicity are nausea, vomiting and confusion.

Since cardiac glycosides are excreted mainly by the kidney, doses have to be adjusted according to renal status of individual patients.

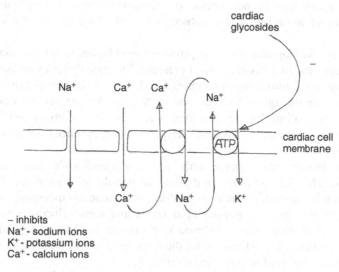

Figure 4.6 Mode of action of cardiac glycosides

Cardiac glycosides interact with many other drugs; some increase the risk of toxicity (for example amiodarone and verapamil) and others reduce it (for example cholestyramine, colestipol and antacids), but also reduce therapeutic effect.

4.3.2   Diuretics

Diuretics are used to treat cardiac failure and hypertension.

Because cardiac failure results in (or is caused by) hypertension and leads to oedema, diuretics are used in both cases because they encourage water loss and therefore produce a reduction in circulating blood volume. This lowers blood pressure, reduces the work the heart has to do and improves oedema by encouraging the movement of fluid from the tissues into the circulation.

Of the different types of diuretics, the ones used in cardiac failure and hypertension are thiazide diuretics and loop diuretics.

Refer to Figure 2.5 for a diagram of the kidney nephron.

4.3.2.1   Thiazide diuretics

Thiazide diuretics work by inhibiting sodium and chloride ion transport in the distal convoluted tubule of the kidney and therefore limit water reabsorption. Thiazides are particularly recommended in older people. A commonly used example of this type of diuretic is bendroflumethiazide.

4.3.2.2   Loop diuretics

Loop diuretics inhibit sodium and chloride ion transport in the ascending limb of the loop of Henlé and therefore prevent water reabsorption there. An example of a loop diuretic is furosemide.

Both thiazides and loop diuretics also produce vasodilation, which reduces peripheral vascular resistance and helps reduce blood pressure. Thiazide diuretics are recommended as an alternative to calcium channel blockers (see below) as first choice drugs in hypertensive patients over the age of 55 years and in those of African origin of any age.

Adverse effects of diuretics are excessive potassium and hydrogen ion loss leading to hypokalaemia and metabolic alkalosis. Hypokalaemia enhances the toxic effects of cardiac glycosides.

Normally potassium ions compete with cardiac glycosides for their site of action, which is the Na^+/K^+ ATPase pump in the cardiac muscle cell membrane. If potassium ion concentration is reduced, the inhibitory effects of digoxin are increased (see page 61).

Diuretics may also increase plasma lipid levels and cause glucose intolerance and insulin resistance, so they have to be used with caution in individuals with hyperlipidaemia and/or diabetes. In addition, most diuretics produce an increase in plasma uric acid levels, which may lead to gout (see Chapter 7).

4.3.3 ACE (angiotensin-converting enzyme) inhibitors

ACE inhibitors are used to treat cardiac failure and hypertension.

Inhibition of ACE prevents formation of angiotensin II, which is a powerful vaso-constrictor (see Figure 4.5). This results in vasodilation and fall in peripheral vascular resistance and therefore reduction in blood pressure. Reduction in blood pressure also improves cardiac output. In addition, ACE inhibitors suppress aldosterone secretion producing a reduction in salt and water retention and improve renal blood flow, which in turn inhibits renin release.

Examples of ACE inhibitors are captopril and lisinopril.

ACE inhibitors have no effect on heart rate or bronchioles, lipid levels or the action of insulin and therefore can be used safely in patients with heart failure, asthma, peripheral vascular disease or diabetes. They are recommended as the drug of first choice for hypertension in white patients below the age of 55 years.

The most common adverse effect is a chronic dry cough; others are taste disturbances and rashes. ACE inhibitors should not be used in pregnancy.

4.3.4 Angiotensin II receptor antagonists

Angiotensin II receptor antagonists are used to treat cardiac failure and hypertension.

Not surprisingly, they have a similar effect to ACE inhibitors but this is brought about by preventing angiotensin II from binding to its receptors. An example is valsartan.

This is a relatively new group of drugs and adverse effects appear to be rare. They provide an alternative for those who cannot tolerate ACE inhibitors because of a persistent dry cough, but should not be used in pregnancy either.

Neither ACE inhibitors nor angiotensin II receptor inhibitors seem to benefit patients of African origin.

4.3.5 Nitrates

Nitrates are used primarily to treat ischaemic heart disease.

These drugs are potent coronary vasodilators and can increase blood flow to the myocardium. However, in many cases of angina, the coronary arteries are partially occluded and blood flow through them does not increase significantly. Nitrates appear to dilate collateral coronary blood vessels allowing partially blocked arteries to be bypassed. In addition, they dilate veins. This brings about a reduction in venous return and reduces the preload on the heart. This decreases the workload of the left ventricle and myocardial oxygen consumption is reduced.

The action of nitrates is likely to be because of the production of nitric oxide, which is a powerful vasodilating agent.

4.3.5.1 Glyceryl trinitrate

Glyceryl trinitrate (GTN) is extremely effective at relieving the pain of angina but it has only a short half-life. Due to extensive first pass metabolism the drug is ineffective when taken orally and is administered sublingually or buccally in either tablet or aerosol form.

Sustained release forms are available to provide prophylactic action against angina attacks; these are formulated into transdermal patches.

4.3.5.2 Isosorbide dinitrate and isosorbide mononitrate

Isosorbide dinitrate and isosorbide mononitrate are more stable forms of nitrate and have a longer lasting effect than GTN making them more suitable for prophylaxis of angina. They are taken orally.

When used prophylactically, tolerance may develop to the vasodilator effects of nitrates. To avoid this, patients have to have drug free periods each day, when they know they are least likely to suffer and angina attack.

Typical adverse effects include headache, postural hypotension and flushing, which often restrict the use of nitrates.

4.3.6 Potassium channel activators

Potassium channel activators are used to treat ischaemic heart disease.

These are a relatively recent development and act by increasing the efflux of potassium ions in smooth muscle cells of blood vessels. This leads to hyperpolarization of vascular smooth muscle thereby reducing the excitability and bringing about vasodilation. The resulting vasodilation in coronary arterioles improves blood flow to the myocardium. This, in combination with a reduction in both afterload (dilation of arteries) and preload (dilation of veins), relieves the angina. Nicorandil is an example of a potassium channel activator.

Adverse reactions are headache, dizziness and hypotension.

4.3.7 Beta blockers (β-blockers)

Beta blockers are used to treat ischaemic heart disease, arrhythmia and hypertension.

These drugs exert their effect in exercise-induced angina by reducing the increase in heart rate and force of contraction in response to exercise, thereby reducing the oxygen demand of the myocardium. β-blockers may be used prophylactically in the treatment of stable angina. They appear to have little effect on the coronary blood vessels, causing slight constriction if anything.

β-blockers are useful in all types of arrhythmia because they block the conducting system in the heart and reduce heart rate.

Along with ACE inhibitors, β-blockers used to be first choice drugs for hypertension (in the absence of asthma/cardiac failure/insulin dependent diabetes) either alone or together with diuretics. However, they are no longer recommended except in younger people who are intolerant to ACE inhibitors and as add-on therapy if other drugs do not reduce blood pressure sufficiently. This is because of their adverse effects in raising blood glucose levels.

The antihypertensive effect of β-blockers is not completely understood. A decrease in heart rate and force of contraction would reduce cardiac output, but this is likely to be compensated for by the baroreceptor reflex.

It is possible that β-blockers work, at least partially, by some central action in the cardiovascular control centres of the brain.

β-blockers also block receptors in the peripheral blood vessels, kidney and bronchi. The main effects of this are vasoconstriction in skeletal muscles causing cold extremities and reduced renin secretion, which would limit the formation of angiotensin II. β-blockers can precipitate asthma and should not be used in people with asthma. They should be used with caution in people with diabetes because the β-blocking effect can mask warning signs of hypoglycaemia.

The original β-blockers did not discriminate between β_1 and β_2 receptors but newer drugs have been developed that are more cardio-selective and there are others that also cause vasodilation.

First generation β-blockers are non-cardio-selective and they can produce bronchoconstriction as an adverse effect. They exacerbate congestive heart failure, adversely affect plasma lipid profiles and reduce exercise tolerance. An example is propranolol.

Second generation β-blockers have greater cardio-selectivity and are less likely to cause bronchoconstriction. Examples are atenolol and bisoprolol.

Third generation β-blockers also cause peripheral vasodilation and do not adversely affect plasma lipid profiles. An example is celiprolol.

4.3.8 Calcium channel blockers

Calcium channel blockers are used to treat ischaemic heart disease, arrhythmia and hypertension.

This is a diverse group of drugs, which fall into three classes. They all block influx of calcium ions during depolarization of muscle cells.

4.3.8.1 Class I

Class I calcium channel blockers work preferentially on calcium ion channels in cardiac muscle cells. This results in a delay in electrical conduction in cardiac muscle, reduced contractility and reduction in the heart rate. Because of these effects, class I calcium channel blockers are used to treat ischaemic heart disease and atrial arrhythmia. An example is verapamil.

Adverse effects are bradycardia, heart block and possible precipitation of heart failure. Class I calcium channel blockers are contraindicated in patients who already have these conditions.

4.3.8.2   Class II

Class II calcium channel blockers act primarily on smooth muscle cells and as such lead to dilation of blood vessels. This in turn reduces peripheral vascular resistance and they are used to reduce blood pressure. Examples are amlodipine and nifedipine.

Nifedipine also has a use in the treatment of Raynaud's disease. In Raynaud's disease, there is inappropriate vasoconstriction in the fingers and toes, usually in response to cold. The condition can be mild or severe leading to ulceration and possibly gangrene of the digits. Raynaud's disease can occur on its own, or as a consequence of systemic sclerosis or systemic lupus erythematosus, both chronic inflammatory diseases (see Chapter 7).

Adverse effects of class II calcium channel blockers are reflex tachycardia, headache, flushing, palpitations and ankle oedema.

4.3.8.3   Class III

Class III calcium channel blockers are intermediate in action affecting calcium ion channels in both cardiac muscle cells and smooth muscle cells in blood vessels. The effect of this is to combine a reduction in heart rate with a fall in blood pressure and they are useful in treating angina and hypertension. An example is diltiazem.

Adverse effects are bradycardia, headache, flushing, palpitations and ankle oedema. There is an increased risk of myopathy if diltiazem is used together with statins (see page 77).

In general, class II and III calcium channel blockers are used to treat hypertension as an alternative to diuretics. In hypertensive patients over 55 years old and those of African origin of any age calcium channels blockers or diuretics are now the drugs of first choice.

4.3.9   Membrane stabilizers

Membrane stabilizers are used to treat arrhythmia.

They act directly on sodium ion channels to restrict the entry of sodium ions into cardiac muscle cells. These drugs therefore act to slow the rate of depolarization, which is more marked in ectopic pacemakers, allowing the SA node to regain control of cardiac rhythm.

4.3.9.1   Lidocaine

Lidocaine is ineffective orally and its use is restricted to preventing and treating ventricular arrhythmia after myocardial infarction. (Lidocaine by virtue of the same mechanism of action is used as a local anaesthetic. See Chapter 12.)

Adverse effects of lidocaine are drowsiness, slurred speech, paraesthesia (pins and needles) and convulsions.

4.3.9.2 Flecainide

Flecainide is an alternative to lidocaine for the treatment of ventricular arrhythmia and can be used orally for disabling paroxysmal atrial fibrillation.

Adverse effects of flecainide are nausea, vomiting and precipitation of arrythmias.

4.3.9.3 Quinidine

Quinidine can suppress all types of arrhythmia, but also cause them. Because of this, it is rarely used without specialist guidance.

Adverse effects of quinidine are nausea, vomiting, diarrhoea, blurred vision, cardiac depression and arrhythmia.

4.3.10 Amiodarone

Amiodarone is used to treat arrhythmia.

This drug prolongs the refractory period in cardiac muscle without affecting the rate of depolarization. This means that the interval required for re-excitation is prolonged and arrhythmias are suppressed. Amiodarone is effective in both atrial and ventricular arrhythmias.

Adverse effects include photosensitivity and skin reactions and the rare but more serious risk of pulmonary fibrosis.

4.3.11 Alpha₁ antagonists (α₁-antagonists)

Alpha₁ antagonists are used to treat hypertension.

They block α_1-adrenergic receptors in veins and arterioles, thereby causing vasodilation and decreased peripheral vascular resistance. As a result, blood pressure is reduced. An example is doxazosin.

Adverse effects include drowsiness, depression and postural hypotension especially in the elderly, which restricts their use to resistant hypertension. α_1 antagonists should not be used in patients with an enlarged prostate or urinary incontinence.

4.3.12 Centrally acting antihypertensives

These drugs are used together with β-blockers and diuretics if necessary in resistant hypertension.

The site of action is in the brain, probably the medulla, and the effect is to reduce sympathetic output.

Examples of centrally acting antihypertensives are methyldopa and clonidine.

Methyldopa becomes converted to methylnoradrenaline, which is a false transmitter with reduced effect on noradrenaline receptors in the brain.

Table 4.4   Summary of drugs used to treat cardiovascular disorders

Disorder Drug group	Cardiac failure	Ischaemic heart disease	Arrhythmia	Hypertension
Cardiac glycosides	×	–	×	–
Diuretics	×	–	–	×
ACE inhibitors	×	–	–	×
Angiotensin II antagonists	×	–	–	×
Nitrates	–	×	–	–
Potassium channel blockers	–	×	–	–
β-blockers	–	×	×	×
Calcium channel blockers	–	×	×	×
Membrane stabilizers	–	–	×	–
Amiodarone	–	–	×	–
α-antagonists	–	–	–	×
Centrally-acting antihypertensives	–	–	–	×
Adrenergic neurone blockers	–	–	–	×

Clonidine stimulates presynaptic receptors on sympathetic neurons in the brain, which reduces the amount of noradrenaline released.

Through both of these mechanisms, there is a decrease in sympathetic output to the peripheral blood vessels. Due to a reduction in vasoconstriction, this leads to a decrease in peripheral vascular resistance and blood pressure falls.

Adverse effects, dry mouth, sedation and drowsiness, can be severe, limiting the usefulness of these two drugs.

4.3.13   Adrenergic neurone blockers

Adrenergic neurone blockers are used rarely to treat hypertension that is resistant to other drugs.

They block release of noradrenaline at sympathetic nerve endings and therefore cause vasodilation and reduced peripheral vascular resistance. An example of an adrenergic neurone blocker is guanethidine.

Adverse effects include postural hypotension, bradycardia and fluid retention.

Table 4.4 summarizes drugs used to treat cardiovascular disorders.

4.4   Blood disorders

Like cardiac disorders and hypertension, disorders of coagulation, anaemia and hyperlipidaemia are all conditions that are more common in older patients. All are treatable with drug therapy and possibly changes in lifestyle as well. Health care professionals, in the course of their professional practice, are sure to come across patients on the drugs commonly used in these conditions. Some of them are of particular significance to podiatric practice.

A basic knowledge of the clotting mechanism, the causes and consequences of anaemia and lipid metabolism is needed in order to appreciate how these conditions can be treated.

4.4.1 Haemostasis

Haemostasis is a protective mechanism to stop bleeding from damaged blood vessels. The normal homeostatic control of bleeding consists of three components. If there is a defect in any of these then bleeding will result.

4.4.1.1 Blood vessel reaction

Damage to a blood vessel causes immediate vasoconstriction due to release of mediators locally. These quickly diffuse to other vessels in the immediate surroundings. The effect is to reduce blood flow to the area and to reduce leakage of blood from the damaged vessel making it easier for other mechanisms to stem the flow of blood.

In a small injury, this alone may be enough to stop the loss of blood.

4.4.1.2 Reaction of platelets

The function of platelets is to plug damaged blood vessels. Damage to the endothelial lining of blood vessels exposes collagen, to which platelets stick providing there is a sufficient supply of Factor VIII (one of the clotting factors – see below).

Platelets release adenosine diphosphate (ADP) and serotonin (5HT, 5-hydroxytryptamine) following adhesion. 5HT is a powerful vasoconstrictor, which reinforces constriction of blood vessels reducing blood flow and loss. ADP attracts more platelets. The platelets fuse together into an inseparable mass. As this happens, large amounts of thromboxane A_2 are released, which further enhances aggregation and is also a potent vasoconstrictor (see Figure 7.1).

Undamaged endothelium produces prostacyclin, which is a potent inhibitor of thromboxane A_2. This prevents clotting in undamaged blood vessels.

4.4.1.3 Blood coagulation

Blood coagulation, or clotting, results from a series of reactions involving clotting factors. There are 13 clotting factors, most of which are synthesized in the liver and vitamin K is necessary for the synthesis of many of them. Clotting factors circulate in an inactive form. Once the reaction is triggered by tissue or blood vessel damage, one factor activates the next by a cascade mechanism leading to the formation of prothrombinase. Prothrombinase is an enzyme that activates prothrombin into thrombin, which in turn activates fibrinogen into fibrin. The insoluble fibres of fibrin enmesh red blood cells and platelets, which forms the clot itself. See Figure 4.7 for a simplified diagram of the clotting cascade.

The process of coagulation is also dependent on the presence of calcium ions.

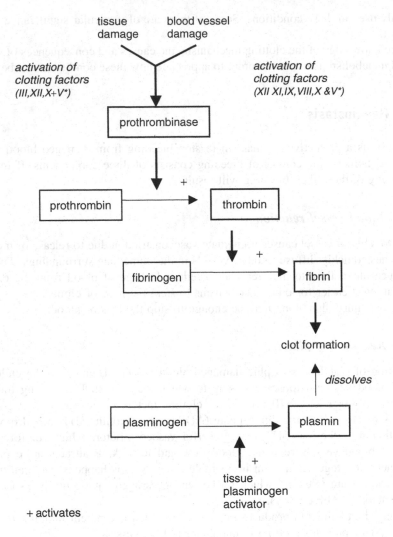

Figure 4.7 Clotting cascade

Small amounts of thrombin are formed all the time but are normally inactivated by formation of a complex with antithrombin III in the presence of heparin.

Once the damage is repaired, the clot is dissolved by the action of a proteolytic enzyme plasmin. Plasmin is normally found circulating in the blood as its inactive precursor plasminogen. It is activated by tissue plasminogen activator as healing takes place.

Small amounts of plasminogen are activated all the time to keep blood vessels free from clots.

4.4.2 Disorders of coagulation

4.4.2.1 Haemophilia

Haemophilia is a sex-linked recessive inherited disorder of coagulation.

There are two types: haemophilia A, or classical haemophilia, is due to lack of Factor VIII and haemophilia B, or Christmas disease, is due to lack of Factor IX.

Both diseases have similar symptoms of excessive bleeding from wounds, easy bruising, bleeding into joints, bleeding into the sheath of nerve trunks resulting in paralysis or muscle spasticity, bleeding into the urinary tract and bleeding into the central nervous system.

Treatment of both conditions is by replacement of the missing factor. Antihaemophilic globulin and Christmas factor complex are prepared from fresh normal plasma. Replacement therapy is used to treat haemorrhage and prophylactically before surgery or dental treatment.

4.4.2.2 Other disorders of coagulation

Coagulation disorder can be secondary to liver disease because most clotting factors are synthesized in the liver.

Vitamin K is essential for successful formation of many clotting factors so lack of vitamin K can cause a coagulation disorder. Normal bile acid production is necessary for absorption of vitamin K from the small intestine.

Although lack of vitamin K is rare, because it is present in green leafy vegetables and synthesized by intestinal bacteria, it may occur with a combination of poor diet and long-term use of antibiotics. Deficiency may also occur because of malabsorption secondary to other conditions such as coeliac disease, or due to lack of bile secretion as in obstructive jaundice.

Deficiency is treated with oral vitamin K or synthetic analogues such as phytomenadione. In the absence of bile, vitamin K replacement has to be given by intramuscular or intravenous injection. Vitamin K can be used to counteract the effects of warfarin in the event of excessive bleeding (see page 72).

4.4.2.3 Thromboembolic disease

A thrombus is a blood clot that forms unnecessarily, usually in response to damage to the lining of blood vessels or disease such as arteriosclerosis. A part of a thrombus that breaks off and circulates in the blood is known as an *embolus*. Thromboembolic disease is the result of thrombus and embolus formation and can take the form of a heart attack or a stroke or blockage of a blood vessel elsewhere in the body.

Thromboses in arteries and veins are slightly different. Venous thrombi form in slow moving blood and contain a lot of fibrin. Arterial thrombi usually occur because of damage to artery walls and contain more platelets than fibrin. Formation of emboli is a risk in either case.

Treatment of thromboembolic disease is with anticoagulants and platelet inhibitors to prevent the formation of thrombi and with fibrinolytic drugs to break down thrombi that have already formed.

4.4.3 Anticoagulants

These are used prophylactically postoperatively and in people who have already had a thrombotic episode (heart attack, stroke, deep vein thrombosis). They are most effective at preventing venous thrombosis.

4.4.3.1 Vitamin K antagonists

Vitamin K antagonists block the activity of vitamin K resulting in the production of inactive clotting factors by the liver. They take a few days to have effect because circulating clotting factors already in existence have to be used up.

Examples of vitamin K antagonists are warfarin (most frequently used) and phenindione (normally used only when patients cannot tolerate warfarin).

Patients on long-term warfarin should have an oral anticoagulant treatment booklet.

Warfarin has many interactions with commonly prescribed drugs. For example, its effects are enhanced by ibuprofen (analgesic), ketoconazole (antifungal) and cimetidine (used to treat stomach ulcers) and reduced by phenytoin (antiepileptic) and alcohol.

4.4.3.2 Heparin

Heparin works by enhancing the formation of a complex of thrombin and antithrombin III. This effectively prevents thrombin from activating fibrinogen. Heparin has a rapid action of short duration and has to be given by injection. It is used to decrease the risk of deep vein thrombosis after surgery and during recovery from heart attack and stroke until warfarin becomes effective.

4.4.3.3 Hirudin

Hirudin is a substance produced by medicinal leeches, which directly inactivates thrombin. Lepirudin is a synthetic version produced by recombinant DNA technology, which has to be given by injection. It is used specifically in patients that have developed hypersensitivity to heparin (see Chapter 3, page 35).

The chief adverse effect of anticoagulants is haemorrhage; hypersensitivity reactions are also possible.

4.4.4 Platelet inhibitors

Platelet inhibitors are most effective in preventing arterial thrombosis.

4.4.4.1 *Aspirin*

Aspirin reduces platelet aggregation by inhibiting the enzyme cyclo-oxygenase, which is necessary for the formation of thromboxane A_2. Its use is recommended prophylactically in patients with heart disease such as angina and arrhythmia as well as those with previous thrombotic disease.

4.4.4.2 *Dipyridamole and clopidrogel*

Dipyridamole has a similar effect on cyclo-oxygenase as aspirin.

Clopidrogel is another platelet inhibitor that works by inhibiting ADP-induced aggregation.

Apart from haemorrhaging, gastrointestinal irritation and bronchospasm are adverse reactions seen with platelet inhibitors.

4.4.5 Fibrinolytic drugs

These drugs are used to destroy thrombi that are occluding blood vessels. They are most effective given soon after the thrombus has formed, for example within 3 hours of a myocardial infarction. All of the drugs in use activate plasminogen resulting in the production of the active enzyme plasmin.

Clinical uses include treatment of acute myocardial infarction, pulmonary embolism and peripheral thromboembolism and in the restoration of circulation through arterial grafts and intravenous catheters.

Examples are streptokinase and alteplase. Streptokinase is naturally produced by streptococcal bacteria whereas alteplase is recombinant tissue plasminogen activator (rTPA). Streptokinase is antigenic and may provoke an allergic reaction, so it should not be used too often.

Adverse reactions are nausea and vomiting and bleeding.

4.5 Anaemias

Anaemia results when, for whatever reason, there is not enough haemoglobin in the blood to carry sufficient oxygen around to the tissues.

Typical symptoms of anaemia are fatigue due to anoxia and intolerance to cold and paleness due to low levels of haemoglobin.

4.5.1 Haemorrhagic anaemia

Haemorrhagic anaemia can be caused by large wounds, stomach ulcers or heavy menstrual bleeding. Sudden great loss of blood causes acute anaemia but slow prolonged bleeding can produce chronic anaemia.

4.5.2   Iron deficiency

Iron deficiency is a common cause of anaemia.

Iron is necessary for haemoglobin production and iron deficiency results in small red blood cells with insufficient haemoglobin.

Iron deficiency and haemorrhagic anaemia are treated with replacement in the form of iron salts, usually ferrous sulfate, and an improved diet. Sometimes a blood transfusion is necessary in the short term.

Adverse effects of oral iron preparations are that they cause gastrointestinal irritation including nausea, epigastric pain, diarrhoea and/or constipation. Parenteral preparations are available if oral therapy cannot be tolerated.

4.5.3   Decreased production of erythrocytes

Decreased production of erythrocytes can be a cause of anaemia.

This can be due to deficiency of vitamin B_{12} or folic acid, both of which are required for formation of erythrocytes.

Vitamin B_{12} deficiency can occur due to lack of intrinsic factor, which is produced in the stomach and is essential for absorption of vitamin B_{12}; after gastrectomy; or because of impaired absorption due to disease of the small intestine.

Oral replacement of vitamin B_{12} is not effective due to malabsorption. Treatment is by intramuscular injection of hydroxycobolamin, a form of vitamin B_{12}.

Folic acid deficiency can occur due to pregnancy, malabsorption syndromes or inadequate diet. Some drugs, for example phenytoin (used in epilepsy), oral contraceptives and isoniazid (used in treating tuberculosis), can cause reduced absorption of folic acid. Oral replacement therapy with folic acid is effective.

Aplastic anaemia is due to loss of haemopoietic tissue in bone marrow and therefore decreased production of red blood cells.

This a common side effect of cytotoxic drugs (cancer chemotherapy, Chapter 10).

Effective haemoglobin synthesis and red blood cell production also depends on the action of a growth factor, erythropoietin, produced by the kidneys. In chronic renal failure lack of this growth factor results in anaemia. Replacement therapy is by intravenous or subcutaneous injection of epoetin.

4.5.4   Increased destruction of erythrocytes

Increased destruction of erythrocytes, or haemolytic anaemia, occurs if erythrocytes are broken down prematurely and can be due to haemoglobin defects, abnormal enzymes, membrane defects, poisons or infections, incompatible blood, autoimmune disease or hereditary disease.

4.5.4.1 Sickle cell disease and thalassaemia

Sickle cell disease and thalassaemia are examples of inherited haemolytic anaemias.

Sickle cell disease is caused by a gene that codes for an abnormal kind of haemoglobin. In low oxygen tension, this type of haemoglobin crystallizes and becomes rigid. As a result, the erythrocytes become distorted and easily ruptured. Sickle cells block blood vessels, cause pain and impair circulation. In young children, this produces hand-foot syndrome, in which there is swelling and pain in wrists and feet.

Sickle cell disease is endemic in parts of Mediterranean Europe, West and East Africa and Asia and is seen in people from those areas in the United Kingdom.

Thalassaemia is a group of inherited disorders in which abnormal haemoglobin is produced. Thalassaemia varies in severity. Thalassaemia is found in people from countries around the Mediterranean.

Pregnancy, infection and malnutrition all make these types of anaemias worse.

There is no specific treatment for sickle cell disease or thalassaemia. Patients are usually given folic acid because there is a high turn over of erythrocytes and they are likely to become deficient in folic acid.

4.6 Lipid metabolism

In normal lipid metabolism, lipids in the diet are converted to triglycerides, which are used in muscle and adipose tissue. Cholesterol is used for the synthesis of steroid hormones and bile acids and by all cells as part of the cell membrane.

Lipids do not dissolve easily in water. Therefore, they have to be transported in blood as lipoproteins. Lipoproteins are formed from a protein envelope, made of apoprotein, containing a variable mix of triglycerides and cholesterol. Lipoproteins vary in size, weight and density. Figure 4.8 shows a simplified version of lipid metabolism.

Chylomicrons are the largest lipoproteins and they transport lipids and cholesterol absorbed from the small intestine to the liver. The liver can synthesize cholesterol if there is not enough in the diet.

Very low-density lipoproteins (VLDLs) are smaller than chylomicrons but still relatively large. They transport lipids and cholesterol from the liver to adipose tissue and muscle where they unload triglycerides and become low-density lipoproteins in the process.

Low-density lipoproteins (LDLs) are medium sized and rich in cholesterol. They transport lipids to any cell that needs them. Cell membranes have LDL receptors to which LDLs attach prior to being taken into the cell. The LDLs are broken down in the cell, the lipids used and the receptors temporarily disappear. Any excess LDLs in the circulation are of a size to deposit cholesterol in the walls of damaged arteries. High levels of circulating LDLs are associated with the development of atherosclerosis.

High-density lipoproteins (HDLs) are the smallest lipoproteins and are formed by the removal of excess cholesterol from cells and possibly from artery walls, which is then

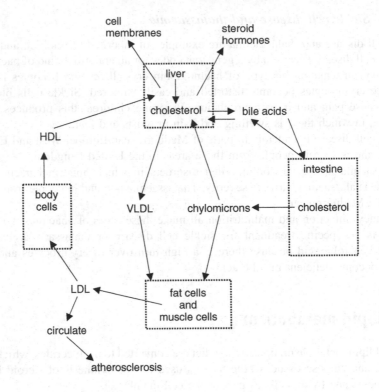

HDL high density lipoprotein
LDL low density lipoprotein
VLDL very low density lipoprotein

Figure 4.8 Lipid metabolism

transported back to the liver. This prevents accumulation of cholesterol in the blood. HDLs are not atherogenic because of their small size.

Total plasma cholesterol (TC) is the sum of cholesterol carried in all forms of lipoproteins, with about 70% of it being in the LDL form.

In the liver, cholesterol can be stored in liver cells, used to form more VLDLs, used to form bile acids or excreted as cholesterol in bile.

4.6.1 Hyperlipidaemias

Hyperlipidaemias are conditions where levels of LDL cholesterol are raised relative to HDL levels. They can be primary or secondary to other conditions such as diabetes or hypothyroidism. About 5% of cases are due to a hereditary condition where there is a deficiency of LDL receptors on cell membranes (familial hyperlipidaemia).

A diet high in saturated fats results in high circulating LDL levels. This is thought to be because saturated fats encourage the reabsorption of bile acids, which means less cholesterol is lost in faeces and because saturated fats can be used in the liver to produce excessive amounts of cholesterol. Polyunsaturated fats are said to encourage cholesterol loss in faeces and its conversion to bile acids.

Hyperlipidaemia is a high risk factor for cardiovascular disease (CVD) and enhances other risk factors such as smoking, excessive alcohol consumption, lack of exercise, diabetes, hypertension, obesity and abdominal distribution of body fat. The risk of CVD over the next 10 years is estimated for an individual using guidelines produced by the British Hypertension Society (or the Joint British Societies 'Cardiac Risk Assessor'), which include age, gender, smoking history, blood pressure and TC: HDL ratios. This estimate is used to determine suitable treatment.

4.7 Lipid-lowering drugs

Drugs can be used to lower lipid levels, but changes in diet and control or elimination of other risk factors play an important role in the management of hyperlipidaemia.

A 10-year risk of CVD of over 30% would indicate treatment with lipid-lowering drugs in addition to dietary and lifestyle advice. If the 10-year risk is between 15 and 30%, changes to lifestyle and diet are recommended and followed by drug therapy if necessary. With a 10-year risk of below 15%, changes in lifestyle and diet may be sufficient. In any case, the aim is to reduce plasma TC levels to below $5 \, \text{mmol} \, l^{-1}$.

Individuals with the hereditary form of hyperlipidaemia will always require drug therapy, possibly with two different types.

4.7.1 Statins

Statins work by reducing the endogenous synthesis of cholesterol in the liver through the inhibition of an enzyme, HMG-CoA reductase (3-hydroxy-3-methylglutaryl-CoA, if you must know). As a result, more LDLs are taken up from the circulation by the liver to provide the cholesterol needed to synthesize bile acids. An example of a statin is atorvastatin and others have similar names.

Statins have few side effects and are now the drug of choice for lowering lipid levels in patients with a high risk of CVD. Some patients have developed myopathy and/or myalgia, which have to be monitored. Statins can interact with many other drugs; some of which can result in increased risk of myopathy (for example imidazole antifungal drugs, Chapter 8; antiviral drugs and some antibiotics, Chapter 9; and calcium channel blockers, page 65).

4.7.2 Ezitimibe

Ezitimibe inhibits absorption of cholesterol from the intestine, probably by interfering with a cholesterol transport system. It is used as an adjunct to therapy with statins in patients with severe hyperlipidaemia, or alone if statins are not tolerated.

Adverse reactions to ezitimibe are gastrointestinal disturbances, headache, fatigue and myalgia.

4.7.3 Omega-3 triglycerides

Derived from fish oils, omega-3 triglycerides are thought to work by inhibiting very low-density lipoprotein (VLDL) synthesis in the liver. Preparations available are omega-3-acid ethyl esters and omega-3-marine triglycerides.

They are recommended as an alternative to a fibrate and in addition to a statin, particularly in familial hyperlipidaemia.

4.7.4 Bile acid sequestrants

These bind bile acids in the intestine, increasing their excretion by preventing entero-hepatic recycling. This in turn causes the liver to use endogenous cholesterol for bile acid synthesis and circulating LDL levels fall. In addition, since bile acids are needed for cholesterol absorption this is reduced as well. Examples of bile acid sequestrants are colestipol and cholestyramine.

Since these drugs are not absorbed, side effects are confined to the intestine and include abdominal discomfort, diarrhoea or constipation. With long-term therapy, fat-soluble vitamin supplements are necessary.

4.7.5 Fibrates

Fibrates suppress endogenous cholesterol synthesis by a mechanism not fully understood but which leads to lower levels of circulating VLDLs and LDLs. There is also an increase in LDL receptors on liver cells, which would encourage removal of LDLs from the circulation by liver cells. Fibrates also increase the activity of lipoprotein lipase, which normally breaks down lipoproteins in muscle cells. An example of a fibrate is bezafibrate.

A serious side effect or fibrates is a myositis-like syndrome, which is made worse by concurrent use of statins. Such a combination is not recommended except in cases of severe familial hyperlipidaemia.

4.7.6 Nicotinic acid derivatives

Nicotinic acid derivatives reduce the release of VLDLs from the liver and the removal of triglycerides from them in adipose tissue. This results in a fall in circulating LDL levels. An example is acipimox.

Because of side effects (flushing, dizziness, palpitations), use of these drugs is now limited to treatment of familial hyperlipidaemia in combination with bile acid sequestrants.

Table 4.5 Lipid lowering drugs

Drug	Mode of action
Statins	Inhibit HMG-CoA reductase
Ezitimibe	Inhibits absorption of cholesterol from the intestine
Omega-3 triglycerides	Inhibit VLDL synthesis in the liver
Bile acid sequestrants	Bind bile acids in the intestine
Fibrates	Lower levels of circulating VLDLs and LDLs by unknown mechanism
Nicotinic acid derivatives	Reduce the release of VLDLs from the liver

HMG-CoA reductase – 3-hydroxy-3-methylglutaryl-CoA reductase

Lipid-lowering drugs are summarized in Table 4.5.

4.8 Summary

There are many disorders of the cardiovascular system and blood. Common cardiovascular disorders are cardiac failure, ischaemic heart disease, arrhythmias and hypertension. Although these conditions cannot be cured by drug therapy, there are many drugs available to help control them. Cardiac glycosides are useful in cardiac failure and arrhythmias because they improve myocardial contractility and slow conduction through the heart.

Diuretics encourage water loss from the body and this helps reduce circulating blood volume. Consequently, blood pressure falls and this reduces the work the heart has to do. Diuretics are used to treat cardiac failure and hypertension.

ACE inhibitors and angiotensin II receptor inhibitors limit the action of angiotensin II and suppress aldosterone production. These effects have benefits in the treatment of cardiac failure and hypertension because the resulting vasodilation and water loss lowers blood pressure.

Ischaemic heart disease can be treated with nitrates, potassium channel activators, β-blockers and calcium channel blockers. All these drugs either improve the blood supply to the myocardium or reduce its metabolic demand for oxygen, or both. Beta blockers are also useful in treating hypertensive patients who do not respond adequately to other antihypertensive drugs, and some types of arrhythmia because they slow the heart rate.

While class I calcium channel blockers are used to treat atrial arrhythmia, class II and class III calcium channel blockers are now considered first-line therapy for many patients with hypertension. Resistant hypertension can be treated with 'add-on' drugs α_1-antagonists, centrally acting antihypertensives and adrenergic neurone blockers.

Arrhythmias can also be treated with membrane stabilizers and amiodarone.

Disorders of coagulation, anaemia and hyperlipidaemia are relatively common, particularly in older patients.

Anticoagulant therapy, chiefly warfarin, and platelet inhibitors, such as aspirin, are used prophylactically to prevent thromboembolic disease while fibrinolytic drugs can be used to destroy thrombi already formed and can be life saving after a myocardial infarction or stroke.

Anaemia is treated according to its cause: iron replacement therapy for iron deficiency; vitamin B_{12} and folic acid for decreased production of erythrocytes and to alleviate the effects of increased destruction.

Hyperlipidaemia predisposes to many CVDs and can be improved by changes to lifestyle and diet. Drug therapy is introduced according to an individual patient's 10-year risk of CVD. The drugs of choice are currently statins, ezitimibe and omega-3 triglycerides with other drugs being added as necessary and especially in familial hyperlipidaemia, which can be difficult to treat.

Useful websites

www.bhsoc.org British Hypertension Society. Hypertension guidelines and 10-year risk of cardiovascular disease estimation.

www.cks.library.nhs.uk National Library for Health Clinical Knowledge Summaries.

http://www.nice.org.uk/CG034 National Institute for Health and Clinical Excellence. Hypertension Guidelines 2006.

http://www.nice.org.uk/CG36 National Institute for Health and Clinical Excellence Atrial Fibrillation Guidelines 2006.

http://www.sign.ac.uk/guidelines/ Scottish intercollegiate Guidelines network. Section 13 Oral anticoagulants.

Case studies

Any health care professional, for whatever reason, might see the first five case studies, which are relatively brief. You should be able to offer general professional advice to patients similar to these. The sixth case study is a patient who might be seen in the podiatry clinic and requires more depth of consideration.

Case study 1

Ms Thomas is a 52-year-old lady of African origin. She has hypertension for which she has been prescribed captopril. Ms Thomas wants to know if she should expect any side effects from this drug. She is particularly worried about her sensitive stomach. What can you tell her about side effects with captopril and is there any other advice you could give her about this medication?

Case study 2

Mrs Jackson is an obese 58-year-old patient who has just been diagnosed with hypertension. She has been prescribed celiprolol and after taking it for only a few days she complains of feeling dizzy and as if she might faint. How can you reassure Mrs Jackson and what advice could you give her to prevent her having a fall?

Mrs Jackson also has type 2 diabetes. Does this have any bearing on the guidance you can give to her?

Case study 3

Mr White is 62 and taking colestipol. He is aware that he has high lipid levels and has decided that he must do something about his lifestyle because he is not keen on taking drugs for the rest of his life.

Mr White asks your advice about his plans. He asks you what would be better – reducing the fat in his diet, increasing his intake of fruit and vegetable or getting more regular exercise (although he says he hates the idea of joining a gym). How would you advise him?

Mr White has noticed that his bowel movements have become irregular lately and he seems to be getting frequent headaches. He wonders if these could be side effects of the colestipol and if they are what he should do. What can you tell him?

Case study 4

Mr Jones is a widower aged 74 who has lived alone since his wife died a few years ago. Up until quite recently, he has been looking after himself very well. Some weeks ago, Mr Jones developed a severe pain in his leg, which was diagnosed as a deep vein thrombosis. He was admitted into hospital for treatment. He was discharged with a prescription for warfarin and told to go to the anticoagulant clinic on a weekly basis. Mr Jones has many questions for you about looking after himself to prevent any complications with his treatment.

He asks you about any potential adverse effects. What should you tell him? He tells you that he has been taking a loop diuretic for his mild hypertension and wonders if this drug might have any effect on his use of warfarin. What should you tell him? Mr Jones also wants to know whether he should modify his diet in any way because of the warfarin. What advice can you give him on diet (and drink)?

Case study 5

You are seeing a patient, Mr Frobisher, who has been taking digoxin for four weeks because he has been diagnosed with heart failure.

What sort of questions could you ask him to ascertain if he is experiencing any side effects of digoxin?

Mr Frobisher volunteers the information that he thinks the digoxin is giving him heartburn. When questioned further he admits he has been taking antacids on a regular basis. Should this be a cause for concern?

Digoxin is a drug with a narrow therapeutic ratio, which means therapeutic doses are very close to those that can cause toxic effects. Would you recognize signs of digoxin toxicity and would you know what to do if Mr Frobisher showed these signs?

Case study 6

This is a case study of a patient who might be seen in a podiatry clinic.

Mr Buckley is 65 and apparently fit and active with the exception of a painful sore on the upper surface of the third toe on his right foot.

On examination, you find that this sore has all the characteristics of an ulcer. Mr Buckley's feet are cold to the touch and he confirms that he often finds it difficult to keep his feet warm. He admits to having hypertension, for which he is being treated, but also says he has not seen his GP for 18 months other than to arrange this visit to the podiatrist. His current drug therapy has not changed for the past five years. There also appears to be significant ankle oedema, which has worsened since Mr Buckley retired 10 weeks ago.

Mr Buckley's current drug therapy is propranolol 240 mg per day.

Discuss the treatment of this patient using the questions below as a guide. They are intended to stimulate discussion not limit debate.

- What are the podiatric consequences of Mr Buckley's current drug treatment?

- Apart from symptomatic treatment, what would you advise the patient to do?

- How would you hope that the drug therapy of this patient could be altered to maximize the natural healing properties of the patient?

Chapter review questions

You should be able to answer these review questions from the material in the preceding chapter.

1. Describe the mode of action of three groups of drugs that are used to treat cardiac failure.

2. Describe the mode of action of four groups of drugs used to treat angina.

3. Describe the mode of action of the following drugs used to treat cardiac arrhythmia: amiodarone, verapamil and lidocaine.

4. Diuretics, ACE inhibitors, β-blockers and calcium channel blockers are all used to treat hypertension. Which of these drugs would be:

    (a) first choice in patients over 55 years of age

    (b) first choice in white patients under 55 years of age

    (c) an unwise choice in patients with asthma or diabetes

    (d) an alternative to calcium channel blockers for patients of African origin?

5. How should you be able to identify a patient who is on anticoagulants?

6. In use to prevent blood clotting, what is the essential difference between warfarin and aspirin?

7. The main aim of lipid-lowering drugs is to reduce plasma LDL levels. How is this achieved by

    (a) statins?

    (b) fibrates?

    (c) ezitimibe?

    (d) omega-3 triglycerides?

5

Respiratory Disorders

5.1 Chapter overview

As with cardiac and blood disorders, respiratory disease is relatively common in many older patients. However, there are common respiratory disorders that affect younger people as well. Although there are probably not as many diseases of the respiratory system as there are of the cardiovascular system, there are still a number of important groups of drugs that are used to treat respiratory disorders and their use can be life saving.

In order to understand diseases of the respiratory system and how they can be treated, it is necessary to have an overview of the normal function of the respiratory system; that is the structures involved, the mechanism of breathing and gas exchange and the control and variation of breathing rate. Air must be able to get in and out of the lungs efficiently and gas exchange must be adequate. Any condition that interferes with these processes will cause disease.

5.1.1 Respiratory physiology

The respiratory system is made up of air passages through the nose, windpipe and into the lungs, which are situated in the thoracic cavity (see Figure 5.1).

The walls of the thoracic cavity are supported by the ribs and its floor is formed by a sheet of muscle called the *diaphragm*.

The inner surface of the thoracic cavity and the outer surface of the lungs are covered by a shiny slippery membrane, the pleural membrane. Pleural fluid is secreted between this double membrane and acts as a lubricant reducing friction as the lungs move against the ribs during breathing. The pleural membranes also ensure that the chest cavity is airtight. This is essential for the lungs to be able to expand as the ribs move.

The hard and soft palates divide the nose from the throat and mouth. Air enters the nose through two openings, the nostrils. The nasal passages are lined with a moist hairy mucous membrane that traps dust, dirt and bacteria breathed in with air. As air passes through the nasal passages it is also warmed and moistened. The pharynx connects the

Pharmacology for the Health Care Professions   Christine M. Thorp
© 2008 John Wiley & Sons, Ltd

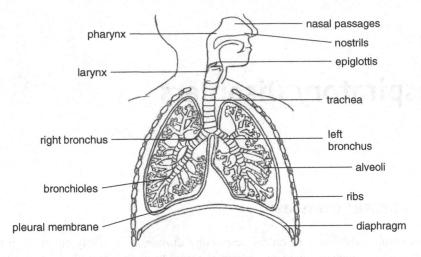

Figure 5.1   Structure of the respiratory system

nasal passages to the trachea. The epiglottis is a flap of cartilage that closes the trachea during swallowing. The larynx is situated in the pharynx at the top of the trachea. It contains membranes called *vocal cords*, which vibrate when air passes over them to produce sounds, thus allowing speech.

The trachea is a short tube supported by incomplete rings of cartilage, which keep the trachea open despite pressure changes that occur during breathing and prevent kinking when the head is turned. The cartilage rings are incomplete to allow for swallowing. The trachea divides into two bronchi, which enter the lungs. The bronchi divide into secondary and tertiary bronchi and further into smaller branches called *bronchioles*. The airways are lined with ciliated mucosal epithelium. The mucus traps dust and particles and the cilia beat upwards to prevent the mucus and trapped particles from going deeper into the lungs. Bronchioles are supported by smooth muscle, which allows their diameter to change. In all, there are about 23 levels of branching until the terminal bronchioles are reached. The terminal bronchioles divide into alveolar ducts at the end of which are tiny thin walled air sacs, the alveoli. The alveoli present an enormous surface area for gas exchange. Surrounding the alveoli are many blood capillaries. Oxygen in the air that is breathed into the lungs diffuses from the alveoli across their thin walls into tissue fluid and then into the blood capillaries. (See Figure 5.2.)

Diffusion of oxygen happens efficiently because there is a concentration gradient from air in the alveoli to blood in the capillaries; the distance through the alveoli walls, tissue fluid and capillary walls is small; and the total surface area of all the alveoli is large. For the same reasons, carbon dioxide diffuses in the opposite direction.

Air moves in and out of the lungs due to changes in volume and pressure inside the chest cavity. Inspiration is brought about by contraction of the diaphragm and the external intercostal muscles (between the ribs). The diaphragm flattens and the ribs are pulled upwards and outwards increasing the volume of the chest cavity. The pressure inside the chest cavity drops and air is drawn in through the trachea. Expiration is normally passive

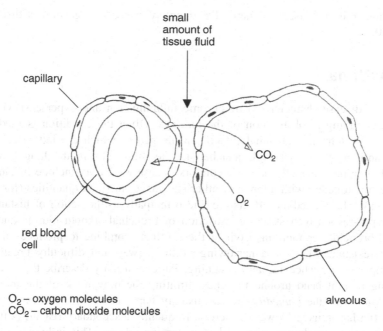

small
amount of
tissue fluid

capillary

CO_2

O_2

red blood
cell

alveolus

O_2 – oxygen molecules
CO_2 – carbon dioxide molecules

Figure 5.2   Gas exchange between alveolus and capillary

and happens when the muscles of inspiration relax and the elastic tissue of the lungs recoils. During exercise, expiration becomes forced involving the internal intercostal muscles and other muscles in the neck and abdomen.

The rhythm of breathing is maintained by the respiratory centre in the brain stem. Regular impulses from the inspiratory area are sent to the muscles of inspiration causing them to contract. During exercise, the expiratory area of the respiratory centre is activated and sends impulses to the muscles of expiration.

The smooth muscle in the walls of the airways allows the bronchioles to both constrict and dilate. Normally the bronchioles are in a state of partial constriction. During exercise, when there is a need for more air to enter the lungs, the bronchioles dilate. At rest, or if there is danger of breathing in noxious fumes, the bronchioles constrict. Dilation and constriction of the bronchioles is controlled by the autonomic nervous system. Partial constriction is the result of parasympathetic stimulation and an increase in this brings about further bronchoconstriction. There is no sympathetic nerve supply to the bronchioles. However, sympathetic stimulation brings about bronchodilation by the action of circulating adrenaline on β_2 receptors in the smooth muscle. In diseases such as asthma and bronchitis, one of the problems is excessive constriction of the bronchioles, which makes breathing difficult.

Bronchial asthma and chronic bronchitis, allergic rhinitis and hay fever are common diseases of the respiratory system and are discussed below together with their treatment.

This is followed by a brief discussion of cystic fibrosis, pneumonia and tuberculosis and their treatment with antibiotics. Use of antibiotics is covered in more detail in Chapter 9.

Lung cancer is not discussed here. Treatment of cancers in general is discussed in Chapter 10.

5.2   Asthma

Asthma is a disease characterized by chronic inflammation and hypersensitivity of the bronchioles leading to obstruction of the airways. Often the condition is mediated by the immune system through inhalation of an antigen to which the individual is allergic. The antigen reacts with IgE attached to the mast cells in the lung causing the release of pharmacologically active mediators. These mediators include histamine and prostaglandins together with a range of other substances in small quantities (for example leukotrienes). The immediate effects are due principally to the action of histamine and include hyper-responsiveness and constriction of bronchial smooth muscle and inflammation of bronchial epithelium. Both of these effects combine to produce an increase in airway resistance, that is, a narrowing of the airway and difficulty breathing out. Other symptoms are wheezing and coughing. Patients usually describe the sensation as like having a tight band around the chest limiting the movement of the thorax. This reaction, known as the *immediate phase*, usually happens within minutes of challenge by the particular antigen. However, there is frequently a secondary reaction often many hours later due to mechanisms other than histamine release. This is known as the *late phase reaction* and is usually less responsive to treatment than the immediate phase reaction.

Not all cases of asthma can be attributed to reaction to an allergen. Other stimuli that cause asthma attacks include exercise, cold air, infections and atmospheric pollutants. In these cases symptoms may be produced through stimulation of irritant receptors and release of mediators from sensory neurones. Treatment of asthma of whatever cause is by use of bronchodilators to reverse the bronchospasm of the immediate phase and by anti-inflammatory drugs to inhibit or prevent the development of the late phase.

5.3   Chronic bronchitis

Chronic bronchitis is characterized by hypertrophy of the glands of the airway and excessive production of mucus. In some cases, the disease is rapidly progressive and may proceed to fatality within five years of symptoms appearing. In most other cases, there is only a slow progression towards respiratory failure. Symptoms start initially with a winter-only productive cough through to a persistent productive cough often associated with frequent respiratory infection. Excessive production of mucus may result in stenosis of the alveoli with obstruction of terminal parts of the airway. The disease appears to have a strong genetic basis but environmental conditions such as smoky, dusty, cold and damp climates seem to accelerate progression of the disease. Treatment of this disease is mainly symptomatic through use of bronchodilators and anti-inflammatory steroids and antibiotics when necessary.

Emphysema often occurs together with, and as a result of, chronic bronchitis. In this condition, the walls of the respiratory bronchioles and alveoli are progressively destroyed leaving fewer but larger alveoli. This effectively cuts down the surface area for gas exchange and results in more air remaining in the lungs during exhalation. Oxygen diffusion is reduced and the patient becomes breathless, initially on exertion and, as the disease progresses, at rest. Whilst there is no specific treatment or cure for emphysema, symptomatic treatment with bronchodilators and anti-inflammatory drugs as used for chronic bronchitis will help. Together chronic bronchitis and emphysema are known as *chronic obstructive pulmonary disease* (COPD).

5.4 Drugs used to treat respiratory disorders

Often the same drugs are used to treat asthma and bronchitis, so they are considered here together. Guidelines on the management of chronic asthma and COPD are produced by the British Thoracic Society and published in the *BNF* and *MIMS*.

Drugs used to treat other respiratory conditions are considered separately.

5.4.1 Bronchodilators

5.4.1.1 Beta$_2$ (β_2) adrenoreceptor stimulants

Beta$_2$ adrenoreceptor stimulants cause bronchodilation by acting on β_2 adrenoreceptors in bronchial smooth muscle. Although they are selective for β_2 receptors these drugs do have some affect on cardiac β_1 receptors. Cardiac side effects are not usually a problem at therapeutic doses, but could be if a patient decided to increase the dose above that recommended.

β_2 adrenoreceptor stimulants are used as first choice for the rapid relief of an acute asthmatic attack and maintenance treatment of chronic asthma. They are more effective in treatment of the immediate phase reaction than the late phase reaction. β_2 adrenoreceptor stimulants are less effective in chronic bronchitis, possibly because in this disease bronchoconstriction is due to reflex stimulation of acetylcholine receptors following stimulation of local irritant receptors in lung tissue.

There are many drugs in this group but little to choose between them. They are most often used as aerosol or dry powder inhalers, but are also available as nebulizer solutions, tablets, syrups, injections and intravenous infusions.

Examples are salbutamol (short-acting for symptomatic use as required) and salmeterol (long-acting for symptom control at night or before exercise).

Side effects of β_2 adrenoreceptor stimulants are rare but include tremor, nervous tension, headache, peripheral vasodilation, tachycardia (increased heart rate), and hypokalaemia (low potassium levels) after high doses, and hypersensitivity reactions.

It is extremely important that patients are warned of the dangers of exceeding the stated dose on inhalers; this would increase the incidence of the above side effects.

5.4.1.2 Antimuscarinic bronchodilators

Antimuscarinic drugs are used to relieve, or at least partially reverse, bronchoconstriction that is refractory to β_2 adrenoreceptor stimulants. These drugs work by blocking acetylcholine receptors (the so-called muscarinic receptors) in the bronchioles. This effectively prevents bronchoconstriction in response to parasympathetic stimulation. They may be used in combination with other bronchodilators. Antimuscarinics are of particular use in chronic bronchitis where the airways seem resistant to β_2 receptor stimulation and because they also reduce mucus secretion. They are most frequently used as aerosol inhalers and in nebulizers.

An example of an antimuscarinic bronchodilator is ipratropium.

Side effects are rare, but patients may get a dry mouth, urinary retention and constipation, all of which are typical effects of parasympathetic inhibition.

High doses of antimuscarinic bronchodilators should not be used if the patient has glaucoma or an enlarged prostate gland, because they would make the glaucoma and urinary retention worse.

5.4.1.3 Xanthine bronchodilators

Xanthines are a group of drugs that directly cause relaxation of bronchial smooth muscle and some central respiratory stimulation. They also have a slight diuretic effect. Xanthines are of principle use in the immediate phase reaction of asthma but also have some effect on the late phase reaction. They are used to treat severe acute asthma attacks and chronic asthma, in particular control of nocturnal asthma and early morning wheezing. Xanthines also have some use in chronic bronchitis.

Formulations of xanthines include intravenous injection, tablets and capsules. Examples of xanthine bronchodilators are theophylline and aminophylline.

Side effects, which become common with high doses, are arrhythmia, hypotension, convulsions, nausea and vomiting, headache and insomnia.

5.4.2 Cromoglicate and related drugs

Drugs in this group are also known as *mast-cell stabilizers* and they probably work by blocking the release of mediators such as histamine and serotonin in the lung, although their exact mode of action is unknown. They can be used prophylactically to reduce the incidence of asthmatic attacks and to allow reduction in the doses of other drugs. They are not used in chronic bronchitis.

Cromoglicate and related drugs do not cause bronchodilation and therefore are of no use in an acute asthma attack. It is important that patients understand this.

These drugs are available as powder and aerosol inhalers for use on a regular basis.

Examples are sodium cromoglicate and nedocromil sodium.

Side effects are rare but dry powder inhalation may provoke bronchospasm. This can be prevented by concurrent use of a bronchodilator.

5.4.3 Leukotriene receptor antagonists

Leukotriene receptor antagonists are a relatively new class of drugs that directly bind to leukotriene receptors in the lung preventing the inflammatory actions of leukotrienes particularly in the late phase of asthma. They are used in cases of asthma where there is a poor response to bronchodilators alone.

Current examples are montelukast and zafirlukast, both of which are given orally, either alone or in combination with corticosteroids. They are not recommended for use in chronic bronchitis.

Side effects of leukotriene receptor antagonists are gastrointestinal disturbances and headache.

5.4.4 Corticosteroids

Corticosteroids are anti-inflammatory drugs that can be used in asthma to reduce airway hyper-responsiveness and to decrease bronchial oedema and mucus secretion. They are effective in the late phase reaction and reduce the intensity of allergic reactions. They are used in emergency treatment of severe acute attacks, for the treatment of mild to moderate attacks and prophylactically to prevent attacks. Corticosteroids can be useful in reducing acute exacerbations of chronic bronchitis.

Examples are beclometasone and budesonide given by inhalation and prednisolone given orally.

Corticosteroids used for inhalation are generally poorly absorbed into the circulation thereby reducing the possibility of systemic side effects.

Oral corticosteroids are sometimes needed in short courses to manage acute exacerbation and longer term when other drugs do not prevent severe attacks of asthma.

Side effects associated with long-term oral therapy and high-dose inhaled therapy are moon face, obesity, purple skin striae, hypertension, osteoporosis, diabetes mellitus, susceptibility to infection and adrenal suppression.

Long-term side effects of corticosteroid use are discussed in Chapter 7.

Regular users of corticosteroids should carry a steroid card.

5.4.5 Mucolytics

These are of doubtful therapeutic value, yet bring relief to many patients particularly those with COPD. They can reduce mucus viscosity and ease expectoration. An example of a mucolytic is carbocisteine.

Mucolytics should be used with caution in patients with a history of peptic ulcer because they may reduce the amount of mucus in the stomach. Steam inhalation may be as effective as a mucolytic in chronic bronchitis.

Drugs used to treat asthma and chronic bronchitis are summarized in Table 5.1.

Table 5.1 Drugs used to treat asthma and chronic bronchitis

Drug group	Action	Example	Use
Bronchodilators			
β_2 adrenoreceptor stimulants	Stimulate β_2 adrenoreceptors in bronchioles	Salbutamol	Asthma Chronic bronchitis
Antimuscarinics	Block muscarinic acetylcholine receptors in bronchioles	Ipratroprium	Asthma Chronic bronchitis
Xanthines	Direct relaxation of bronchial smooth muscle	Aminophylline	Asthma Chronic bronchitis
Prophylactics			
Cromoglicates	Mast cell stabilizers	Sodium cromoglicate	Asthma
Anti-inflammatories			
Leukotriene receptor antagonists	Block leukotriene receptors in the lung	Montelukast	Asthma
Corticosteroids	Reduce inflammation, bronchial oedema and mucus secretion	Beclometasone	Asthma Chronic bronchitis

5.4.6   Severe acute asthma (status asthmaticus)

Although asthma is considered to be a reversible condition, severe acute attacks can cause obstruction that can take days to reverse and in some cases is not reversible at all. Such attacks need to be treated as a medical emergency requiring hospital treatment. Treatment includes oxygen, inhalation of salbutamol in oxygen, intravenous hydrocortisone and oral prednisolone. Sometimes inhaled antimuscarinics are also used and intravenous salbutamol and aminophylline plus antibiotics if there is infection as well.

5.5   Treatment of other respiratory conditions

5.5.1   Allergic rhinitis and hay fever

These two conditions very similar and are described by their symptoms of either a runny nose or nasal congestion together with sneezing and itching of the nose and eyes.

Treatment is largely symptomatic and often consists of the use of antihistamines. However, cromoglicate can be used prophylactically by inhalation. Corticosteroids in the form of nasal drops may be used to treat local inflammation. Sympathomimetic drugs such as xylometazoline can be used as decongestants.

5.5.1.1 Antihistamines

Antihistamines are used in the treatment of allergic rhinitis and hay fever. They reduce symptoms such as runny nose and sneezing, but are less effective for nasal congestion.

There are many antihistamines available and they differ in their duration of action, anticholinergic effects (dry mouth, blurred vision, urinary retention and constipation) and incidence of drowsiness. Older antihistamines are the most likely to cause drowsiness. Newer ones, for example acrivastine, are preferred because they penetrate the blood-brain barrier less easily and cause less sedation. Nevertheless, all antihistamines should be labelled with a warning that they may cause drowsiness and patients should not drive or operate machinery when taking them.

Antihistamines should be used with caution in people with epilepsy, enlargement of the prostate, urinary retention, glaucoma and hepatic disease. Children and older people are more susceptible to the adverse effects.

5.5.2 Cystic fibrosis

Cystic fibrosis is an inherited disorder characterized by abnormally viscous secretions of the exocrine glands including those in the pancreas and the mucus secreting glands of the respiratory system and the gastrointestinal tract. The liver, salivary glands, sweat glands and glands of the reproductive organs also produce abnormal secretions.

Respiratory symptoms are obstruction of bronchioles with viscous mucus and recurrent infection. Respiratory infections require intensive antibiotic therapy and together with lung damage are the commonest causes of death in people with cystic fibrosis. Viscous secretion is treated with vigorous physiotherapy and the use of mucolytics.

Infections in cystic fibrosis are often due to *Staphylococcus aureus*, *Haemophilus influenzae* or *Pseudomonas aeruginosa*. Infecting organisms need to be identified so that the most appropriate antibiotics can be used. Antibiotics used to treat respiratory infections in cystic fibrosis commonly include ciprofloxacin, erythromycin, flucloxacillin and amoxicillin. However, specialist individual therapy is essential for maximum benefit to the patient and avoidance of the development of resistant strains of bacteria.

General use of antibiotics and their mechanisms of action and adverse reactions are discussed in Chapter 9.

5.5.3 Pneumonia

Pneumonia is inflammation of the lower parts of the lungs with accumulation of fluid in or around the alveoli. It can be caused either by viruses, bacteria and fungi, or by accidental inhalation of food or vomit.

Bacterial pneumonia can be treated with antibiotics such as the ones given above under cystic fibrosis, depending on the infecting organism.

Viral pneumonia is uncommon and not usually serious unless it occurs in the very young, the very old, those with chronic lung disease or the immunocompromized. Treatment may simply be rest and increased intake of fluids, but there are antiviral drugs available for serious viral pneumonia, for example amantadine, oseltamivir or zanamivir.

Side effects of amantadine and oseltamivir are gastrointestinal disturbances, insomnia and dizziness. Zanamivir should be used with caution in asthma and COPD because of the risk of bronchospasm.

Antiviral drugs are considered in Chapter 9.

Fungal pneumonia is most likely in the immunocompromized as an opportunistic infection with aspergillus or histoplasma and can be treated with amphotericin, itraconazole or voriconazole. Amphotericin has to be given intravenously and is toxic to the kidneys; itraconazole is given orally or by intravenous infusion and can produce liver toxicity and heart failure; voriconazole is given orally or by intravenous infusion and can be toxic to both liver and kidneys.

Antifungal drugs are considered in Chapter 9.

5.5.4 Tuberculosis

Tuberculosis is caused by infection with *Mycobacterium tuberculosis* or *Mycobacterium bovis*. Although the lungs are the major site of infection, the organism can also infect other tissues such as bones, joints and the brain. In the lungs, mycobacteria infect alveolar macrophages and this initiates an inflammatory reaction that involves many of the cells of the immune system. Infected cells form granulomas enclosing the organism and replacing healthy tissue. In the centre of a granuloma is an area of necrotic tissue, referred to as caseation, which can become fibrotic and calcified. The granulomas contain bacteria lying dormant but capable of becoming reactivated later in the patient's life if they become ill from some other cause or become immunosuppressed.

Treatment for tuberculosis is lengthy taking up to two years and requiring combinations of at least three different antibiotics.

First line drugs used to treat tuberculosis include isoniazid, rifampicin and pyrazinamide with streptomycin and ethambutol as secondary drugs.

Antibiotics for tuberculosis are discussed in Chapter 9.

5.6 Summary

Common respiratory disorders are asthma and chronic bronchitis. Asthma is often, but not always, caused by an allergic response and bronchitis is caused by chronic irritation of the lungs. Both conditions can be treated with inhaled bronchodilators. Although generally asthma responds best to β_2 adrenoreceptor stimulants and bronchitis responds to antimuscarinic drugs, both types of drugs may be needed. If these do not control symptoms, xanthine bronchodilators may be required in addition to one or both of the other two. Attacks of asthma can be prevented by prophylactic use of cromoglicate, although this drug is of no use in an acute attack or in chronic bronchitis. Leukotriene receptor antagonists can be used to reduce the inflammation of asthma particularly in the late phase. Inhaled corticosteroids are useful for treatment and prophylaxis of asthma and for reducing acute exacerbations of bronchitis. In severe cases oral corticosteroids might be needed, although long-term use is best avoided. Severe acute asthma is treated as an emergency with oxygen, inhalation of salbutamol and possibly inhaled antimuscarinics and intravenous and oral corticosteroids, intravenous salbutamol, aminophylline and antibiotics if necessary.

Less serious allergic rhinitis and hay fever are treated mainly with antihistamines and decongestants.

Cystic fibrosis is a serious inherited disorder where abnormally viscous mucus can cause obstruction of bronchioles and lead to frequent infection. Management of cystic fibrosis is through a combination of physiotherapy and antibiotics. It is important to identify infecting micro-organisms for effective individual treatment.

Pneumonia can result from infection by bacteria, viruses or fungi. Bacterial pneumonia is treated with antibiotics; viral pneumonia may not need specific drug treatment, but in serious cases antiviral drugs can be used. Fungal pneumonia usually only occurs in immunocompromized patients and must be treated with oral antifungal drugs.

Tuberculosis is a serious lung disease that can be difficult to treat. Successful therapy requires the prolonged use of a combination of at least three different antibiotics.

Treatment of other lung diseases, for example lung cancer, is beyond the scope of this book.

Useful websites

www.brit-thoracic.org.uk British Thoracic Society. Management guidelines.
www.cftrust.org.uk Cystic Fibrosis Trust. General information.

Case studies

Case study 1

The following case study is of a patient who might be seen in a podiatry clinic and who might be a suitable case for supplementary prescribing under a clinical management plan.

You may need to refer to Chapter 14 for details of supplementary prescribing and the use of clinical management plans.

Mrs Singleton is an 85-year-old woman with a long history of bronchial asthma. Despite this, she is fit and active for her age and visits the podiatrist every few months for the trimming of one corn. She appears to have no other problems with her feet with the exception of the occasional swelling of one ankle.

Mrs Singleton seems to be on quite a complex treatment regime:

Budesonide 400 µg one puff twice daily

Combivent ipratroprium 20 µg and salbutamol 100 µg two puffs four times daily

Terbutaline turbohaler 500 µg as required but no more than four times daily

Occasional antibiotics and oral corticosteroids

Mrs Singleton is quite chatty and tells you she gets breathless if she misses any of her inhalers and has regular 12-month reviews of her condition with her GP. She sees the practice nurse in between who does the breathing test with her but has to get the doctor to sign the repeat prescription. She confides in you that she very much does not like taking steroids because she has heard bad things about them.

Discuss the treatment of this patient using the questions below as a guide. They are intended to stimulate discussion not limit debate.

- Are there any podiatric consequences of Mrs Singleton's current drug treatment?

- Is there anything else about her case that might concern you?

- What can you deduce about Mrs Singleton's condition from the multiple use of inhalers and occasional prescription of oral corticosteroids and antibiotics?

- Given Mrs Singleton's aversion to taking steroids, what could you say about their use to reassure her?

- Do you think this patient would be suitable for supplementary prescribing with a clinical management plan and why?

Case study 2

This is a patient who could be seen by any health care professional for different reasons. You are asked to discuss how you would answer general questions about the patient's condition and treatment. You may find the British Thoracic Society Guidelines (see web site) useful for this case study.

Mrs Xavier has been prescribed montelukast for prevention of her asthma symptoms. When you speak to her, she complains that the drug does not seem to be working. After some gentle probing, you discover that Mrs Xavier has lost her reading glasses and cannot remember how or when she is supposed to take montelukast. Discuss how you would explain to her exactly what montelukast is for, how it works and when she should take it.

Although Mrs Xavier has not mentioned any side effects of this drug, are there any you should warn her about, and are there any other drugs that should not be taken with montelukast?

Mrs Xavier tells you that her granddaughter, who is one year old, also has asthma and recently whilst visiting her grandma had an attack. Mrs Xavier wonders if it would be alright to give her granddaughter montelukast if she has an asthma attack again. What should you tell her?

Chapter review questions

You should be able to answer these review questions using the information provided in this chapter. You may need to refer to a physiology book for physiology of the autonomic nervous system.

1. What are the main differences in the pathology of asthma and bronchitis?

2. Receptors for sympathetic stimulation can be divided into α and β receptors.

    – Where are β_1 receptors found and what happens when they are stimulated?
    – Where are β_2 receptors found and what happens when they are stimulated?

3. Why are β_2 stimulant bronchodilators less effective in bronchitis than in asthma?

4. Acetylcholine receptors come in two varieties: nicotinic and muscarinic.

    – Explain why antimuscarinic bronchodilators are likely to be more effective in chronic bronchitis rather than asthma.
    – Explain why antimuscarinic drugs should not be used in patients with benign prostatic enlargement.

5. Why is sodium cromoglicate of no use in acute attacks of asthma?

6. Why are inhaled steroids generally preferred to oral steroids for the long-term therapy of asthma?

7. Why do patients with cystic fibrosis suffer from recurrent respiratory infections?

8. How does identification of infecting micro-organisms ensure effective antibiotic therapy for individual patients?

9. Define pneumonia and explain why it can be serious.

10. Why is tuberculosis so difficult to treat?

6

Disorders of the endocrine system

6.1   Chapter overview

The endocrine system includes the endocrine glands and the hormones that they produce. (See Figure 6.1)

It is a varied and complex system involving many glands and many hormones that work in balance via interconnected feedback control systems. Hormones are chemicals that are released into the blood stream where they travel to distant tissues and their sites of action.

As with any body system, there is the possibility of disease and malfunction: glands can secrete too much (hypersecretion) or too little (hyposecretion) of their hormones; cells in tissues can have altered sensitivity to hormones; feedback control systems can break down. In these cases, hormones or their synthetic analogues are used as drugs. In other cases, hormones are used as drugs in ways that differ from their physiological role.

This chapter gives a general overview of the endocrine system. Some detail is given of the thyroid gland, the adrenal glands, the pancreas and disorders of them because health care professionals are likely to come across patients with such disorders.

6.2   Pituitary gland

The pituitary gland plays a very important role in the overall control of the endocrine system.

The gland consists of the posterior pituitary (neurohypophysis) and the anterior pituitary (adenohypophysis). The functions of the two are not related but both are controlled by the hypothalamus. (See Figure 6.2.)

6.2.1   Posterior pituitary gland

The posterior pituitary gland secretes two hormones, antidiuretic hormone (ADH) and oxytocin. They are both synthesized in the hypothalamus and reach the posterior pituitary via a neural connection. See Figure 6.2.

Pharmacology for the Health Care Professions   Christine M. Thorp
© 2008 John Wiley & Sons, Ltd

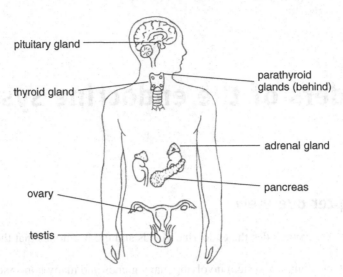

Figure 6.1 The endocrine system

6.2.1.1 Antidiuretic hormone

ADH is secreted in response to an increase in plasma sodium chloride concentration or a decrease in circulating blood volume. ADH acts on kidney collecting duct cells to increase water reabsorption. In this way plasma sodium chloride concentration and/or blood volume are restored to normal. ADH also acts as a vasoconstrictor, which helps to maintain blood pressure if circulating blood volume has fallen.

Lack of ADH causes diabetes insipidus, which is characterized by the production of large volumes of dilute urine. It is treated by replacement therapy with ADH or an analogue, desmopressin.

6.2.1.2 Oxytocin

Oxytocin is secreted in response to cervical dilation and uterine contractions during labour. By a positive feedback mechanism, oxytocin brings about further contraction of the uterus until parturition is complete. Oxytocin is used clinically to induce or augment labour.

6.2.2 Anterior pituitary gland

The anterior pituitary gland receives releasing factors from the hypothalamus in blood via a venous portal system. The releasing factors are hormones that regulate the secretions of the anterior pituitary gland. See Table 6.1 for actions and clinical uses of hypothalamic releasing factors.

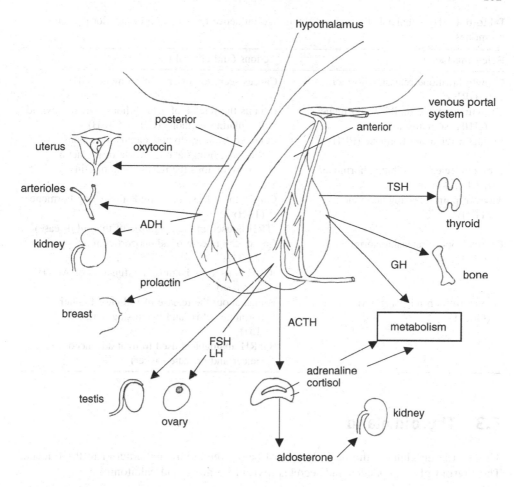

ADH – antidiuretic hormone
FSH – follicle stimulating hormone
LH – luteinising hormone
ACTH – adrenocorticotrophic hormone
GH – growth hormone
TSH – thyroid stimulating hormone

Figure 6.2 The pituitary gland

Many of the hormones of the anterior pituitary gland in turn regulate the activity of other endocrine glands and are known as trophic hormones or trophins. Table 6.2 summarizes the hormones produced by the anterior pituitary gland and their main actions and clinical uses. See also Figure 6.2.

Table 6.1 Hypothalamic hormones Factors that influence the secretion of anterior pituitary hormones

Releasing factor	Actions (and clinical use)
Growth hormone releasing hormone (GHRH)	Causes secretion of growth hormone (GH)
Growth hormone inhibiting hormone (GHIH, somatostatin)	Inhibits the release of growth hormone and thyroid stimulating hormone (GH and TSH)
Prolactin releasing hormone (PRH)	Released from the hypothalamus in response to suckling; brings about release of prolactin.
Prolactin release inhibiting hormone (PRIH, dopamine)	Normally inhibits the release of prolactin
Thyrotrophin releasing hormone (TRH)	Causes the release of thyroid stimulating hormone (TSH) (TRH is used in the diagnosis of thyroid disease)
Corticotrophin releasing hormone (CRH)	Causes the release of adrenocorticotrophic hormone (ACTH) (CRH is used to diagnose malfunction of ACTH secretion)
Gonadotrophin releasing hormone (GnRH)	Brings about the release of follicle stimulating hormone (FSH) and luteinising hormone (LH) (GnRH analogue is used to treat advanced breast cancer and prostate cancer)

6.3   Thyroid gland

This is a large gland in the neck, lying just below the larynx and anterior to the trachea. The thyroid gland produces and secretes thyroid hormone and calcitonin.

6.3.1   Thyroid hormone

The thyroid is the only gland that stores large amounts of its hormones – enough for about 100 days supply.

Under the influence of thyroid stimulating hormone (TSH) thyroid hormones are made by attaching iodine to tyrosine. The source of tyrosine is thyroglobulin, a protein synthesized in the thyroid gland. There are two hormones, thyroxine and triiodothyronine, referred to together as thyroid hormone.

An increased secretion of thyroid hormone results in an overall increase in metabolic rate in all tissues. There is a general increase in metabolism of carbohydrates, fats and proteins together with an increase in oxygen consumption and heat production.

Thyroid hormone also has an influence on growth by potentiating the effects of growth hormone and on skeletal development by affecting the action of parathyroid hormone and calcitonin.

Table 6.2   Anterior pituitary gland hormones Actions and clinical uses

Hormone	Actions	Clinical use
Growth hormone (GH)	Acting together with thyroxine, sex hormones and corticosteroids, GH controls normal growth	GH deficiency
Prolactin	Secretion normally inhibited by prolactin release-inhibiting hormone (PRIH) Suckling by an infant brings about a reflex release of prolactin Prolactin regulates milk production	
Thyroid stimulating hormone (TSH)	TSH stimulates the thyroid gland to release thyroid hormones	To check successful thyroidectomy
Adrenocorticotrophic hormone (ACTH)	ACTH controls synthesis and secretion of glucocorticosteroids from the adrenal cortex	To test adrenocortical function
Follicle stimulating hormone (FSH)	In the female, FSH stimulates development of follicles in the ovary and the secretion of oestrogen by the maturing follicle In the male, FSH is involved in the development of sperm	Infertility treatment
Luteinising hormone (LH)	In the female, LH is responsible for ovulation and the secretion of progesterone by the corpus luteum In the male, LH stimulates the interstitial cells of the testes to secrete testosterone	Infertility treatment

6.3.2   Calcitonin

Calcitonin is secreted by the thyroid gland in response to high levels of plasma calcium (see Section 6.4).

6.3.3   Thyroid disorders

6.3.3.1   Hypersecretion

Hypersecretion of thyroid hormone (*thyrotoxicosis*) occurs in Graves's disease due to autoimmune stimulation of the thyroid gland mimicking the effects of TSH. Symptoms are high metabolic rate, increased temperature and sweating, nervousness, tremor, tachycardia, increased appetite and loss of weight, goitre and protrusion of the eyeballs (*exophthalmia*).

Treatment can be by surgery or with drugs, for example carbimazole, to reduce the amount of thyroid hormone being produced.

The main adverse effect of carbimazole is granulocytopenia; this is rare and reversible on stopping drug therapy.

Radioactive iodine is also used to treat thyrotoxicosis and to assess thyroid function. Eventually it causes hypothyroidism, which has to be treated by thyroid hormone replacement.

6.3.3.2 *Hyposecretion*

Hyposecretion of thyroid hormone due to absence or incomplete development of the thyroid in early life causes cretinism, which is characterized by stunted growth and mental deficiency.

Hyposecretion in adults results in myxoedema, a condition where there is a characteristic thickening of the skin together with low metabolic rate, slow speech, poor appetite, weight gain, lethargy, bradycardia, sensitivity to cold and mental impairment. One form is Hashimoto's thyroiditis, an autoimmune condition where antibodies are produced against thyroglobulin.

Treatment is by replacement therapy with thyroxine.

Adverse effects of thyroxine therapy are due to overdose and are essentially those seen with hyperthyroidism together with the risk of angina, cardiac arrhythmia or cardiac failure.

Hypothyroidism with goitre can result (rarely) from deficiency of iodine in the diet; in which case treatment is with iodine supplementation.

6.4   Parathyroid glands

The parathyroid glands are four tiny glands embedded in the back of the thyroid gland.

6.4.1 Parathyroid hormone

The parathyroid glands secrete parathyroid hormone (PTH) when calcium levels in the body are low.

PTH, together with vitamin D, raises plasma calcium levels by increasing activity of osteoclasts, which results in release of calcium from bone; by increasing reabsorption of calcium by the kidney and by increasing absorption of calcium from the intestine.

Deficiency of PTH is treated with injections of calcium and calcitriol (a vitamin D analogue).

6.4.2 Calcitonin

High levels of calcium in blood plasma cause the release of calcitonin from the thyroid gland, which inhibits the reabsorption of calcium from bone by inhibiting osteoclasts.

Calcitonin is used in the treatment of Paget's disease and osteoporosis (see Chapter 7).

6.5 Adrenal glands

The adrenal glands are situated one above each kidney.

They have two separate parts; an inner medulla derived from nerve tissue and an outer cortex, which is typically endocrine in structure.

6.5.1 Adrenal medulla

The adrenal medulla secretes adrenaline (80%) and noradrenaline (20%). They have similar physiological effects, which augment activation of the sympathetic nervous system.

6.5.2 Adrenal cortex

The adrenal cortex is made up of three distinct layers, each of which secretes a different type of steroid hormone.

6.5.2.1 Aldosterone

The outermost layer of the cortex produces aldosterone. Aldosterone is released in response to fall in blood volume and/or blood pressure and it increases the rate of sodium and water reabsorption by the kidney distal tubules (see Figure 4.5). This increases blood volume and restores blood pressure.

6.5.2.2 Glucocorticosteroids

The middle and thickest layer of the cortex produces glucocorticosteroids, the main one being cortisol. Secretion of cortisol is mediated through feedback pathways involving adrenocorticotrophic hormone (ACTH) from the anterior pituitary and corticotrophin releasing hormone (CRH) from the hypothalamus as well as stimulation from higher brain centres. Psychological factors can affect the release of CRH, as can stimuli such as trauma, injury, infection or extremes of heat and cold. In this way, the adrenal glands are activated in response to a threatening situation.

Cortisol has many target tissues and has a variety of metabolic actions:

- increase in circulating blood glucose due to decrease in uptake by tissues other than the brain and increase in gluconeogenesis;
- increase protein breakdown, particularly in muscle, making amino acids available for repair and gluconeogenesis;
- lipolysis providing fatty acids for alternative energy supply.

Cortisol also has anti-inflammatory and immunosuppressive actions, which under physiological conditions help to limit response to injury and infection. When used pharmacologically, these effects are useful in treating chronic inflammatory conditions (see Chapter 5, use in asthma and Chapter 7, use in rheumatic diseases).

6.5.2.3 Androgens

The inner layer of the cortex produces adrenal androgens, the main one being testosterone. Adrenal androgens are produced in both sexes and are responsible for the development of secondary sexual characteristics that occur at puberty.

In males, the testes also secrete testosterone.

6.5.3 Disorders of the adrenal glands

Disorders of all parts of the adrenal glands are possible. Table 6.3 shows some of them together with their treatment.

6.6 Pancreas

The pancreas is a large organ containing exocrine and endocrine tissue. The islets of Langerhans make up the endocrine part and contain cells that produce insulin, glucagon and somatostatin. Cells that produce glucagon (α-cells) and somatostatin (D-cells) are found peripherally in the islets while the predominant insulin producing cells (β-cells) are found in the centre.

Insulin is essential for the metabolism of glucose and control of blood glucose levels, but it also has important effects on fat and protein metabolism.

In the liver, insulin reduces *glycogenolysis* (breakdown of glycogen) and *gluconeogenesis* (formation of glucose) and increases glycogen synthesis. In skeletal muscle, the action of insulin increases uptake and use of glucose and increases synthesis of glycogen. In fat tissue, insulin increases glucose uptake and the use of glucose to form lipids and suppresses *lipolysis* (lipid breakdown). Insulin also stimulates the uptake of amino acids and formation of protein in skeletal muscle and suppresses protein breakdown in the liver.

Secretion of insulin rises in response to increased blood glucose concentration. When glucose concentration falls to normal insulin secretion drops back to basal levels. A summary of the actions of insulin on glucose, protein and lipid metabolism is given in Table 6.4.

6.6.1 Diabetes mellitus

Diabetes mellitus is a chronic metabolic disorder caused by deficiency of insulin. Hyperglycaemia results because cells are unable to take up and use glucose. When the renal

Table 6.3 Disorders of the adrenal glands and treatment

Adrenal gland	Disorder	Cause	Treatment
Medulla	Phaeochromocytoma	Excessive adrenaline secretion	Surgical removal of adrenal gland
Cortex	Conn's syndrome Increased water retention, hypertension and disturbed sodium and potassium balance	Hypersecretion of aldosterone	Spironolactone (an aldosterone antagonist)
	Cushing's syndrome Obesity and abnormal fat deposition, muscle wasting, thin skin, poor wound healing and diabetes	Hypersecretion of cortisol	Irradiation/surgery
	Addison's disease Low blood pressure, muscular weakness, anorexia, loss of weight, hypoglycaemia and depression	Hyposecretion of aldosterone and cortisol	Replacement therapy with fludrocortisone and cortisol
	Androgenital syndrome Precocious puberty in males; masculinization in females	Hypersecretion of adrenal androgens	Irradiation/surgery

threshold for glucose is reached, glucose appears in the urine. This *glycosuria* causes osmotic diuresis and *polyuria* (production of large amounts of urine). Loss of fluid causes dehydration and increased thirst and *polydipsia* (excessive drinking). Other metabolic disturbances are seen, including protein wasting and ketosis due to increased lipid breakdown.

Long-term complications develop, many due to disease of blood vessels. Atheroma is common and this increases the risk of heart disease and stroke in diabetics. Damage to small blood vessels (*micro-angiopathy*) causes a number of problems. Diabetic *retinopathy* can lead to blindness and damage to blood vessels in the kidneys predisposes to hypertension, which in turn leads to further kidney disease and eventual kidney failure. Nerve damage (*neuropathy*) possibly due to fluctuating glucose levels over many years, can result in pain in hands, feet, thighs or face; digestive problems; bladder or bowel control problems; loss of sensation; muscle weakness; and impotence.

Table 6.4   Summary of actions of insulin

Metabolism of:	Effect of insulin
Glucose	Reduced glycogenolysis and gluconeogenesis and increased glycogen synthesis in liver
	Increased uptake of glucose and synthesis of glycogen in skeletal muscle
	Uptake of glucose and synthesis of lipid in fat tissue
Protein	Uptake of amino acids and synthesis of protein in skeletal muscle
	Decreased protein breakdown in liver
Lipid	Glucose converted to lipid in fat tissue
	Reduced lipolysis in fat tissue

Excessive blood glucose increases the risk of infections of the mouth and gums, lungs, skin, feet, genital areas and cuts. Neuropathy may allow damage to go unnoticed and untreated until a major infection develops.

There are two types of diabetes mellitus: insulin dependent diabetes mellitus (IDDM, Type 1), where there is absolute deficiency of insulin and non-insulin dependent diabetes mellitus (NIDDM, Type 2), where there is insulin resistance and impaired insulin secretion.

IDDM must be treated by insulin replacement or the patient will die from diabetic ketoacidosis.

NIDDM patients are usually obese and treatment starts with dietary control followed by oral hypoglycaemic drugs and insulin if necessary.

6.7   Treatment of diabetes mellites

6.7.1   Insulins

Bovine, porcine and recombinant human insulin preparations are currently available, although bovine insulin is rarely used. There are three types of insulin preparations: rapid-acting; intermediate-acting; and long-acting. A mixture of soluble (rapid and short duration of action) and isophane (slower onset and longer duration) human insulin twice a day is the most commonly used regime. However, since the duration of action of insulin preparations varies from patient to patient, combinations of different preparations have to be determined for individual patients. The aim of insulin therapy is to achieve the best possible control of blood glucose concentration, without precipitating hypoglycaemic episodes, to avoid complications.

Table 6.5 shows examples of some of the many insulin preparations with onset of action, peak effects and duration of action.

Insulin is ineffective orally and has to be administered by injection. The usual route is subcutaneously, though the soluble rapid-acting preparations can be given intravenously in emergencies.

Table 6.5 Insulin preparations and their properties

Type of preparation	Onset of action	Peak effects	Duration of action
Rapid-acting			
Insulin lispro	15 min	30–90 min	2–5 h
Soluble insulin	30 min	60–120 min	6–8 h
Intermediate-acting			
Isophane insulin	2 h	4–12 h	18–26 h
Insulin zinc suspension	2–4 h	6–12 h	18–26 h
Long-acting			
Extended insulin zinc suspension	4 h	10–30 h	36 h
Protamine zinc insulin suspension	4–8 h	14–24 h	28–36 h

All insulin preparations in the United Kingdom are produced as 100 units/ml.

(Insulin is a natural product and its concentration was originally measured by bioassay and given as international units. This tradition has persisted. One unit is inconveniently equivalent to 3.6 mg/ml.)

6.7.2   Oral hypoglycaemics

These drugs lower blood glucose by a variety of mechanisms. They are of no use in IDDM, but do have a place in the treatment of NIDDM. There are two main groups of oral hypoglycaemics: the sulfonylureas and the biguanides. There are some newer drugs recently added or in development.

6.7.2.1   Sulfonylureas

Sulfonylureas work by stimulating remaining functional β-cells in the islets of the pancreas so that insulin is secreted in response to normal stimuli. They also seem to make tissues more sensitive to insulin by an unknown mechanism.

Examples of sulfonylureas are gliclazide, glipizide and glibenclamide.

Adverse effects are increase in appetite and weight gain, which is serious in already obese patients. Hypoglycaemia can occur with any of these drugs and is particularly dangerous in the elderly. Gastrointestinal upsets and allergic skin rashes have been reported, as has bone marrow damage, which is rare but nevertheless severe.

6.7.2.2   Biguanides

Biguanides do not affect insulin production but they do require some functioning β-cells to work.

They seem to increase glucose uptake by peripheral muscle cells and reduce gluconeogenesis and intestinal glucose absorption. Only one, metformin, is available in the United Kingdom. It does not stimulate appetite and therefore is useful in obese patients.

Metformin also reduces plasma concentrations of cholesterol [in the form of low density lipoproteins (LDLs) and very low density lipoproteins (VLDLs)], which could help reduce the risk of atheroma, and it is unlikely to cause hypoglycaemia. It can be used with sulfonylureas.

Adverse effects of metformin are few, mainly gastrointestinal upsets and a rare but potentially fatal lactic acidosis, which is why other drugs in this class are unavailable. Metformin should never be prescribed for patients with renal disease or severe pulmonary or cardiac disease.

Metformin is contraindicated with concurrent use of iodine containing contrast agents. Patients have to stop metformin prior to radiological examination and must not use it again until renal function has returned to normal (see Chapter 13).

6.7.2.3 Drug interactions

Drug interactions with sulfonylureas and biguanides are common: non-steroidal anti-inflammatory drugs, warfarin, alcohol, monoamine oxidase inhibitors, some uricosurics, some antibacterials and some antifungals can interact with them. All increase the risk of hypoglycaemia. The mechanism is probably competition for metabolizing enzymes or displacement from plasma protein binding sites.

Drugs that decrease the effectiveness of sulfonylureas include diuretics and corticosteroids.

6.7.3 Other drugs

There are three other groups of drugs licensed in the United Kingdom for the treatment of NIDDM.

6.7.3.1 Acarbose

Acarbose is an alpha glucosidase inhibitor. This intestinal enzyme is responsible for the breakdown of complex sugars into monosaccharides to enable absorption. Without this enzyme, only glucose can be absorbed effectively from the intestine. The effect of acarbose is to reduce the postprandial rise in blood glucose levels.

Adverse effects of acarbose are abdominal discomfort and flatulence.

6.7.3.2 Prandial glucose regulators

Prandial glucose regulators are relatively new oral hypoglycaemics. Examples are nateglinide and repaglinide. They stimulate insulin release from β-cells in the pancreas. They are conveniently taken just before each main meal and have a rapid onset and short duration of action. They should be used together with metformin if glucose control is not sufficient with metformin alone.

Adverse effects are hypoglycaemia and hypersensitivity rashes.

6.7.3.3 Glitazones

Pioglitazone and rosiglitazone are similarly relatively new. They reduce insulin resistance by enhancing the uptake of glucose in the liver, adipose tissue and skeletal muscle. They can be used with metformin or sulfonylureas or alone but only if those drugs cannot be tolerated or are contraindicated.

Adverse effects are gastrointestinal disturbances and weight gain.

Glitazones can cause oedema and are contraindicated in people with cardiac failure.

6.7.4 Lifestyle changes

In addition to drug therapy, with insulin or oral hypoglycaemics and keeping as tight a control on blood glucose concentration as possible, changes in lifestyle can reduce the risk of diabetic complications. Losing weight, exercising, limiting alcohol intake, stopping smoking and eating a healthy diet with lots of whole grains, fruits and vegetables and a moderate amount of protein and reduced salt are all recommended. Many diabetics have high blood pressure and if lifestyle changes do not reduce it then anti-hypertensive therapy should be used.

Care should be taken to avoid skin infections. The feet in particular should be examined regularly for cuts, calluses and corns. Diabetics should visit the podiatrist regularly.

Good oral hygiene and regular visits to the dentist can prevent gum disease. Women should be aware of increased susceptibility to vaginal yeast infections.

Regular general health checks with a General Practioner (GP) are important, including glycosylated haemoglobin levels (this test gives a measure of how well blood glucose concentration is being controlled), cholesterol levels (including high-density lipoproteins [HDL]) and kidney function.

Eyes should be examined for signs of retinal disease annually so that treatment can begin before damage to vision occurs.

6.8 Summary

The endocrine system is complex, involving many glands and hormones, which are inter-connected by feedback mechanisms. The pituitary gland and the hypothalamus produce hormones that affect the activity of many other glands. There are many disorders of the endocrine glands.

Disorders of the thyroid gland are relatively common and can be due to either hyper- or hyposecretion of thyroid hormone. Hypersecretion is treated by surgery or with carbamazepine or radioactive iodine. Hyposecretion is managed with thyroxine replacement therapy.

The parathyroids are small, but important, glands involved in regulation of blood calcium levels together with calcitonin secreted by the thyroid gland and vitamin D. Calcitonin has a use in treating Paget's disease and osteoporosis (see Chapter 7).

The adrenal glands produce a variety of hormones. Adrenaline and noradrenaline are secreted from the adrenal medulla in response to sympathetic stimulation.

The adrenal cortex secretes aldosterone, cortisol and testosterone from different layers. Aldosterone causes salt and water reabsorption in the kidney if blood pressure or blood volume is low. Cortisol has a complex role in the body, being able to increase circulating glucose levels and mobilize amino acids and fatty acids in response to stress and trauma. Cortisol derived drugs are used therapeutically for their anti-inflammatory and immunosuppressant actions.

There are several disorders of the adrenal glands. Hypersecretion of aldosterone (Conn's syndrome) is treated with an aldosterone antagonist. Hypersecretion of cortisol (Cushing's syndrome) and testosterone (androgenital syndrome) are managed with surgery or irradiation. Hyposecretion of aldosterone and cortisol (Addison's disease) is life threatening and must be treated with replacement therapy of both hormones.

Diabetes mellitus is due to either absolute deficiency of insulin (Type 1) or insulin resistance and reduced insulin secretion (Type 2). In both types metabolic disturbances of carbohydrates, proteins and lipids occur with serious long-term consequences including atheroma, micro-angiopathy (which can lead to blindness and kidney damage) and neuropathy.

Careful control of glucose levels either with insulin replacement by injection or with oral hypoglycaemics together with lifestyle and dietary changes can help avoid long-term complications.

Useful websites

http://www.biomedcentral.com/bmcendocrdisord Biomedical Central Endocrine Disorders ejournal.

http://www.diabetes.org.uk/professionals Diabetes UK Healthcare Professionals and scientists page.

Case studies

Case study 1

This is a hypothetical patient, who might be seen by any health care professional for whatever reason. You are asked to discuss how you would answer the patient's questions about his condition, general lifestyle and medication.

Mr Qureshi, aged 68, has recently been diagnosed with Type 2 diabetes and has been prescribed an oral hypoglycaemic drug, gliclazide, by his GP. He has been told to take two 30 mg tablets each day with breakfast.

Mr Qureshi tells you he has a friend who has been diabetic for a long time and he has to inject insulin several times a day. He wonders why he has not been prescribed insulin. Discuss how you would explain to Mr Qureshi why he does not need insulin at this stage. What other advice could you give Mr Qureshi regarding his lifestyle and other things he can do to maximize his diabetic control?

Mr Qureshi mentions that he has experienced nausea and occasional vomiting since he started on gliclazide and wonders if this is a side effect of the drug and if he should be worried about it. What would you advise him about side effects of gliclazide?

Case study 2

The following patient is being investigated for recurrent urinary tract infection. It is planned that the patient's urinary tract will be imaged using intravenous iodine contrast.

Mrs Thompson is 55 years old and of normal weight. Apart from the recurrent infection and a skin condition, she appears to be well. You take a careful patient history and discover that Mrs Thompson is taking the following medication:

Levothyroxine sodium 100 mg before breakfast

Trimethoprim 100 mg at night

Discuss the patient's management using the following questions as a guide:

- What condition would levothyroxine be prescribed for?

- What could the skin condition be?

- Would it be safe to give this patient intravenous iodine contrast agent?

- What is trimethoprim and what is it used for?

Chapter review questions

You should be able to answer these review questions using the information covered in the preceding chapter.

1. What are the symptoms of Graves's disease?

2. What is the treatment for hyposecretion of thyroid hormone?

3. Explain how it is possible for autoimmune disease to cause hyposecretion of thyroid hormone in one condition and hypersecretion in another.

4. Describe the actions of calcitonin and why it is used to treat Paget's disease and osteoporosis. You may need to refer to Chapter 7.

5. What are the physiological actions of cortisol?

6. What is the essential difference between Type 1 and Type 2 diabetes?

7. What are the long-term complications of uncontrolled diabetes and what can a diabetic do to reduce the risk of getting them?

8. Can you explain why there is no one standard insulin therapy for all diabetics?

9. Discuss the use of oral hypoglycaemics in Type 2 diabetes.

7
Disorders of the musculoskeletal system

7.1 Chapter overview

There are many diseases and conditions affecting the musculoskeletal system; many of them are rare and certainly too numerous to include here. The following diseases have been chosen, either because they are relatively common or because they are of particular interest or relevance to podiatrists, physiotherapists or radiographers.

Rheumatic diseases are chronic inflammatory diseases primarily affecting the joints, but with involvement of other tissues and organs. Gout is included as a chronic inflammatory joint disease and osteoarthritis as a chronic degenerative joint disease.

Paget's disease, osteoporosis and osteomalacia are chronic diseases of bone.

Myasthenia gravis is a disease of the neuromuscular junction that affects skeletal muscle function.

Multiple sclerosis and motor neuron disease are diseases of the nervous system, but are included here because major effects of these diseases are on skeletal muscle function.

There are no cures for any of these diseases but many drugs are available to reduce inflammation and alleviate symptoms.

7.2 Rheumatic diseases

Rheumatic diseases are a group of chronic inflammatory diseases that include rheumatoid arthritis, juvenile rheumatoid arthritis, ankylosing spondylitis, systemic lupus erythematosus and psoriatic arthritis. As a group they are characterized by increased activity of phagocytes and the release of inflammatory mediators, which cause tissue damage. In diseases that primarily affect the joints there is inflammation and proliferation of cells of the synovial membrane and damage to cartilage and bone, but other connective tissues and organs in the body can also be damaged.

Pharmacology for the Health Care Professions   Christine M. Thorp
© 2008 John Wiley & Sons, Ltd

7.2.1   Rheumatoid arthritis

Rheumatoid arthritis is a chronic inflammatory autoimmune disease of largely unknown aetiology affecting about 1% of the adult population and is more common in women than men. It is diagnosed after persistent symmetrical multiple joint inflammation of more than six weeks duration. There is elevation of 'acute phase' plasma proteins, which are released during inflammation and often vasculitis and damage to heart muscle. The serum of most patients contains rheumatoid factors, which are autoantibodies that induce complement and macrophage activation and contribute to inflammation. The disease progression varies in different individuals.

Juvenile rheumatoid arthritis occurs before the age of 16. There are several subtypes and the disease can adversely affect growth in children.

7.2.2   Ankylosing spondylitis

Ankylosing spondylitis particularly affects the iliosacral and vertebral joints which eventually become fused. Due to ossification of ligaments connecting the vertebrae, the spinal column becomes curved and rotational motion is restricted. Extra-articular involvement includes conjunctivitis, psoriatic inflammation of the skin and nails and lung fibrosis.

7.2.3   Systemic lupus erythematosus

Systemic lupus erythematosus is a systemic inflammatory disease with formation of autoantibodies against DNA, wide spread immune complex deposition and complement activation. Symptoms include arthritis, a characteristic butterfly rash on the face, myalgia, glomerulonephritis, vasculitis, pericarditis and inflamed and fibrosed lungs.

7.2.4   Psoriatic arthritis

Psoriatic arthritis occurs in approximately 15% of patients with psoriasis. The cause is unknown. Skin lesions are minimal with a polyarthritis affecting particularly the small joints of the hand in an asymmetrical pattern. Usually the disease is mild but chronic. Sometimes the joint involvement is similar to that seen in rheumatoid arthritis, that is, symmetrical polyarthritis. Rarely, the disease is rapidly progressing with joint destruction. The serum is rheumatoid factor negative.

7.3   Drugs used to treat rheumatic diseases

Most rheumatic diseases require symptomatic treatment to relieve pain and increase joint movement using non-steroidal anti-inflammatory drugs (NSAIDs) and anti-inflammatory corticosteroids to limit or stop the inflammation. In addition treatment aims to prevent destruction of joints by the use of disease-modifying antirheumatic drugs (DMARDs). Indeed, use of DMARDs early on in the disease process is now recommended. The British Society for Rheumatology produces clinical guidelines for treatment of rheumatic diseases (see web site).

Many of the same types of drugs are used to treat different rheumatic diseases.

7.3.1 Non-steroidal anti-inflammatory drugs (NSAIDs)

NSAIDs are used in the treatment of chronic inflammation because they have a lasting analgesic action and an anti-inflammatory effect. See Chapter 12 for their use as analgesics.

Both of these effects are due to the inhibition of an enzyme, cyclo-oxygenase (COX). This enzyme normally converts arachidonic acid to prostaglandins, thromboxanes and prostacyclin. (See Figure 7.1.)

Arachidonic acid is an unsaturated long chain fatty acid found in all cell membranes, which is released as a result of tissue damage.

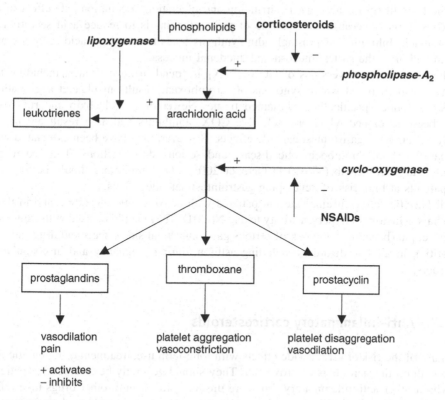

Figure 7.1 Prostaglandin synthesis: Site of action of NSAIDs and anti-inflammatory corticosteroids

Prostaglandins play a role in inflammation by causing vasodilation and increased vascular permeability and in pain by sensitizing sensory nerve endings to the effects of other mediators such as bradykinin.

Thromboxanes promote platelet aggregation.

Prostacyclin is produced from arachidonic acid in undamaged endothelium of blood vessels and plays a role in preventing unnecessary blood clotting. (See Chapter 4.)

Inhibition of COX by NSAIDs reduces rather than abolishes inflammation because the drugs do not inhibit production of other mediators of inflammation. Most patients experience some relief from pain, stiffness and swelling, but these drugs do not alter the course of the disease, prevent tissue destruction or produce remission.

7.3.1.1   Individual NSAIDs

There are many drugs belonging to the NSAIDs group and the main differences between them are the incidence and type of side effects. Aspirin used to be the obvious first choice anti-inflammatory analgesic, but its use has been superseded by less the toxic drugs. Aspirin is more likely to be used as a short-term analgesic than an anti-inflammatory drug. Ibuprofen, fenbrufen and naproxen are now the drugs of first choice because they are the least likely to cause gastric irritation and ulceration, a common side effect of all NSAIDs. This is because another role of prostaglandins is to reduce acid secretion in the stomach. Inhibition of prostaglandin synthesis leads to excessive acid secretion and inflammation of the gastric mucosa and duodenal mucosa.

There are different versions of COX. COX_1 is found in most tissues, including the stomach and in blood where synthesis of thromboxane results in platelet aggregation. COX_2 is found specifically in inflamed tissue. Newer drugs celecoxib and rofecoxib have been developed which are selective COX_2 inhibitors and as a result are not as likely to cause the gastrointestinal side effects. However, they have been associated with increased risk of thromboembolic disease and serious skin reactions. This led to the withdrawal of rofecoxib (Vioxx) in October 2004. COX_2 inhibitors should be reserved for patients at high risk of developing gastrointestinal side effects.

All NSAIDs are contraindicated in patients with active peptic ulceration and in those who have a history of hypersensitivity to any NSAID. They should be used with caution in older people (because of the risk of serious gastrointestinal side effects and drug-induced hepatitis), in allergic disorders including asthma, during pregnancy and in coagulation disorders.

7.3.2   Anti-inflammatory corticosteroids

Because of the risk of serious side effects with long-term use, treatment of rheumatic disease with corticosteroids is controversial. They should generally be reserved for patients in whom other anti-inflammatory drugs are unsuccessful or until other drugs take effect (see DMARDs below). However, the use of corticosteroids does depend on individual circumstances.

In acute exacerbations, which could be life threatening, a high initial dose of corticosteroid can be given to induce remission. The dose should then gradually be reduced
to the lowest maintenance level that will control the disease or, if possible, discontinued
altogether. This approach carries the risk that relapse may occur as the dose is reduced,
particularly if this is done too quickly. In order to avoid this relapse, the dose is often
increased again and maintained. The danger then is that the patient becomes dependent
on corticosteroids. Sometimes pulse doses (large doses on three consecutive days) of
corticosteroids are used to suppress inflammation while longer term and slower acting
DMARDs are being started.

Corticosteroids suppress all phases of the inflammatory response, including the early
swelling, redness and pain and the later stages seen in chronic inflammation. Circulating
lymphocytes and macrophages are reduced in number and the formation of prostaglandins
and leukotrienes is inhibited via inhibition of phospholipase A_2. Phospholipase A_2 is the
enzyme that converts cell membrane phospholipids into arachidonic acid. (See Figure 7.1.)

7.3.2.1 Prednisolone

Prednisolone is used for most rheumatoid disease. It is not the most potent of the corticosteroids available, but it has the advantage of allowing fine adjustment of doses. To
minimize side effects the maintenance dose of prednisolone should be kept as low as
possible.

In moderate to severe rheumatoid arthritis of less than two years duration, low dose
prednisolone may reduce the rate of joint destruction. Nevertheless, it is recommended
that maintenance therapy should be for two to four years only and then gradually reduced
to avoid possible long-term adverse effects.

7.3.2.2 Local corticosteroid injection

In some cases where there is involvement of only a few joints, corticosteroids (usually
triamcinolone or hydrocortisone) can be administered locally by intra-articular injection
to relieve pain, increase mobility and reduce deformity.

Methylprednisolone has recently (November 2006) been added to the list of drugs
that registered podiatrists can administer by injection into soft tissue injury. Registered
physiotherapists can administer corticosteroids by injection according to patient group
directions.

See Chapter 14 for legislation applicable to professional use of drugs in podiatry and
use of patient group directions by health care professionals.

7.3.2.3 Disadvantages of corticosteroids

Disadvantages of systemic corticosteroids are widely recognized and perhaps the most
serious is suppression of the pituitary-adrenal axis. The addition of exogenous corticosteroids to the systemic circulation is detected by the pituitary gland. In response
the pituitary gland reduces its own release of adrenocorticotrophic hormone (ACTH).
ACTH normally stimulates the adrenal glands to release their natural corticosteroids, but
with reduced levels of ACTH this does not happen (see Chapter 6). With prolonged

use of anti-inflammatory corticosteroids, the adrenal glands atrophy and lose their ability to synthesize their natural corticosteroids. If the corticosteroid drugs are stopped and the adrenal glands are no longer able to produce sufficient natural corticosteroids, the body will be unable to withstand every day stresses, a situation that can be life threatening.

Anti-inflammatory corticosteroids depress the immune system. Response to infection can be severely impaired as a result of inhibition of the activity of leukocytes and reduction in their numbers. Infections such as tuberculosis can spread extensively before they are discovered.

Prolonged use of anti-inflammatory corticosteroids can also impair the function of fibroblasts, which reduces the ability of wounds to heal.

In addition, changes in the synthesis of collagen occur resulting in thinning and loss of the mechanical resistance in the skin. On cessation of corticosteroid use, the structure of the skin may recover to a degree but there may be permanent damage in the form of striae (stretch marks) particularly at sites of fat accumulation or oedema.

There are other metabolic affects associated with chronic corticosteroid use including: redistribution of fat giving the typical moon face and buffalo hump; hyperglycaemia and possible diabetes; protein loss from skeletal muscles with wasting and weakness; increased bone metabolism leading to osteoporosis; and growth inhibition in children due to early closure of the epiphyseal plates in long bones.

Adverse reactions to anti-inflammatory corticosteroids are summarized in Table 7.1.

7.3.3   Drugs that suppress the rheumatic disease process

This is a diverse group of drugs that do not inhibit COX; neither do they have analgesic or direct anti-inflammatory activity. They appear to suppress the disease process in rheumatoid arthritis and other chronic inflammatory conditions and may cause remission

Table 7.1   Adverse reactions to corticosteroids

Adverse reaction	Effects
Suppression of the pituitary-adrenal axis	Inhibition of ACTH release
	Atrophy of adrenal glands; reduced synthesis of natural corticosteroids
	Body unable to withstand stress if corticosteroid withdrawn – life threatening
Immunosuppression	Leukopenia; reduced activity of leukocytes
	Increased susceptibility to infection
Impaired healing	Reduced activity of fibroblasts
Skin atrophy	Reduced collagen production; thin skin easily damaged
Metabolic effects	Redistribution of fat; moon face and buffalo hump
	Hyperglycaemia; possible diabetes
	Protein wasting; muscle weakness Osteoporosis

of disease. These are the disease-modifying antirheumatic drugs (DMARDs) (see above). Unlike NSAIDs and corticosteroids, they do not produce an immediate therapeutic effect but require four to six months of treatment for a full response. However, if one drug produces no objective benefit within six months, then it should be discontinued and replaced with an alternative.

DMARDs can be used in rheumatoid conditions when treatment with NSAIDs alone does not give adequate benefit, or in patients who are taking excessive doses of corticosteroids. Since DMARDs are toxic and the course of rheumatoid arthritis can be unpredictable, their use may be delayed in order to ascertain that severe symptoms are persisting. However, treatment with DMARDs is now recommended as early as possible. This is because it is now known that joint damage occurs within the first 18 months of onset of symptoms.

In use DMARDs reduce the symptoms and signs of inflammatory joint disease and also improve systemic effects such as vasculitis. In addition, serum markers of disease progression (erythrocyte sedimentation rate and rheumatoid factor concentration) are reduced.

For many of these drugs the mode of action is not fully understood, but they all interfere with the activity of cells of the immune system in one way or another.

7.3.3.1 Gold salts

Gold salts accumulate in phagocytic cells thereby reducing their activity and inhibit the migration of lymphocytes into inflamed tissue. An example is sodium aurothiomalate, which is given by intramuscular injection.

They are toxic drugs capable of causing skin reactions ranging from erythema to exfoliating dermatitis, inflammation of mucous membranes and blood dyscrasias.

7.3.3.2 Penicillamine

Penicillamine may inhibit production of inflammatory mediators and can delay the progress of disease.

It has severe side effects; in particular bone marrow depression, and precipitation of myasthenia gravis and myositis.

7.3.3.3 Sulfasalazine

Sulfasalazine is split by bacteria in the colon to produce aminosalicylic acid and sulfapyridine, the latter of which is probably the active drug. Its mode of action is unknown.

Toxic effects of sulfasalazine include gastrointestinal disturbances, skin rash and leukopenia.

7.3.3.4 Chloroquine and hydroxychloroquine

Chloroquine and hydroxychloroquine are normally used to treat malaria. In inflammatory disease their exact mode of action is unknown but they inhibit the migration of leukocytes and proliferation of lymphocytes in rheumatoid arthritis. They also reduce release of enzymes in phagocytic cells and inhibit phospholipase A_2 and therefore reduce prostaglandin and leukotriene synthesis. Chloroquine and hydroxychloroquine are sometimes used to treat systemic and discoid lupus erythematosus, but should not be used for psoriatic arthritis.

Because of the risk of retinopathy, long-term therapy should be reserved for cases in which all other DMARDs have failed.

7.3.3.5 Immunosuppressants

Immunosuppressants, such as azathioprine, ciclosporin and methotrexate, can be used to treat severe inflammatory disease. Their use depends on their ability to inhibit the activity and proliferation of lymphocytes and other leukocytes and because of this they are very toxic to the bone marrow. Other adverse effects are nausea, leukopenia, blurred vision, rashes and hair loss.

7.3.3.6 Cytokine inhibitors

Cytokines are chemicals released by damaged cells and cells involved in inflammation and repair. Drugs have been developed against two of them, tumour necrosis factor α (TNF-α) and interleukin 1 (IL-1). These two cytokines are released from macrophages and together they are key players in the inflammatory response. Levels of both are raised in rheumatoid arthritis and other autoimmune diseases.

Cytokine inhibitors are relatively new and they must only be prescribed under specialist supervision. Their use is still being monitored by the Medicines and Healthcare Products Regulatory Agency (MHRA). Treatment should be withdrawn if there are severe side effects or no obvious improvement after three months. None of these drugs is recommended for the treatment of systemic lupus erythematosus, because they can make the condition worse.

7.3.3.6.1 Inhibitors of TNF-α Inhibitors of TNF-α are used in highly reactive rheumatoid arthritis, severe ankylosing spondylitis and active progressive psoriatic arthritis that have not responded to at least two DMARDs. Administration regimes vary. Etanercept can be used either alone or with methotrexate and is given by sub-cutaneous injection twice weekly. Infliximab and adalimumab should both be used together with methotrexate. Infliximab is given by intravenous infusion every eight weeks. Adalimumab is given by sub-cutaneous injection on alternate weeks and should only be used for moderate to severe rheumatoid arthritis and active progressive psoriatic arthritis.

TNF-α inhibitors are associated with increased risk of severe infections, for example with TB, and septicaemia. Adverse effects are many and include nausea, abdominal pain, worsening heart failure, hypersensitivity reactions, headache, depression and blood disorders.

7.3.3.6.2 Inhibitors of IL-1 Anakinra inhibits the activity of IL-1. It is licensed for use in combination with methotrexate for rheumatoid arthritis that is unresponsive to methotrexate alone. Anakinra is given by sub-cutaneous injection once weekly. National

Table 7.2 Summary of antirheumatic drugs

Drug group	Examples
Non-steroidal anti-inflammatory drugs (NSAIDs)	Ibuprofen, celecoxib
Anti-inflammatory corticosteroids	Prednisolone
Drugs that suppress the rheumatic process (DMARDs)	Gold salts, penicillamine, sulfasalazine, chloroquine
Immunosuppressants	Azathioprine, ciclosporin
Cytokine inhibitors	Etanercept, infliximab, adalimumab, anakinra
Monoclonal antibodies	Rituximab

Institute for Health and Clinical Excellence (NICE) guidelines currently recommend it is only used in controlled long-term clinical studies.

7.3.3.7   *Monoclonal antibody*

Rituximab is a monoclonal antibody that destroys B lymphocytes. It was previously licensed only for the treatment of particular types of leukaemia (see Chapter 10). Since November 2006, rituximab has been licensed for treatment of refractory rheumatoid arthritis, where other treatments, including anti TNF-α have failed to produce an adequate response. NICE guidelines were produced in 2007.

A summary of drugs used to treat rheumatic diseases is given in Table 7.2.

7.4 Gout

Strictly speaking, gout is not a rheumatic disease, but as it involves inflammation of joints it is often included. The pain and swelling of gout is caused by the deposition of uric acid crystals in the joint. Uric acid is a substance that forms when the liver metabolizes purines (from DNA and RNA). It is usually in solution in the blood and filters through the kidneys into the urine. In people with gout, the uric acid level in the blood exceeds its solubility and crystals are deposited in joints and other tissues. In joints this causes the synovium to become inflamed. Initially, episodes of inflammation last only a week or so with no symptoms in between. However, untreated acute attacks occur more frequently and chronic inflammation results in damage to affected joints leading to stiffness and limited mobility.

Attacks of gout usually develop very quickly. The first attack often occurs in the middle of the night with the sufferer waking up with extreme joint pain. Typically the first metatarsal-phalangeal joint of the big toe is affected, although the finger joints and the joint at the base of the thumb may also be involved.

An episode of gout can be triggered by excessive alcohol consumption, a diet high in purines (meat and fish), surgery, sudden severe illness, injury to a joint, use of diuretics or chemotherapy.

Over several years, uric acid crystals can build up in other tissues to form large deposits, called *tophi*, under the skin. Tophi are often found in or near severely affected joints, on or near the elbow, over the fingers and toes and in the outer edge of the ear. Uric acid crystals can form stones in the kidneys, in the ureters or in the bladder.

Gout can occur at any age, but the first attack often affects men between the ages of 40 and 50. Gout does affect women as well, particularly after the menopause.

Almost all people with gout have excess uric acid in their blood, a condition called *hyperuricaemia*. However, there are many people who have hyperuricaemia but do not suffer from gout. Hyperuricaemia can result from the kidneys being unable to excrete uric acid fast enough or from the body making too much uric acid.

7.5 Drugs used to treat gout

Approaches to management of gout include treatment of acute attacks with anti-inflammatory drugs and medication to control uric acid levels long term, either by increasing elimination of uric acid or by reducing the amount of uric acid being produced.

7.5.1 Treatment of acute attacks

7.5.1.1 Colchicine

Colchicine has been used to treat gout for over 2000 years. It relieves the pain and swelling of acute attacks. It works best if taken during the first two days of an attack. Colchicine interferes with the formation of tubulin necessary for cell division and cell motility and as such inhibits the formation and migration of leukocytes into the inflamed joint. It is not, however, useful for other types of inflammation.

Colchicine can cause diarrhoea, nausea, and abdominal cramps and bone marrow depression, which limits its long-term use.

7.5.1.2 Non-steroidal anti-inflammatory drugs

NSAIDs are used to relieve the pain and swelling of an acute attack (see page 117 above). Aspirin should not be used because it competes with uric acid for tubular secretion in the kidney.

7.5.2 Control of uric acid levels

Long-term treatment with drugs that control uric acid levels prevents future episodes of gout and helps resolve tophi. However, these drugs do not relieve the pain and inflammation of an acute attack and may take many months to have an effect. Concurrent treatment with colchicine or a NSAID is necessary for the first three to six months.

Table 7.3 Drugs used to treat gout

Treatment of acute attacks	Action
Colchicine	Inhibits formation and migration of leukocytes
Non-steroidal anti-inflammatory drugs	Anti-inflammatory and analgesic

Control of uric acid levels	Action
Allopurinol	Inhibits xanthine oxidase and reduces purine metabolism into uric acid
Probenecid and sulfinpyrazone	Reduce reabsorption of uric acid in the kidney

7.5.2.1 Allopurinol

Allopurinol reduces the amount of uric acid in blood and urine by slowing the rate at which the liver produces uric acid. It does this by inhibiting an enzyme, xanthine oxidase. This enzyme normally catalyses the later stages of metabolism of purines to uric acid. As a result of xanthine oxidase inhibition less uric acid is produced. Intermediate metabolites xanthine and hypoxanthine are produced in larger amounts, but these are freely water soluble and eliminated by the kidneys.

Side effects of allopurinol include skin rash and stomach upset and in rare cases a severe allergic reaction.

7.5.2.2 Probenecid and sulfinpyrazone

Probenecid (available on a named-patient basis only) and sulfinpyrazone lower the uric acid level in blood by increasing the amount of uric acid passed in the urine. They do this by competing for the transport mechanism responsible for tubular reabsorption of uric acid in the kidney.

Side effects of these two drugs include nausea, skin rash, stomach upset or headaches.

A drawback to the use of probenecid or sulfinpyrazone is that because of increased excretion of other organic acids that use the same transport system, the urine becomes more acidic than normal. Acidic urine may encourage the precipitation of uric acid and the formation of kidney stones. To avoid this patients should be treated with allopurinol first.

Sulfinpyrazone enhances the effects of warfarin.

Table 7.3 summarizes drugs available to treat gout.

7.6 Osteoarthritis

Osteoarthritis is the most common form of arthritis worldwide. It occurs in about 50% of people over 60 years of age and is more common in women. Disease progression usually starts slowly with individual joints. Involvement of weight-bearing joints is the greatest cause of disability in older people. Osteoarthritis is primarily a degenerative disease of cartilage. With breakdown of cartilage inflammation of the synovium can occur. This is followed by more cartilage destruction and bone remodelling as an attempt at repair.

Unlike rheumatoid arthritis, there is no systemic disease associated with osteoarthritis.
Treatment of osteoarthritis is with NSAIDs for their analgesic and anti-inflammatory effects. Corticosteroids are not recommended and disease-modifying drugs are not effective in osteoarthritis.

7.7   Paget's disease

Paget's disease is a disorder of bone turnover of unknown aetiology producing characteristic bone deformities. The disease is rarely seen before the age of 40 and occurs in approximately 3% of the population over 40 and 10% of those over 70. This is a disease that is often without symptoms and progresses slowly over many years, being discovered during other investigations in approximately one-fifth of those diagnosed. In Paget's disease, osteoclasts, the cells that digest bone during the normal bone remodelling process, become hyperactive and osteoblasts, the cells that form new bone are unable to replace bone at the same rate. Instead affected bones develop fibrous tissue and become enlarged, which may cause pressure on nerves and result in bone pain and neurological complications. The fibrous tissue reduces the strength of the bone and may lead to stress fractures, which are notoriously difficult to heal. Deformity of the bone(s) is common and may result in an alteration in gait. There is an increased risk of bone tumour (osteosarcoma), which occurs in approximately 7% of patients with Paget's disease. In patients over the age of 50 with osteosarcoma about 20% have Paget's disease.

7.8   Treatment of Paget's disease

The majority of patients require no treatment for the disease. If necessary, pain can be treated with NSAIDs. If bone pain is severe or if the patient has neurological symptoms, or fractures that will not heal, calcitonin or bisphosphonates may be indicated.

7.8.1   Calcitonin

Calcitonin is the natural hormone that inhibits the activity of osteoclasts (see Chapter 6). Produced either from pigs or salmon or more recently from human sources by recombinant DNA technology, it is given daily by sub-cutaneous injections. Calcitonin is the preferred treatment in severe Paget's disease.
Side effects of calcitonin are nausea, vomiting and diarrhoea together with flushing, paraesthesia and a peculiar taste in the mouth.

7.8.2   Bisphosphonates

The mechanism of action of bisphosphonates is not completely understood. They are deposited in place of normal bone and may slow bone turnover because they are resistant to degradation by enzymes produced by osteoclasts or they may inhibit the activity of osteoclasts. Bisphosphonates are given orally and about 50% of the oral dose concentrates in bone. They should not be taken at the same time as milk because it impairs their absorption. An example is sodium etidronate.

Side effects of bisphosphonates are bone pain and gastrointestinal disturbances.

7.9　Osteoporosis

Osteoporosis, literally meaning porous bone, involves a loss of total bone mass and a change in the microstructure of bone. This is usually an age-related process that is accelerated in post-menopausal women due to oestrogen deficiency. Oestrogen normally plays a role in the inhibition of bone digestion and this protection is lost after the menopause. Osteoporosis may also occur due to hormone imbalances or chronic hormone therapy (for example with corticosteroids or thyroxine). Bone mass begins to decline naturally after the age of 25. In later life this increases the risk of bone fractures and vertebral collapse. Specialized radiographic imaging techniques or bone biopsy are used to diagnose and monitor the condition. Where the condition occurs in the young it is secondary to another cause, for example osteogenesis imperfecta (brittle bone disease), Cushing's disease, Crohn's disease or prolonged immobility.

Osteoporosis can be prevented or reduced by regular exercise and adequate dietary intake of calcium and vitamin D throughout life but especially in adolescence.

Treatment of osteoporosis includes the use of drugs that decrease the rate of bone reabsorption and (as a last resort) replacement of oestrogen in post-menopausal women.

7.10　Drugs used to treat osteoporosis

Treatment of osteoporosis is essentially the same regardless of whether the cause is post-menopausal or corticosteroid-induced or simply due to old age.

7.10.1　Bisphosphonates and calcitonin

Bisphosphonates and calcitonin can be used in osteoporosis, when they have an action similar to that described under Paget's disease (see page 126). Calcium salt supplements can be added to calcitonin therapy if the diet is low in calcium or if the patient is very old.

7.10.2　Other drugs

There have been some relatively new developments in the treatment of osteoporosis. Selective oestrogen receptor modulators, for example raloxifene mimic the inhibitory effects of oestrogen on osteoclasts. Teriparatide, which is a recombinant fragment of parathyroid hormone (PTH) paradoxically stimulates the activity of osteoblasts. These drugs are recommended for use in women over 65 with low bone mineral density and a history of fractures.

7.10.3 Hormone replacement therapy

Hormone replacement therapy (HRT) with oestrogen and progesterone is no longer recommended and should not be used as first line treatment in post-menopausal women for osteoporosis. This is because of the increased risk of breast, endometrial and ovarian cancer with HRT. Its use should be reserved for patients in whom other drugs are contraindicated, not tolerated or ineffective. HRT is most effective if started early in the menopause and continued for up to five years (after which osteoporosis will return, possibly at an accelerated rate).

7.11 Osteomalacia

Osteomalacia is caused by deficiency of vitamin D. Vitamin D regulates calcium homeostasis in the body by facilitating absorption of calcium from the intestine and, together with PTH, by enhancing calcium mobilization from bone and by reducing excretion of calcium by the kidney. Deficiency of vitamin D leads to inadequate absorption of calcium. Low levels of calcium stimulate the release of PTH, which in turn causes release of calcium from bone and failure to mineralize newly formed bone.

Osteomalacia in childhood is known as *rickets* and leads to soft bones with characteristic deformities. In adults there is generalized reduction in mineralization of bone matrix and symptoms of bone pain and tenderness.

Nutritional rickets results from inadequate exposure to sunlight or deficiency of vitamin D.

Metabolic rickets and osteomalacia result from abnormality in synthesis of or response to calcitriol, the active form of vitamin D.

Treatment of osteomalacia and rickets is by vitamin D supplements.

Some drugs, for example anticonvulsants phenytoin, phenobarbital and corticosteroids can lead to osteomalacia and rickets by depressing vitamin D dependent calcium uptake in the intestine.

7.12 Myasthenia gravis

Myasthenia gravis is an autoimmune disorder, which is diagnosed in about 1 in 20 000 people every year in the United Kingdom, mainly affecting women between 20 and 40 and men over 50. In myasthenia gravis, normal communication between nerve and

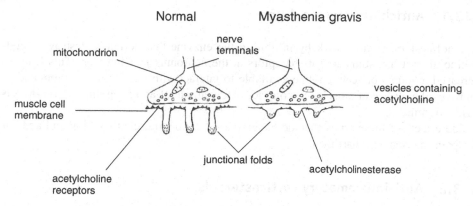

Figure 7.2   Destruction of acetylcholine receptors in myasthenia gravis

skeletal muscle is disrupted. Normally when impulses travel down a neuron to skeletal muscle the nerve endings release acetylcholine, which crosses the neuromuscular junction and brings about the activation of muscle contraction.

In myasthenia gravis the receptors for acetylcholine on the muscle cell membrane are progressively destroyed by antibodies. The antibodies are produced by the patient's own immune system resulting in an autoimmune reaction, which causes structural changes at the synapse. (See Figure 7.2.)

The disease is characterized by fluctuating weakness and easy fatigability of certain skeletal muscles. Patients may present with double vision or a drooping eyelid at the end of the day. Other facial muscles may be affected, resulting in difficulty in chewing, an abnormal smile, or nasal speech. In severe cases there may be difficulty in swallowing with the risk of aspirating food into the lungs and difficulty in breathing, due to weakness of the respiratory muscles, which can be life threatening if untreated. Not all muscle groups are affected in all patients, and the disease has a variable progression.

7.13   Treatment of myasthenia gravis

In some cases, the first approach to treatment is surgical removal of the thymus gland, providing the patient has reached puberty and is younger than 60. Many patients with myasthenia gravis have an abnormality of the thymus. Normally the thymus gland is involved in the maturation of T lymphocytes, but in myasthenia gravis it appears to be the site of production of antibodies against the acetylcholine receptors. Thymectomy increases the chances of a patient going into remission, although it may take 6–18 months for the benefits to be seen.

Drug therapy has to be tailored to individual patients and includes anticholinesterase drugs, anti-inflammatory corticosteroids and immunosuppressant drugs.

7.13.1 Anticholinesterase drugs

Anticholinesterase drugs work by inhibiting the enzyme that normally destroys acetylcholine after it has stimulated its receptors at the neuromuscular junction. This leads to increased amounts of acetylcholine available to interact with remaining receptors and so improves the ability of muscles to contract. An example of an anticholinesterase drug is pyridostigmine.

Side effects of these drugs include excessive salivation, muscle twitching and abdominal pain, nausea and diarrhoea.

7.13.2 Anti-inflammatory corticosteroids

Anti-inflammatory corticosteroids, for example prednisolone, can be used to suppress the antibody formation in myasthenia gravis. They should be used together with anticholinesterase drugs and once an improvement is seen the dose should be decreased.

Long-term use of corticosteroids leads to serious side effects, including suppression of the pituitary-adrenal axis, immunosuppression, muscle wasting, osteoporosis and impaired wound healing (see page 119 and Table 7.1).

7.13.3 Immunosuppressants

Immunosuppressants such as azathioprine are used when other forms of treatment fail to control the progression of myasthenia gravis.

They have severe side effects including bone marrow suppression, gastrointestinal disturbances and hair loss (see page 122 and use in cancer, Chapter 10).

7.14 Motor neuron disease and multiple sclerosis

Strictly speaking, motor neuron disease and multiple sclerosis are diseases of the nervous system, rather than the musculoskeletal system. However, although these conditions are relatively rare, they are included here because the consequences of them affect skeletal muscle function and because health care professionals are likely to be involved in the care of patients with these two conditions.

7.14.1 Motor Neuron Disease

Motor neuron disease is also known as *amyotrophic lateral sclerosis*. It is a condition in which there is loss of spinal motor neurons and neurons of descending motor pathways from pyramidal cells of the cerebral cortex. The disease results in rapidly progressing muscular weakness, muscle atrophy, fasciculations, spasticity, difficulty speaking, difficulty swallowing and difficulty breathing. Usually the disease is progressive and fatal with most patients dying from respiratory failure and pneumonia within two to three years from onset.

(A notable exception to this generalization is Professor Stephen Hawking who was diagnosed with motor neuron disease at the age of 21 and is still alive some 40 years later, albeit paralysed for most of that time.)

The cause of motor neuron disease is unknown.

There is one drug that can be used to slow the progression of motor neuron disease and extend life or prolong the time to mechanical ventilation. Riluzole inhibits the release and postsynaptic action of glutamate in the motor pathways. It is said to be neuroprotective and slows deterioration of nerve function. This could be because excessive amounts of glutamate are released in motor neuron disease and this is toxic to neurons. Riluzole must be used under specialist supervision.

Side effects of riluzole are nausea, vomiting, tachycardia, headache, dizziness and vertigo.

7.14.2 Multiple sclerosis

Multiple sclerosis is an autoimmune disease of the central nervous system in which the cells that produce the myelin sheath of neurons are progressively destroyed. Particular areas of the central nervous system affected are the optic nerves, brain stem and cerebellum, corticospinal tracts of the spinal cord and areas around the ventricles. Demyelination leads to variable symptoms depending on where the lesions are. Double vision and vertigo are common presenting signs of brain stem involvement. Difficulty in walking with sensory disturbances is a sign of spinal cord lesions. Typically the disease affects young people between 20 and 35 and is more common in women and follows a course of relapses interspersed with variable periods of remission.

The cause of multiple sclerosis is unknown and there is no cure.

Interferon β has been used for patients with relapsing intermittent multiple sclerosis who can walk unaided. However, this drug is no longer recommended by NICE for this purpose.

Short courses of corticosteroids are used during exacerbations and seem to reduce the effect of a relapse.

Spasticity is a common symptom in multiple sclerosis. It is defined as an increase in muscle tone characterized by initial resistance to passive movement followed by sudden relaxation. Spasticity can be treated with a number of drugs.

Dantrolene has a direct action on skeletal muscle to cause relaxation; baclofen inhibits transmission at the spinal cord by acting on inhibitory presynaptic gamma-aminobutyric acid B (GABA$_B$) receptors; benzodiazepines cause muscle relaxation by some central action; and tizanidine is an α_2 adrenoreceptor agonist presumably with a central action.

Side effects of dantrolene, baclofen and tizanidine are similar, being motor incoordination, drowsiness, dizziness, weakness, fatigue and gastrointestinal disturbances.

7.15 Summary

Rheumatic diseases are a group of chronic inflammatory conditions such as rheumatoid arthritis, ankylosing spondylitis, systemic lupus erythematosus and psoriatic arthritis. In these diseases joints become inflamed and damaged, although there are also systemic effects such as skin, lung and muscle involvement. Management can be complex and individual. Early use of disease-modifying antirheumatic drugs can prevent disease progression and limit joint damage. This is a diverse group of drugs. NSAIDs are often needed for pain control and corticosteroids can be useful for reducing exacerbations of inflammation. Corticosteroids should not be used long term because of adverse effects, the most serious being suppression of the pituitary-adrenal axis. Newer drugs are cytokine inhibitors and monoclonal antibodies. These are reserved for severe active rheumatic disease (but not systemic lupus erythematosus) that is unresponsive to other drugs and must only be used under close medical supervision.

Gout is an inflammatory joint disease caused by deposition of uric acid crystals in the joints. Treatment of gout aims to control acute attacks with analgesics and colchicine and in the longer term to reduce circulating uric acid levels either by inhibiting its production with allopurinol or by encouraging its excretion by the kidneys with probenecid or sulfinpyrazone.

Osteoarthritis is primarily a degenerative joint disease, but as joint destruction occurs this creates local inflammation. Treatment of osteoarthritis is with NSAIDs.

Paget's disease and osteoporosis are conditions where there is loss of normal bone tissue due to overactivity of osteoclasts. In Paget's disease, bone is replaced with fibrous tissue; in osteoporosis, there is a loss of bone mass and altered microstructure of bone. Both conditions result in increased risk of fracture and can be treated with calcitonin and bisphosphonates. New developments in the treatment of osteoporosis include selective oestrogen receptor modulators and PTH fragments. Classic oestrogen and progesterone replacement therapy is no longer recommended for post-menopausal osteoporosis prevention because of the increased risk of inducing cancers of the breast, uterus and ovaries.

Osteomalacia is a deficiency disease. Insufficient vitamin D leads to low plasma calcium levels. This stimulates the secretion of PTH, which causes release of calcium from bone and failure to mineralize new bone. Characteristic bone deformities result in children and bone pain and tenderness in affected adults. Treatment is by vitamin D replacement.

Myasthenia gravis is an autoimmune disease where acetylcholine receptors at the neuromuscular junction are progressively destroyed. This interferes with muscle contraction causing muscle weakness and fatigue. In many patients, removal of the thymus gland can induce remission of the disease. Drug therapy includes the use of anticholinesterase drugs, anti-inflammatory corticosteroids and immunosuppressant drugs in resistant cases.

Motor neuron disease and multiple sclerosis are diseases of the nervous system that affect muscle function. Both these diseases are progressive and there are no cures for them. Riluzole may slow progression of motor neuron disease and dantrolene can alleviate the spasticity of multiple sclerosis.

Useful websites

www.arc.org.uk/ Arthritis Research Campaign. Information for medical professionals.

www.mgauk.org/ Myasthenia Gravis Association. Medical guides for patients.

www.mndassociation.org/ Motor Neuron Disease Association. Section for professionals.

www.mssociety.org.uk/ Multiple Sclerosis Society. Professional resources.

http://www.nice.org.uk/TA125 National Institute for Health and Clinical Excellence. Psoriatic Arthritis (moderate to severe) – Adalimumab Guidance 2007.

http://www.nice.org.uk/TA126 National Institute for Health and Clinical Excellence. Rheumatoid Arthritis (refractory) – Rituximab Guidance 2007.

http://www.nos.org.uk/professionals.htm National Osteoporosis Society. Web page for health care professionals.

http://www.paget.org.uk/guidelines.pdf National Association for the Relief of Paget's Disease. Guidelines for the management of Paget's disease of bone.

www.rheumatology.org.uk British Society for Rheumatology. Clinical guidelines.

Case studies

The first two case studies are about hypothetical patients who might be seen in the podiatry clinic. However, other health care professionals might see similar patients for other reasons. The third case study is a patient who any health care professional might see for different reasons.

Case study 1

Mrs Charles is a middle-aged patient who has an ingrowing toenail and requires a TNA (total nail avulsion). She is aware that she suffers from a disorder with a long name, which makes her muscles weak. Mrs Charles currently requires a wheelchair in which she is pushed around by her carer.

Mrs Charles's drug therapy is pyridostigmine bromide 60 mg per day and prednisolone 10 mg per day.

Discuss the treatment of this patient using the questions below as a guide. They are intended to stimulate discussion not limit debate.

- What disease do you gather that the patient suffers from?

- What problems may the treatment of this patient pose?

- Can this patient be treated safely in the community?

- What can you deduce about the healing potential of this patient?

- Would the aftercare of this patient vary from that of a normal healthy adult?

Case study 2

The following patient was initially referred to the podiatrist because he has fungally infected toenails. He now sees the podiatrist every two months and on this visit, he is complaining of a rash on the plantar surface of his feet, which is red and itchy.

Mr Lewis is 68 years old. He has a chronic rheumatoid disease called *Wegener's granulomatosis*. In this condition, there is polyarthritis and vasculitis of small and medium sized blood vessels throughout the body. Mr Lewis also has type 2 diabetes.

Mr Lewis is on a complex regime of drug therapy (and you may need to consult the *BNF*):

Metformin 500 mg three times a day with meals

Prednisolone 5 mg four times a day

Asasantin (aspirin 25 mg and dipyridamole 200 mg) one capsule twice daily

Omeprazole 20 mg once a day

Sertraline 50 mg once a day

Temazepam 10 mg at night

Salbutamol 100 μg metered dose inhalator four times daily

Glyceryl trinitrate 0.4 mg metered dose at onset of attack

Amlodipine 5 mg once a day

Perindopril 2 mg once a day

Acetazolamide 250 mg twice a day

Erythromycin 250 mg four times a day

Discuss the treatment of this patient using the questions below as a guide. They are intended to stimulate discussion not limit debate.

- What can you deduce about Mr Lewis' general health from the number of drugs he is currently taking?

- Identify the drugs that are being used to treat the Wegener's granulomatosis

- Apart from type 2 diabetes. What other major disease does Mr Lewis appear to have?

- Given the conditions that Mr Lewis is suffering from, how should his fungal nail infection be managed?

- What is the itchy red plantar rash likely to be? How can you tell?

Case study 3

In the following case study, you must discuss how you would answer the patient's questions and offer advice about her condition and medication.

Mrs Begum has been prescribed methotrexate by her consultant rheumatologist for rheumatoid arthritis that has not responded to other therapies. Mrs Begum has some preconceived ideas about methotrexate and is nervous about starting with it. She tells you that she knows someone who is being treated with methotrexate for cancer and has heard that it can cause many serious adverse reactions. She is not sure why she is being asked to take methotrexate at all and she thinks it very strange that she has been told to take the drug once a week only (7.5 mg). Discuss how you could reassure Mrs Begum about the need to take methotrexate in her case, about the once weekly dosage and about the side effects and how likely they are to affect her. You could also consider alternatives if she proves intolerant of methotrexate.

If you were a physiotherapist, would you consider this patient suitable for supplementary prescribing under a clinical management plan?

Chapter review questions

You should be able to answer these questions by referring to the material in the preceding chapter.

1. Discuss the advantages and disadvantages of using NSAIDs to treat rheumatoid disease

2. Discuss the rationale for using DMARDs early on in rheumatoid arthritis.

3. How long should treatment be continued?

4. Describe the circumstances when it would be appropriate to use corticosteroids in the treatment of rheumatoid arthritis.

5. Describe the disadvantages of prolonged corticosteroid therapy.

6. Discuss typical symptoms and cause of gout and factors that can trigger an episode of gout.

7. Discuss approaches to treatment of gout.

8. Describe the treatment of osteoarthritis.

9. Discuss the use of bisphosphonates and calcitonin in the treatment of Paget's disease and osteoporosis.

10. Describe the underlying pathology of myasthenia gravis and explain how using anticholinesterase drugs can help.

11. There is no cure for either multiple sclerosis or motor neuron disease. What can be done to improve the outlook for patients with these diseases?

8

Disorders of the skin

8.1 Chapter overview

This chapter should be of interest to all health care professions but some of it may be of more relevance to podiatrists since they are likely to see and treat patients with skin disorders of the lower limb.

The skin is the largest organ of the body, being about 16% of body weight and covering about $1.8\,m^2$ of the body surface. The skin has many functions, important ones being to act as a barrier against the environment; to contribute to regulation of body temperature; to receive sensory stimulation; and to synthesize vitamin D.

Common disorders of the skin include eczema, psoriasis and infections with viruses, bacteria and fungi. Treatment of skin conditions is usually topical because the skin is uniquely accessible to this mode of administration. However, systemic treatment is sometimes necessary.

Proprietary names of preparations commonly available as general sales list medicines are given for treatment of viral warts and fungal infections of the skin.

8.1.1 Structure of the skin

A basic knowledge of the structure of the skin is required in order to understand how disorders arise and how they can be treated.

The skin consists of three layers: the epidermis, the dermis and the subcutis (hypodermis). (See Figure 8.1.)

Cells on the border between the dermis and the epidermis are constantly dividing and moving towards the skin surface. These cells are known as *keratinocytes* because they produce and contain keratin. Keratin is a tough fibrous protein.

Keratinocytes of the epidermis are arranged into four or five layers. The deepest layer is the *stratum basale*, which is a single layer of dividing cells. The *stratum spinosum* is next and consists of 8–10 layers of polyhedral keratinocytes. Above this layer is the *stratum granulosum*, composed of three to five layers of keratinocytes that have lost their

Pharmacology for the Health Care Professions   Christine M. Thorp
© 2008 John Wiley & Sons, Ltd

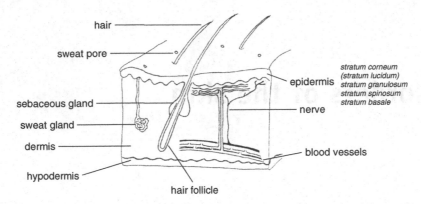

Figure 8.1   Structure of the skin

nuclei and are beginning to die. Cells in this layer contain lamellar granules that release a thick lipid secretion that sticks the cells together. Finally, the *stratum corneum* containing 25–30 layers of dead flat keratinocytes stuck together with lamellar lipid provides a relatively tough and waterproof barrier. These cells are continuously lost from the surface and replaced with cells from below. Where skin is thicker on the palms of the hands and the soles of the feet there may be a fifth layer of the epidermis between the *stratum corneum* and *stratum granulosum*. This layer is known as the *stratum lucidum* and consists of three to five layers of clear, flat keratinocytes that contain large amounts of keratin.

Three other types of cell are found in the epidermis. Merkel cells and melanocytes are found in the *stratum basale*. Merkel cells are attached to a sensory neuron and function to detect the sensation of touch. Melanocytes produce the pigment melanin, which contributes to skin colour and provides some protection against sunlight. Langerhans cells are found in the *stratum spinosum*. They are part of the immune system and help protect the skin from microbial infection.

Hair follicles, sebaceous glands and sweat glands are part of the epidermis, but they penetrate into the dermis. The dermis consists of loose connective tissue containing fibroblasts that produce collagen and elastic fibres. The dermis also contains lymphocytes, mast cells and macrophages, which respond to injury, infection, allergens and harmful chemicals by releasing inflammatory substances.

The dermis has a good blood supply and variation in blood flow through the dermis and the production of sweat are important in body temperature control. Sensory nerve endings found in the dermis are sensitive to touch, pressure, pain and change in temperature.

Subcutaneous fat provides insulation and an energy store.

8.2   Eczema

Eczema is a localized inflammatory reaction in the skin that has several possible causes. It can be a reaction to contact with chemicals including some drugs, metals or allergens, in which case it is known as *contact dermatitis*. In some cases, eczema occurs without an obvious cause and is known as *atopic eczema*.

Characteristics of eczema are red itchy epidermis and oedema of the dermis. Most often, the skin is dry and scaly, but sometimes in the early stages of eczema, the lesions weep and can become infected.

8.3   Treatment of eczema

If the cause of eczema is known, every effort should be made to avoid contact with the cause. Otherwise treatment is symptomatic, there being no cure for eczema.

8.3.1   Emollients and astringents

If the eczema is dry, emollients can be used to soothe the skin and restore hydration. Emollients are made from liquid paraffin, soft paraffin and various fats and oils formulated into creams, ointments and bath oils with or without antimicrobials.

If the eczema is wet, mild astringents such as dilute potassium permanganate may be useful in the short term.

8.3.2   Topical corticosteroids

In many cases of eczema, the use of emollients or astringents may not be sufficient to relieve the symptoms. Topical corticosteroids may then be required to reduce the inflammation. Corticosteroids come in different potencies and the potency of the preparation chosen should be appropriate to the severity of the condition. One percent hydrocortisone cream or ointment, which is available as a pharmacy medicine is usually effective. Long-term application of topical steroids can cause damage to the skin. The aim is to use the lowest effective concentration of corticosteroid for the shortest period. However, it is preferable to use a high potency steroid for a short period rather than a low potency steroid for a longer period. This minimizes the damage to the skin structure and reduces the possibility of systemic adverse effects. (See Chapter 7; page 119 and Table 7.1 for adverse effects of corticosteroids.)

8.3.3   Antihistamines

Oral antihistamines that have a sedative effect may be useful at night to reduce nocturnal itching, which can cause skin damage and secondary infection. Antihistamines are discussed in Chapter 5.

8.3.4 Immunosuppression

In very severe eczema, topical corticosteroids may not be sufficient to relieve the inflammation.

In these cases, specialist treatment with drugs that suppress the immune system may be required. These drugs are normally used to prevent transplant rejection. For severe eczema refractory to other treatment, ciclosporin is licensed for oral use and tacrolimus is licensed for topical use.

Adverse effects of cyclosporin are toxicity to bone marrow, nausea, leukopenia, blurred vision and rashes.

Adverse effects of topical application of tacrolimus are rash, irritation, pain and paraesthesia.

8.4 Psoriasis

Psoriasis is a chronic inflammatory skin disease affecting about 1% of the population. The condition is characterized by hyperproliferation of keratinocytes. Cell turnover in the epidermis is increased with cells reaching the surface in as few as 3 days instead of the normal 21–28 days. This results in a thickened epidermis with immature keratinocytes containing nuclei piling up on the surface. The keratinocytes produce excessive amounts of keratin and form typical silvery scales. There are dilated blood vessels in the upper dermis and the skin bleeds easily if scratched. Leucocytes infiltrate into the epidermis and can form microabsesses.

All parts of the skin can be involved including mucous membranes and nails, but more usually isolated patches of skin are affected, particularly at the elbows and knees with clear demarcation between the plaques of psoriasis and normal skin. Plaque psoriasis is the most common form, accounting for 80% of cases.

Psoriasis occurs in several other forms. A form particularly difficult to treat is palmoplantar psoriasis (PPP), which manifests as pus-filled spots (pustules) on the hands and feet. Pustular and erythrodermic psoriasis are two rare forms of the disease, which present as medical emergencies. Both types cover the entire body, as either pus-filled spots (*pustular*) or red patches with skin shedding (*erythrodermic*).

(The 'Singing Detective' in the Dennis Potter play had particularly severe erythrodermic psoriasis.)

The disease appears to have a genetic basis with environmental factors, such as infections, emotional trauma or mechanical trauma, causing outbreaks in susceptible individuals. Psoriasis should be regarded as a chronic systemic disease, which can be associated with other chronic diseases such as heart disease, hypertension and diabetes and may reduce life expectancy. About 15% of people with psoriasis also have psoriatic arthritis (see Chapter 7, page 173).

Sometimes psoriasis can occur in response to drugs such as lithium (Chapter 11), chloroquine and NSAIDs (Chapter 7), beta-blockers and ACE inhibitors (Chapter 4).

8.5   Treatment of psoriasis

There is no cure for psoriasis but there are many therapies and treatment depends on the extent and severity of the disease.

8.5.1   Emollients

As with eczema, emollients can have beneficial effects on dry, cracked skin and may also reduce hyperproliferation in mild psoriasis.

They are also useful as adjuncts to other more specific treatments.

8.5.2   Calcipotriol

Calcipotriol is a vitamin D analogue. Its use in psoriasis came about relatively recently following the observation that an oral vitamin D derivative used for osteoporosis coincidentally seemed to improve psoriasis.

The mode of action of calcipotriol in psoriasis is apparently unconnected to the role of vitamin D in calcium metabolism.

In the skin, the drug combines with an intracellular vitamin D receptor to form a drug-receptor complex, which then binds to a specific sequence of DNA in the nucleus that modulates and controls transcription. (See Chapter 3, Type 4 receptors.) As a result, the action of calcipotriol is to reduce proliferation and allow differentiation and maturation of keratinocytes.

Calcipotriol is available only on prescription as a cream, an ointment and a solution. It is normally applied twice daily directly to the affected areas of skin. The majority of patients see some improvement within one or two weeks and maximum response is produced within six to eight weeks. Approximately 15% of patients have complete remission of symptoms, although maintenance therapy is usually necessary.

Calcipotriol should not be used on the face or in skin folds because it may produce irritation due to increased absorption. Since vitamin D normally enhances calcium absorption from the intestines and its release from bone, hypercalcaemia is a potential side effect. However, only rarely is sufficient amounts of drug absorbed to affect calcium homeostasis in this way.

Another derivative of vitamin D is talcacitrol, which has a similar pharmacological profile.

8.5.3   Dithranol

Dithranol was developed from a natural remedy called *Goa powder*. It is a prescription-only medicine, but preparations containing 1% or less dithranol can be sold to the public over the counter.

Dithranol is effective but has to be applied carefully to plaques of psoriasis avoiding normal skin, because it stains the skin violet-brown, leaving discoloration of healed areas. For this reason, and because it also stains clothing and bedding, patient compliance is poor with this drug. Its use is more successful with hospital inpatients and it has relatively few other side effects.

It is not clear exactly how dithranol exerts its action but it appears to interact with DNA and directly inhibit cell proliferation.

Dithranol can cause inflammation, so initially concentrations no greater than 0.1% should be applied for 5 minutes to 1 hour, after which the preparation should be washed off (known as *short-contact treatment*). If this is tolerated, higher concentrations can be used to provide optimum response.

Since dithranol irritates normal skin, it should not be used in flexures, on the face or in skin folds.

8.5.4   Coal tar preparations

Coal tar preparations are complex and variable mixtures of hydrocarbons that have cytostatic or antimitotic actions. Crude coal tar is the most effective, but few outpatients tolerate the mess and smell. In hospitals, coal tar is often combined with ultraviolet B (UVB) radiation. There are many proprietary preparations, which are more refined and therefore more acceptable to patients, but they are less effective. They are used as a creams, lotions, ointments and bath oils in concentrations from 1 to 12%.

Coal tar preparations are less irritating to normal skin than dithranol, but can cause folliculitis.

8.5.5   Keratolytics

Keratolytics break down keratin and soften skin, which improves penetration of other treatments. Salicylic acid ointment is most frequently used either on its own or in combination with coal tar or dithranol.

8.5.6   Phototherapy and photochemotherapy

Phototherapy is the use of ultraviolet light, which produces improvement in most patients with psoriasis but should only be done under supervision of a dermatologist. UVB is useful if patients have extensive small plaques of psoriasis resistant to topical treatment. Ultraviolet A (UVA) requires more specialized equipment and prior administration of an oral or topical photosensitizing drug called *psoralen*, in which case the treatment is known as *photochemotherapy*.

The combination of psoralen with UVA is known as *PUVA*. After absorption, psoralen is distributed throughout the body and is activated in the skin by absorbing the energy of UVA radiation.

Activated psoralen probably acts by locating between base pairs in the DNA helix and thereby inhibiting replication. PUVA has also been shown to cause alteration in the immune response, which may play a role in reducing inflammation. PUVA treatment is usually reserved for severe, resistant psoriasis, when there is a high success rate, often as high as 90%. Treatment is given three or four times a week until the maximum response is seen. However, relapse is very common, usually within six months of stopping therapy. Various types of maintenance regimes have been devised with variable success.

Acute side effects of PUVA treatment are nausea from the psoralen and blistering and erythema from the UVA.

UVA is principally responsible for photo-ageing and penetrates tissues far more deeply than UVB. Long-term risks of PUVA include accelerated skin ageing and an increased incidence of skin cancer.

Patients must not be taking other photosensitizing drugs, for example phenothiazines and benzodiazepines (Chapter 12), NSAIDs (Chapter 7) and some antibiotics (Chapter 9).

Psoralens are naturally occurring plant extracts found in the parsnip family and the technique has a long history, being used by Egyptians and Indians to treat vitiligo as long ago as 1500 BC. Vitiligo is a condition where there is patchy loss of pigmentation in the skin and photochemotherapy is still used to treat it. Psoralen causes stimulation of melanocytes and some repigmentation in vitiligo.

8.5.7 Topical corticosteroids

Topical corticosteroids should not be used for long periods or on extensive areas of psoriasis because their withdrawal can produce a rebound reaction and long-term changes in skin structure. However, corticosteroids remain useful for small lesions. Combining corticosteroids with other topical treatment is effective in some cases.

8.5.8 Systemic antipsoriatic drugs

These drugs are reserved for patients with the most severe forms of psoriasis that do not respond to, or cannot tolerate, topical or phototherapy.

Use of systemic antipsoriatic drugs is generally under supervision of a dermatologist in a hospital setting.

8.5.8.1 Methotrexate

Methotrexate is the most common treatment for resistant and widespread psoriasis. Its main actions are cytotoxic and immunosuppressant. Methotrexate inhibits an enzyme (dihydrofolate reductase) necessary for the synthesis of purine nucleotides (adenine and guanine), which are components of DNA. This results in a reduction in DNA synthesis and particularly affects rapidly dividing cells.

Methotrexate is prescribed for oral administration once a week and it is extremely important the patients understand this. Bone marrow depression and hepatotoxicity are the main complications. Blood counts should be checked every two to three months and liver biopsies taken every one to two years to monitor treatment.

8.5.8.2 *Ciclosporin*

Ciclosporin is an immunosuppressant normally used to prevent transplant rejection. Its mode of action in psoriasis is not completely understood, but it has anti-inflammatory actions. This highly toxic drug is reserved for psoriasis that is resistant to other systemic treatment and it should only be used under supervision of a dermatologist.

8.5.8.3 *Efalizumab*

Efalizumab is a monoclonal antibody that inhibits the activation of T lymphocytes. It is licensed for severe plaque psoriasis resistant to other systemic treatment and phototherapy. Presumably, efalizumab works by inhibiting the inflammatory component of this type of psoriasis.

8.5.8.4 *Retinoids*

Retinoids are synthetic vitamin A derivatives. Vitamin A is known to affect normal cell differentiation. Retinoids bind directly to retinoic acid receptors in the cell nucleus and reduce cell proliferation in the skin.

Acitretin is currently the only retinoid licensed in the United Kingdom for the treatment of psoriasis and is only available for treatment of patients in hospitals. It is given orally and has anti-inflammatory and cytostatic actions. This is useful in inflammatory psoriasis. The half-life of acitretin is about two days. However, there is a high risk of teratogenesis and women must use adequate contraception and stop treatment for two years before conception.

Side effects of acitretin include dryness of mucous membranes and skin with localized exfoliation of the palms and plantar surface of the feet, skin and nail erosion and transient thinning of hair. These effects are dose-dependent and reversible. Longer-term adverse effects include ossification of ligaments and raised plasma triglycerides and cholesterol levels. Acitretin should not be taken in combination with methotrexate because of toxicity to the liver.

A summary of treatment available for psoriasis is given in Table 8.1.

8.6 Warts

Warts are caused by the human papilloma virus (HPV) of which around 100 subtypes is known. Warts can appear on any part of the skin, but are most common on the hands and feet.

Warts on the sole of the foot are generally known as *verrucas* or *plantar warts*. The HPV-1 group most commonly causes verrucas although HPV-2, HPV-60 and HPV-63 can also cause them.

Table 8.1 Treatment of psoriasis

Topical drugs	Comments
Calcipotriol	Vitamin D analogue
Dithranol	Inhibits cell proliferation
Coal tar	Many preparations
Keratolytics	Used with coal tar/dithranol
Phototherapy	UVB
Photochemotherapy	UVA plus psoralen
Corticosteroids	Limited use
Systemic drugs	Comments
Methotrexate	Cytotoxic, immunosuppressant
Ciclosporin	Immunosuppressant
Efalizumab	Monoclonal antibody
Retinoids	Vitamin A derivatives

Warts are very common in childhood and are spread by direct contact or autoinoculation. If a wart is scratched, the viral particles may be spread to another area of skin. It may take as long as 12 months for the wart to appear.

In children, warts disappear without treatment within six months to two years. They tend to be more persistent in adults, some lasting five to seven years, but always go eventually due to the actions of the immune system.

Anogenital warts require specialist diagnosis and screening for other sexually transmitted diseases. Treatment is with prescription-only medications and is not considered here.

8.7 Treatment of warts

Treatments are often painful and lengthy with poor results. Warts have a very good blood supply and this may be why they are able to withstand prolonged treatment with topical keratolytics.

With no treatment, a wart will eventually disappear. However, warts may be painful, often look ugly, and cause embarrassment in which case treatment may be appropriate.

8.7.1 Chemical treatments

Chemical treatments include wart paints, of which there are many available as over-the-counter preparations (that is, they are general sales list items). As such, they can be accessed and supplied by qualified, registered podiatrists (see Chapter 14).

Wart preparations contain keratolytics such as salicylic acid, formaldehyde or glutaraldehyde, which work by removing the dead skin cells at the surface of the wart.

Table 8.2   Treatment of viral warts

Treatment	Method	Success rate (%)
Cryotherapy	Use of liquid nitrogen for 10–30 s to freeze and destroy warts. Multiple treatment over several months.	70
Electrosurgery	Under local anaesthetic, the wart is pared away and the base burned by diathermy or cautery. Two weeks to heal.	80
Excision	Dermatologists rarely recommend excision because this type of surgery leaves a permanent scar.	80
Bleomycin injections	Under local anaesthetic, cytotoxic drug is injected into warts. Treatment repeated at three-week intervals. Painful.	
Laser	Carbon dioxide laser used to vaporize persistent warts. Repeated treatment is often necessary and can be painful.	

One preparation contains podophyllum (Posalfilin), which is a cytotoxic drug and must not be used in patients who are pregnant or considering pregnancy or breast-feeding. Podophyllum inhibits cell division in virally infected cells.

Treatment with wart paints is usually once or twice a day with abrasion between applications. Perseverance with treatment is essential and surrounding skin should be protected with plasters or Vaseline.

8.7.2   Other treatments

There are many other treatments of warts. Brief details of some of these are given in Table 8.2. Because such treatments can be lengthy and painful, they are reserved for multiple or unsightly warts.

8.8   Other viral infections of the skin

Herpes simplex of the skin (cold sores) can be treated with acyclovir. Treatment should be started as early as possible, when the first tingling sensation is felt.

Side effects of topical use of acyclovir are transient stinging or burning of the skin. Antiviral drugs are discussed in Chapter 9.

8.9   Fungal infections of the skin and nails

Organisms called *dermatophytes* cause the majority of superficial fungal infections of the skin, hair and nails. The most common dermatophytes that infect humans belong to the trichophyton, epidermophyton and microsporum families. These fungi are commonly occurring in the environment and they thrive in tissues that contain keratin. Such fungal infections of the skin are known as *tinea*: Tinea pedis is infection of the foot (athlete's foot); tinea corporis is infection on the body (ringworm) and tinea capitis is infection of the scalp (ringworm). Fungal infection of the nails is known as *tinea unguium* or *onychomycosis*.

Superficial fungal infections can usually be treated with topical preparations, but systemic antifungal drugs can be necessary for nail and scalp infections, or if skin infection is extensive and unresponsive to topical therapy.

8.10 Drugs used to treat fungal infection of the skin and nails

There are many topical treatments available of varying efficacy.

Most of them are available as over-the-counter or pharmacy medicines but some are available only on prescription. Proprietary names are given where appropriate. Because podiatrists are likely to be treating fungal infections of the feet, some prescription-only antifungal drugs have been added to the list of exemptions for qualified podiatrists registered with the Health Professions Council (see Chapter 14).

The first three of the following drugs are available in many preparations as over-the-counter medicines. An example of each is given.

8.10.1 Undecenoates

Undecenoates are mixtures of similar fatty acids with antifungal activity. Preparations should be protected from light as this denatures them. They have a peculiar odour (similar to sweat) that is difficult to mask. Preparations of undecenoates are available as over-the-counter medicines, for example Mycota.

8.10.2 Borotannic complex

This is the product of a reaction between tannic and boric acids. Tannic acid is an astringent; boric acid is an antifungal. The only preparation is a complex one combined with salicylic acid, available over the counter as Phytex.

8.10.3 Tolnaftate

Tolnaftate is not available on prescription, but it can be bought over the counter as a 1% cream (for example Mycil).

In some patients, it may produce skin reactions such as irritation, puritis or allergic contact dermatitis.

8.10.4　Benzoic acid

Benzoic acid is used as an antifungal in combination with salicylic acid, which is kera-tolytic. It may irritate sensitive areas.

It is prepared as Whitfield's ointment and available as a pharmacy medicine.

8.10.5　Imidazoles

This group of drugs is effective against many fungi. Imidazoles are often the drugs of first choice for treating tinea because they are more effective than other topical drugs described above. The mode of action of imidazoles is to make fungal cells leaky by inhibiting an enzyme necessary for ergosterol synthesis. Ergosterol is an essential component of fungal cell membranes.

Examples of imidazoles available without prescription are clotrimazole, available as 1% cream (for example Canestan) and miconazole, available as 2% cream (for example Daktarin).

Ketoconazole and tioconazole are currently prescription-only medicines, but qualified podiatrists registered with the Health Professions Council are able to access and supply tioconazole 28% (as from November 2006).

Ketoconazole is not available on the NHS for tinea infections, but tubes of maximum 15 g ketoconazole 2% can be sold over the counter and as such could be supplied by registered podiatrists. See Chapter 14 for more information on how the law applies to sale and supply of medicines by podiatrists.

The most common side effects of imidazoles are irritation and itching of the skin.

Some imidazoles (miconazole, fluconazole) are known to interact with oral hypogly-caemic drugs (Chapter 6) by enhancing their activity and they may increase the risk of myopathy with lipid lowering drugs statins (Chapter 4). Some imidazoles enhance the anticoagulant properties of warfarin. However, these interactions are unlikely with topical application. Systemic administration of imidazoles is often necessary for tinea of the nail. Systemic use of antifungal drugs is considered in Chapter 9.

8.10.6　Terbinafine

Terbinafine displays a broad spectrum of antifungal activity at low concentrations.

Its mechanism of action is by interfering with the synthesis of ergosterol in the fungal cell membrane. Terbinafine inhibits the enzyme that converts a substance called *squalene* to *ergosterol*. This results in a build up of squalene and leakage of cellular contents.

Terbinafine is a prescription-only antifungal drug. It is available as a 1% cream (Lamisil cream), a 1% spray (Lamisil spray) and as 250 mg tablets (Lamisil) for oral therapy. Lamisil cream has recently (November 2006) been added to the list of drugs registered podiatrists are allowed to access and supply and no longer has to be prescribed through a GP. In addition, tubes of maximum 15 g cream and 15 ml spray can be sold over the counter to the public.

Podiatrists are not allowed to supply Lamisil tablets (see Chapter 14).

Mild localized infections of the skin respond to topical application of terbinafine cream for one to two weeks.

Side effects of terbinafine cream include local redness, itching or stinging which the patient may tolerate.

Oral therapy may necessary to treat severe and extensive skin infection and particularly onychomycosis.

When given orally, terbinafine binds strongly to plasma proteins and diffuses rapidly through the dermis concentrating in the *stratum corneum*. It is excreted in the sebum and achieves high concentration in hair follicles and hair. Within two weeks of oral treatment, the drug is distributed into the nail plate.

Skin infections should respond within six weeks and onychomycoses within 6–12 weeks. Infections of the large toenail may take six months or longer to respond.

Terbinafine tablets may produce mild to moderate gastrointestinal disturbance, for example anorexia, loss of taste, nausea and diarrhoea, and occasionally urticaria.

Cimetidine (histamine antagonist) interacts with terbinafine to increase its plasma concentration. Adjustment of the dose of oral terbinafine may be necessary if the two drugs are taken together.

8.10.7 Amorolfine

Amorolfine is a novel topical antifungal drug. It has a similar action to terbinafine affecting synthesis of fungal cell membrane ergosterol and is generally fungicidal.

Amorolfine is available as a 0.25% cream (Loceryl cream) and as a 5% lacquer (Loceryl lacquer). The cream is indicated to treat all fungal infections of the skin and the lacquer is indicated for the treatment of onychomycoses. Both preparations are classified as prescription-only medicines but are included in the list of drugs that registered Podiatrists can access and supply. See Chapter 14.

Amorolfine readily enters the *stratum corneum* and the nail plate but systemic absorption is minimal from topical administration.

Skin infections may require up to six weeks treatment with a minimum of two to three weeks.

Toenail infections will typically require 6–12 months of treatment with 3-monthly reviews of progress.

Side effects are, rarely, pruritis or erythema and a burning sensation in the area of treatment. Hypersensitivity may require withdrawal of treatment.

8.10.8 Griseofulvin

Griseofulvin is an antifungal drug that works by interacting with microtubules to stop mitosis in fungal cells. The drug is used to treat fungal infections of the skin and nails. It is available as a 400 μg metered spray (Grisol) and as prescription-only 500 mg tablets.

Table 8.3 Topical antifungal drugs

Topical antifungals
Undecenoates
Borotannic complex
Tolnaftate
Benzoic acid
Imidazoles
Terbinafine
Amorolfine
Griseofulvin

Griseofulvin spray has recently (November 2006) been added to the list of medicines that registered podiatrists can access and supply.

Oral griseofulvin is prescription-only and is indicated where topical therapy is ineffective but it has largely been replaced by the newer imidazole antifungals. Griseofulvin should be taken with or after fatty food for adequate absorption.

Following oral administration, griseofulvin is sequestered into the keratin of skin and nails.

Griseofulvin has to be taken for at least four weeks and because the drug may be teratogenic, it must not be used in during pregnancy, or in the month before conception or by prospective fathers six months prior to conception.

Common adverse effects of oral griseofulvin are diarrhoea, nausea, vomiting, headache and photosensitivity.

Griseofulvin enhances the effects of alcohol and reduces the effectiveness of oral contraceptive pills.

Griseofulvin should not be used in patients with systemic lupus erythematosus or with liver disease, because it can worsen both of these conditions.

A summary of topical antifungal drugs is given in Table 8.3.

8.11 Bacterial infection of the skin

Bacterial infection of the skin and subcutaneous tissue can lead to rapidly spreading inflammation known as *cellulitis*, which requires systemic antibacterial treatment. Treatment is usually with benzylpenicillin and flucloxacillin or erythromycin.

Impetigo on small areas of skin can be treated with topical application of fusidic acid. Extensive impetigo needs treating with a systemic antibacterial such as flucloxacillin.

There are many topical antibacterials available. In order to minimize the development of resistant bacteria, their use should be limited to those that are not used systemically.

Silver sulfadiazine, 1% cream, was recently added to the list of prescription-only medicines that qualified, registered podiatrists can access and supply. Silver sulfadiazine is used in the prophylaxis and treatment of infection in wounds, and as such can be used to prevent and treat infection following nail avulsions and ablations and after verrucae treatment. Caution has to be used when treating large areas because systemic concentrations can rise sufficiently to cause side effects. Severe skin and blood disorders are possible.

See Chapter 9 for general information about antibiotics.

8.12 Summary

The skin is the largest organ of the body and forms a barrier between the external environment and the internal organs of the body. The skin is subject to trauma and damage. Common skin disorders are eczema, psoriasis and infection.

Eczema is inflammation of the epidermis and is best managed by avoiding the cause, if known. Mild eczema can be treated with emollients if dry, or astringents if wet. Topical corticosteroids are effective in most cases of persistent eczema. With very severe resistant eczema, immunosuppressant drugs may be needed.

Psoriasis is a chronic inflammatory skin disease with hyperproliferation of keratinocytes, which leads to thickening of the epidermis and typical silvery scales on the surface. Topical treatments include drugs that reduce cell division, such as calcipotriol, dithranol and coal tar, and keratolytics and emollients. Phototherapy using UVB and photochemotherapy using UVA plus an oral photosensitizing drug, psoralen, can be successful in severe resistant psoriasis, although relapse may occur. For patients who do not respond to topical therapy, systemic drugs such as methotrexate, ciclosporin, efalizumab and acitretin can be used. These drugs all have serious side effects and must be used under specialist supervision.

Viral warts are common, particularly in children. Without treatment, warts will eventually disappear due to actions of the immune system. There are treatments available, but their use can be tedious, painful and prolonged. Many chemical treatments containing keratolytics are sold as over-the-counter preparations. Other more specialized treatments are cryotherapy, using liquid nitrogen, electrosurgery, surgical excision, bleomycin injections and carbon dioxide laser.

Fungal infection of the skin and nails will be familiar to podiatrists. There are many topical antifungal drugs available over the counter and of varying efficacy. Drugs of choice are imidazoles, terbinafine and amorolfine. Some of the preparations of these drugs are prescription-only medicines but they can be accessed and supplied by podiatrists. In severe extensive infection, oral terbinafine or griseofulvin may be needed. Even so, infection may take weeks or months to clear, particularly infection of the nails.

Bacterial infection of the skin causes a condition known as cellulitis, which requires systemic antibiotics. As does impetigo if it is extensive, although small patches can be treated topically. Silver sulfadiazine is used prophylactically to prevent infection in burns and other wounds and is now on the list of antibiotics that qualified registered podiatrists can access and supply.

Useful websites

www.cks.library.nhs.uk/ National Library for Health. Clinical knowledge summaries.

www.dermatology.co.uk/ Leeds General Infirmary Dermatology Department. Information about eczema, psoriasis, warts, skin infections, and their treatment.

www.feetforlife.org Society of Chiropodists and Podiatrists

http://www.merck.com/mmhe/sec18.html The Merck Manuals Online Medical Library. Section on skin disorders.

Case studies

The first of the following case studies is a patient who might be seen in the podiatry clinic, although other health care professionals might see similar patients for other reasons. The second case study is a patient who could be seen by any health care professional.

Case study 1

Mr Jacobson is a widower of 64, who lives alone. This patient has serious fungal infection of most of his toenails. The use of topical antifungals has not resolved the problem – probably because of poor compliance by the patient. Mr Jacobson is reluctant to consider total nail avulsion and so are you now because the infection has spread to so many toenails. You decide to refer the patient to his GP. You tell Mr Jacobson you will write to his GP and your recommendation is oral therapy with fluconazole.

Mr Jacobson's current therapy is:

   Warfarin 6 mg per day

   Simvastatin 40 mg two at night

Discuss the treatment of this patient and your suggestion of oral fluconazole. You should consider his current medication and take his other medical conditions into account. From his medication, what other conditions could the patient have? Is there any reason for not recommending oral fluconazole for this patient?

Case study 2

Mrs Shabnam is a 32-year-old patient who you are seeing for treatment following an injury. You take a medical history and note that she is taking an antianxiety drug, diazepam. Mrs Shabnam has quite severe and extensive psoriasis. She tells you that the psoriasis causes her a lot of anxiety and that this is the reason why the doctor has put her on diazepam. She is waiting for an appointment at the hospital for what she calls 'light therapy'. She asks you if you know anything about this therapy and if you think it will help improve her psoriasis. What can you tell her about it? Is there anything about Mrs Shabnam's current therapy that might have to be reviewed if she has the therapy for her psoriasis? You would like to recommend a NSAID for Mrs Shabnam's injury. Does she have any contraindications that might affect your choice of drug to recommend?

Chapter review questions

You should be able to answer these review questions using the information in the preceding chapter.

1. Describe eczema and approaches to its treatment.

2. Describe what happens in the epidermis to give the typical appearance of psoriasis.

3. Discuss the advantages and disadvantages of topical treatments available for psoriasis.

4. Explain the difference between phototherapy and photochemotherapy.

5. Discuss the use of systemic antipsoriatic drugs.

6. There are many treatments for warts. Review their practicality and usefulness.

7. Discuss the use of imidazoles in the treatment of tinea.

8. Discuss the use of terbinafine, amorolfine and griseofulvin in the treatment of tinea.

9

Chemotherapy of infectious diseases

9.1 Chapter overview

The use of drugs to eradicate micro-organisms and parasitic worms in the body is called *chemotherapy*. Drugs that are used to treat infections with micro-organisms are known as *antimicrobial drugs* or *antibiotics*. The first antibiotics were naturally produced by micro-organisms; nowadays many are produced synthetically. Infestations with parasitic worms are treated with drugs called *antihelmintics*.

The success of chemotherapy depends on the drugs being relatively more toxic to the infecting organism than to the cells of the human host. This is known as *selective toxicity*. Chemotherapy also applies to treatment of cancers. Cancer cells are similar to normal cells and most anticancer drugs show little selective toxicity and therefore produce serious adverse effects. (See Chapter 10.)

Conventionally, micro-organisms include bacteria, viruses and fungi; protozoa and parasitic worms are parasites. However, bacteria, viruses and fungi are also parasitic, in that they live in or on other living organisms. The term parasite can be applied to any organism that causes disease.

A basic knowledge of the characteristics of these organisms is needed in order to understand how they differ from human cells and how these differences can be exploited in chemotherapy.

9.2 Bacteria

Bacteria are single cell organisms, which can be rod-shaped or spherical. Spherical bacteria are known as *cocci* (as in staphylococcus), and they can be single or double, in strings or in clumps. Bacteria have a cell wall, DNA and the means to synthesize proteins, but no nucleus. Bacteria have many biochemical differences compared with human cells, and

Pharmacology for the Health Care Professions Christine M. Thorp
© 2008 John Wiley & Sons, Ltd

some antibiotics are strikingly non-toxic to humans. Others, however, can cause adverse effects as a result of their effects on normal healthy cells. Most antibiotics are effective against rapidly dividing cells.

Bacteria are classified by gram staining into gram-positive and gram-negative bacteria. The cell wall of gram-positive bacteria contains a mucopolysaccharide that takes up the gram stain. The cell wall of gram-negative bacteria does not contain as much of this mucopolysaccharide and they do not take up the gram stain. The cell wall of gram-negative bacteria is complex and includes an outer double lipid layer, which limits susceptibility to some antibiotics. Broad-spectrum antibiotics are effective against most bacteria, whereas narrow spectrum antibiotics may be effective against only some. The use of broad-spectrum antibiotics can reduce populations of normal flora (the resident harmless bacteria that we all have), which can cause adverse effects and superinfections of fungi and other drug-resistant micro-organisms.

9.2.1　Antibiotics

Bacteriostatic antibiotics inhibit bacterial growth and proliferation, while bactericidal antibiotics actually kill bacteria. Many antibiotics are bacteriostatic at low concentrations and bactericidal at higher concentrations. This distinction is often not important clinically. If bacteria are prevented from multiplying, they will eventually be destroyed by the normal immune reaction of the host. Infections in immunocompromized individuals (for example, those with human immunodeficiency virus (HIV) infection and those on systemic corticosteroid, anticancer or immunosuppressant therapy) have to be treated with potent bactericidal drugs.

9.2.2　Bacterial resistance

The first drugs to be effectively used against bacterial infections were the sulphonamides introduced in the 1930s and followed in the 1940s by the penicillins. More recently, new drugs have been necessary because of the problem of resistance. This is best exemplified by the staphylococcus strains now called *MRSA* (meticillin resistant *Staphylococcus aureus*, or multi-resistant *Staphylococcus aureus*. Meticillin is now unavailable in the United Kingdom).

Resistance to antimicrobial drugs can be innate or acquired.

With innate resistance an entire bacterial species or a certain percentage of a population are naturally resistant to a drug. For example, *Pseudomonas aeruginosa* has always been resistant to flucloxacillin.

Acquired resistance happens when bacteria that were once sensitive to a drug become resistant. The gene that codes for resistance is transferred from one bacterium to another. This can occur by conjugation between two bacteria or by viral transduction.

Resistance to antibiotics can develop by the following mechanisms:

- enzymes produced by bacteria destroy the drug (for example, resistance to penicillin);

- the bacterial cell wall becomes impermeable to a drug or the drug is rapidly removed (for example, resistance to tetracycline);

- drug binding sites within bacteria become altered (for example, resistance to aminoglycosides);

- alternative metabolic pathways develop in bacteria using enzymes not affected by the drug (for example, resistance to sulphonamides).

The development of resistance can be reduced by less indiscriminate use of antibiotics, using them only where there is a distinct clinical need and ensuring that patients complete the entire course of antibiotics. This reduces the risk of survival and reproduction of resistant strains. Drug combinations are used to minimize resistance in certain infections, for example tuberculosis (see page 161).

9.3 Antibiotic drugs

There are so many antibiotics it is helpful to consider them in groups according to their mode of action. Examples are given for each major group.

9.3.1 Inhibitors of nucleic acid synthesis

See Table 9.1 for examples of antibiotics in this group.

The sulphonamides were the first drugs found to be effective in the treatment of systemic infections and although they do still have some specific uses, for example treatment of toxoplasma infections in patients with HIV (see page 170), they are not much used nowadays because of bacterial resistance and the development of more effective drugs, which are less toxic. Silver sulfadiazine is used topically to treat infected wounds (see Chapter 8).

Sulphonamides inhibit an enzyme, dihydropteroate synthatase, used in the production of folic acid in many bacterial cells. Folic acid is the starting point for a cofactor necessary for DNA synthesis in both bacterial and human cells. However, in susceptible bacterial cells, folic acid must be synthesized, whereas in humans it is provided preformed in the diet.

Sulphonamides interact with warfarin (anticoagulant) and sulphonylureas (oral hypoglycaemics) by enhancing the actions of these drugs.

The most important adverse effects of sulphonamides are rashes (common), renal failure and various blood disorders.

Table 9.1 Inhibitors of nucleic acid synthesis

Antibiotic group	Examples
Sulphonamides	Sulfadiazine
Trimethoprim	Trimethoprim
Quinolones	Ciprofloxacin

Trimethoprim inhibits another enzyme, dihydrofolate reductase, in the same folic acid metabolic pathway. Folic acid is converted into folate, which then has to be converted into an activated form by dihydrofolate reductase. In this way, trimethoprim interferes with the conversion of folate into its activated form, which is a cofactor in the synthesis of bacterial DNA. Dihydrofolate reductase also occurs in host cells, but it is less sensitive to trimethoprim. Trimethoprim is used to treat urinary tract infections. It is also formulated in combination with a sulphonamide, when it is known as *co-trimoxazole*, to treat pneumonia in patients with HIV (see page 170). Due to the synergistic effect of the two drugs, this combination is more effective than either drug alone.

Adverse effects on the production of blood cells occur with prolonged use of trimethoprim.

The quinolones inhibit a bacterial enzyme, DNA gyrase. This enzyme is involved in the unwinding of DNA prior to replication. Inhibition of this enzyme prevents DNA replication and bacterial cells cannot survive. Human cells contain a different form of DNA gyrase and so are unaffected by these drugs.

Ciprofloxacin is a broad-spectrum antibiotic, active against both Gram-positive and Gram-negative bacteria. So far, bacterial resistance to quinolones is uncommon.

Adverse reactions to quinolones are infrequent but include nausea, vomiting, rashes, dizziness and headache.

9.3.2 Inhibitors of cell wall synthesis

See Table 9.2 for examples of antibiotics in this group.

Penicillins were first used in the 1940s; they remain the most important group of antibiotics despite the production of many other antibiotics since, and new penicillin derivatives continue to be developed.

Penicillins bind to the cell wall of bacteria and prevent cross-linking of proteoglycan chains that stabilize the cell wall. They are bactericidal and most effective on rapidly dividing bacteria. Human cells have membranes rather than cell walls and so are unaffected by penicillins.

Many bacteria produce an enzyme called *penicillinase*, which destroys some penicillins. Flucloxacillin is not affected by penicillinase and is used for staphylococcus infection.

Resistance to penicillins also occurs by reduction in entry of the drug into the bacterial cell and by alteration of penicillin binding protein.

Table 9.2   Inhibitors of cell wall synthesis

Antibiotic group	Examples
Penicillins	Flucloxacillin
Cephalosporins	Cefadroxil
Glycopeptides	Vancomycin

Adverse reactions to penicillins are rare; diarrhoea can occur due to alteration in normal gastrointestinal bacteria. In fact, penicillins are probably the least toxic drugs known. Hypersensitivity reactions to penicillins occur in up to 10% of patients and vary from mild skin rashes to exfoliative dermatitis and Stevens-Johnson syndrome (immune vasculitis with arthritis, nephritis, central nervous system abnormalities and myocarditis) and from bronchoconstriction to life-threatening anaphylactic shock.

Flucloxacillin and amoxicillin have recently (November 2006) been added to the list of drugs that can be accessed and supplied by podiatrists registered with the Health Professions Council (see Chapter 14 for details of how the law applies to sale and supply of medicines by podiatrists).

There are many drugs in the cephalosporin group. They are closely related to penicillins and have the same mechanism of action and potential for hypersensitivity reactions. Resistance to them can occur by several mechanisms.

Vancomycin also inhibits cell wall synthesis by a slightly different mechanism at an earlier stage. This antibiotic is still effective against most strains of MRSA and has to be given intravenously because it is not absorbed well enough to be effective orally.

It is useful in patients who are allergic to penicillin.

Adverse reactions to vancomycin are fever, rashes and rarely ear and kidney toxicity.

9.3.3 Inhibitors of protein synthesis

See Table 9.3 for examples of antibiotics in this group.

Many antibiotics inhibit protein synthesis in bacteria. They do so by binding to ribosomes, which are different to those in human cells. Ribosomes are essential for protein synthesis because they read the code on messenger RNA. This ensures that amino acids are assembled in the correct order to make a protein. Drugs that bind to ribosomes inhibit cell growth and are therefore bacteriostatic.

Bacterial ribosomes have two subunits, known as *30s* and *50s* according to their sedimentation coefficient (which is a measure of how different fractions sediment during centrifugation).

Aminoglycosides bind to the 30s subunit of ribosomes. They are not well absorbed orally and have to be given intravenously.

Resistance to aminoglycosides occurs by the action of bacterial enzymes, reduced uptake of the drug into bacterial cells and by alteration of their binding protein.

Table 9.3 Inhibitors of protein synthesis

Antibiotic group	Examples
Aminoglycosides	Gentamicin
Tetracyclines	Oxytetracycline
Macrolides	Erythromycin
Chloramphenicol	Chloramphenicol

Adverse effects of aminoglycosides are deafness, kidney damage and allergy. People with myasthenia gravis (see Chapter 7) should not take them because they can impair neuromuscular transmission.

Streptomycin is an aminoglycoside, which is reserved for use in treating tuberculosis (see below).

Tetracyclines also bind to the 30s subunit of ribosomes. They are active orally, but uptake is decreased by the calcium ions in dairy products, so they should not be taken together.

Tetracyclines accumulate in tooth enamel and bone and are not recommended for use in children (or during pregnancy) because they cause discolouration of the teeth and stunt bone growth.

Resistance to tetracyclines is common due to decreased accumulation in bacterial cells. Because of this, broad-spectrum use of tetracyclines has declined and they are now used mainly to treat chlamydia infections, Lyme disease and acne.

Adverse effects of tetracyclines include nausea, vomiting, diarrhoea and liver toxicity. People with myasthenia gravis or systemic lupus erythematosus (see Chapter 7) should not use tetracyclines because they can exacerbate these diseases.

Macrolides bind to the 50s subunit of ribosomes. They are effective against a wide range of bacteria and are active orally. Erythromycin in particular is an alternative in individuals with penicillin hypersensitivity.

Erythromycin has recently (November 2006) been added to the list of drugs that can be accessed and supplied by podiatrists registered with the Health Professions Council (see Chapter 14 for details of how the law applies to sale and supply of medicines by podiatrists).

Resistance is a major problem with macrolides, due to reduced uptake into bacterial cells and alteration of macrolide binding proteins.

Macrolides can cause drug interactions, through inactivation of liver enzymes, with antiarrhythmic drugs, anticoagulants, antipsychotic drugs, anxiolytics and antiepileptic drugs. Erythromycin in particular can increase the risk of myopathy if taken in combination with lipid-lowering statins. Such combinations should be avoided.

Adverse effects of macrolides are nausea, vomiting, diarrhoea, deafness (with high doses but reversible) and jaundice.

Chloramphenicol also binds to the 50s subunit of ribosomes.

Resistance to chloramphenicol occurs due to activity of bacterial enzymes.

Because it is extremely toxic to bone marrow, its use is reserved for life-threatening infections where no other antibiotic is effective.

9.4   Treatment of tuberculosis

Tuberculosis is caused by *Mycobacterium tuberculosis* and *Mycobacterium bovis*. Both organisms can cause tuberculosis in man and cattle. A related organism is *Mycobacterium leprae*, which is the cause of leprosy.

Despite vaccination programmes in many parts of the world, tuberculosis remains the largest single cause of death by infectious disease, causing three million deaths worldwide annually. The problem is compounded by the emergence of resistant strains of the organism and the return of infection to parts of the world where the disease had been eradicated.

Mycobacteria take a long time to duplicate themselves (18–24 hours compared with 30 minutes for *Escherichia coli*). This means it can take a long time to culture enough organisms for identification and there is a long period between infection and clinical manifestation of disease.

Mycobacteria are unusual organisms in that the cell wall has several outer waxy lipid layers, which makes drug penetration very difficult. They do not take up the gram stain, but can be characterized by an acid-resistant stain. Mycobacteria are intracellular organisms and infect lung tissue mainly, but other tissues such as bones, joints and the brain can be infected.

Treatment for tuberculosis is a long and difficult process, taking up to two years and requiring combinations of at least three different antibiotics.

Drugs used to treat tuberculosis include isoniazid, rifampicin and pyrazinamide primarily with streptomycin and ethambutol as secondary drugs.

9.5 Viruses

Viruses are made up of a protein capsule, some with a lipoprotein envelope around it, containing genetic material (either DNA or RNA), with maybe a few enzymes but very little else and hence they are not considered to be cells, but, rather, infectious particles. Viruses are able to bind to cell membranes, penetrate and infect cells of other organisms. Viruses are intracellular parasites; they cannot replicate themselves without using the contents of the host cell to synthesize the cellular components necessary for their reproduction. This makes the development of effective antiviral drugs that do not damage healthy host cells very difficult.

See Table 9.4 for some examples of viruses and the diseases they cause.

Table 9.4 Types of viruses and diseases caused

DNA viruses	Diseases
Poxviruses	Small pox
Herpesviruses	Chicken pox, shingles, cold sores
Adenoviruses	Throat infection, conjunctivitis
Papillomaviruses	Warts
RNA viruses	Diseases
Orthomixoviruses	Influenza
Paramixoviruses	Measles, mumps
Rubella virus	German measles
Picornaviruses	Colds, meningitis, poliomyelitis
Retroviruses	AIDS

9.6   Antiviral drugs

Historically, the development of antiviral drugs has been slow when compared with the development of other antimicrobial drugs. This was probably because the majority of viral infections are eliminated by the host's immune system. With the advent of the HIV and acquired immunodeficiency syndrome (AIDS) that follows infection with HIV, many new antiviral drugs have been introduced since the end of the 1990s. As more is known about how viruses replicate inside cells and how they affect cell function, then no doubt more antiviral drugs will be developed.

Antiviral drugs have been developed that work by the following mechanisms:

- inhibition of the synthesis of viral DNA, RNA and proteins (for example, nucleoside analogues, reverse transcriptase inhibitors, protease inhibitors);

- reduction in release of viral genetic material once inside the host cell (for example, amantadine);

- interference with penetration of virus through the cell membrane (for example, neuraminidase inhibitors);

- interference with attachment of the virus to the host cell membrane (for example, vaccines, immunoglobulins and interferons).

9.6.1   Resistance to antiviral drugs

Resistance can occur through a number of consequences of viral mutation. There may be reduced phosphorylation of the drug (for example with nucleoside analogues), alteration of viral enzyme targets (for example with reverse transcriptase inhibitors), or alteration of target channel protein (for example with amantadine).

9.6.2   Nucleoside analogues

These synthetic compounds resemble the natural nucleotides of viral DNA or RNA. They are phosphorylated in infected cells and then compete with the normal nucleotides for incorporation into viral DNA or RNA. This results in irreversible inhibition of viral polymerase (the enzyme that sticks nucleotides together to form DNA/RNA).

(Note: a nucleoside is a base [either purine or pyrimidine] plus a sugar molecule; a nucleotide is a base plus a phosphorylated sugar molecule. See Figure 10.2, page 180 for a diagram of a nucleotide.)

Aciclovir is an analogue of guanosine nucleoside. It is used to treat herpes infections.

Side effects of topical acyclovir are rare. Use of the drug intravenously to treat systemic herpes can cause nausea and headache and possibly kidney damage.

Ganciclovir is similar in mode of action. Adverse reactions can be severe so this drug is reserved for serious viral infections such as cytomegalovirus in immunocompromised patients.

Adverse effects include bone marrow suppression, liver and kidney toxicity and central nervous system damage.

9.6.3 Reverse transcriptase inhibitors

Some RNA viruses are known as *retroviruses*. These viruses use an enzyme called *reverse transcriptase* to copy viral RNA to produce viral DNA. The viral DNA is then incorporated into the host cell's DNA where it can lie dormant before being used as a template for the production of more viral RNA. Drugs have been developed to inhibit reverse transcriptase and are used particularly to treat HIV. Some drugs in this group are also nucleoside analogues; others are not.

Zidovudine is a thymidine (a nucleoside) analogue and was the first drug introduced that inhibits reverse transcriptase. This action terminates the growing viral DNA strand. Because of HIV resistance, several similar drugs have been developed, including abacavir and lamivudine.

All drugs in this group are toxic, common side effects being gastrointestinal disturbances, anorexia, pancreatitis and liver damage.

Non-nucleoside reverse transcriptase inhibitors such as efavirenz and nevirapine are similar and have a similar profile of adverse effects.

9.6.4 Protease inhibitors

Viral protease is essential in retroviruses for the production of mature viral particles. Inhibition of the enzyme slows spread of virus from cell to cell because immature particles cannot infect host cells.

There are now a number of protease inhibitors available; examples are indinavir and amprenavir.

Adverse effects of protease inhibitors are similar to those seen with reverse transcriptase inhibitors. In addition, this group of drugs causes metabolic disturbances, particularly insulin resistance and hyperglycaemia, and fat redistribution leading to raised plasma lipid levels, which increases the risk of heart disease. These effects are collectively known as *lipodystrophy syndrome*, which appears to be similar to what happens with long-term corticosteroid use.

All drugs in this group inhibit liver enzymes and cause interactions with many other drugs. In particular, they increase the risk of myopathy with statins (Chapter 4).

9.6.5 Inhibition of attachment to, or penetration of, host cells

Some antiviral drugs in this group act by blocking an ion channel that allows the release of genetic material from viruses once they are inside the host cell and which allows viral particles to leave the infected host cell to infect other cells. Amantadine is such a drug, but its use in influenza A is no longer recommended by the National Institute for Health

Table 9.5 Antiviral drugs and their uses

Drug group	Examples	Use
Nucleoside analogues	Aciclovir	Herpes
Protease inhibitors	Indinavir	HIV
Neuraminidase inhibitors	Zanamivir	Influenza
Interferons	Interferon	Hepatitis

and Clinical Excellence (NICE). It has an unrelated use in the treatment of Parkinson's disease (see Chapter 11).

Neuraminidase inhibitors inhibit an enzyme, neuraminidase, that is produced by virus infected cells and which helps viruses to leave infected host cells. Zanamivir and oseltamivir are examples, which are used to treat influenza A and B in vulnerable individuals (those over 65 years or those with chronic respiratory disease, cardiovascular disease, renal disease, immunosuppression or diabetes). They must be used within a few hours and up to 48 hours of onset of symptoms to be effective. However, vaccination against influenza is considered the most effective way of preventing influenza and drugs should not be seen as a substitute. There are NICE guidelines on the use of zanamivir and oseltamivir.

Adverse effects of neuraminidase inhibitors are bronchospasm with zanamivir, which is inhaled, and gastrointestinal disturbances with oseltamivir.

Interferons are cytokines that are produced naturally by virally infected cells. They act to prevent infection of other cells by increasing the synthesis of enzymes that can destroy viral RNA. This interferes with the synthesis of viral proteins and reduces the production of virus particles. Interferon is used to treat chronic hepatitis C infection.

Adverse effects of interferon are fever, lethargy, headache and myalgia.

A summary of antiviral drugs and their uses is given in Table 9.5.

9.6.6 HIV infection and acquired immunodeficiency syndrome (AIDS)

AIDS is caused by infection with HIV. It is a virus that causes impairment or deficiency of the immune system.

The virus is small, fragile and of relatively simple construction. The virus binds to cells that have a particular surface antigen called *CD4*. CD4 is found on T helper cells, monocytes and macrophages.

Once bound, the viral RNA enters the cell and using the cell's components makes a DNA copy of itself by reverse transcription. The newly formed DNA then inserts itself into the host DNA. It may cause cell death immediately or lie dormant until the cell is activated by some other infection. Such latency can last for months or years. Once activated, the cell starts reading its DNA to make proteins. This includes making more viral RNA and viral proteins, which are assembled into new virus particles. The host cell dies as hundreds of virus particles leave the cell in this way and go on to infect other cells.

Monocytes and macrophages are infected but not necessarily destroyed, so they serve as a reservoir of infection.

In addition, macrophages present virus to T lymphocytes, so there is direct spread from cell to cell without the virus entering plasma where it would be attacked by antibodies.

HIV infection leads to AIDS when sufficient T lymphocytes have been destroyed and it is diagnosed when there is the presence of opportunistic infections. These are many and include atypical tuberculosis, herpes, chicken pox and shingles, cytomegalovirus, candida and other fungal infections and protozoan infections including pneumocystis and toxoplasma.

9.7 Treatment of HIV infection

There is no cure for HIV infection. The aim of drug therapy is to slow or stop disease progression by using a combination of different antiviral drugs. Treatment should start as early as possible, before the immune system is irreversibly damaged. However, this aim has to be weighed up against the toxicity of the drugs used and the fact that treatment must be continued over many years. Recommended treatment combines two nucleoside reverse transcriptase inhibitors with either a non-nucleoside reverse transcriptase inhibitor or a protease inhibitor. Such combinations of drugs reduce the development of drug resistance in the virus.

9.8 Fungi

Fungi are multicellular or single cell organisms abundant in the environment; many live in association with humans without causing disease but some of them can be pathogenic in certain circumstances. Multicellular fungi are classed as moulds and single cell fungi are yeasts.

Fungal infections are known as *mycoses* and can be either superficial infections of the skin and nails or systemic infection of the internal organs, particularly the lungs. Both types of infection have become more common since the 1970s due to overuse of broad-spectrum antibiotics and an increase in numbers of immunocompromized individuals because of HIV infection and the use of immunosuppression and cancer chemotherapy. Overuse of broad-spectrum antibiotics leads to a reduction in the body's normal flora (harmless bacteria) that compete with fungi, allowing the fungi to overgrow.

Immunocompromized individuals are susceptible to opportunistic infection with fungi that normally would not be pathogenic, or would easily be eliminated with antifungal drugs.

Fungal skin infections are considered in Chapter 8 and are of particular interest to podiatrists.

Systemic fungal infections are very serious and can be difficult to treat. Examples of fungal infections are candidiasis, aspergillosis and cryptococcus. In the immunocompromized, lung infections are common because fungi produce spores that can be inhaled, but systemic infection of other internal organs is possible including infection of the blood, heart, kidneys and brain.

9.9 Antifungal drugs

The cells of fungi have a nucleus, the usual cellular contents, a cell membrane and a cell wall.

The fungal cell membrane contains a sterol called *ergosterol*, which keeps the membrane stable. In human cell membranes, the sterol is cholesterol. This difference in cell membrane structure allows some selective toxicity of antifungal drugs and most antifungal drugs work by interfering with ergosterol production. Others prevent division of fungal cells. Antifungal drugs can be either fungicidal or fungistatic.

9.9.1 Amphotericin

Amphotericin is a macrolide antibiotic effective against fungi. It binds to ergosterol in the fungal cell membrane and this makes the cell leaky. Loss of intracellular potassium ions in particular leads to cell death.

Amphotericin is used to treat systemic fungal infections. It is not absorbed orally so, unless it is being used to treat gastrointestinal infections, it has to be given by intravenous injection.

Renal toxicity is the most important adverse effect of amphotericin, although patients can also suffer low potassium levels.

9.9.2 Nystatin

Nystatin is similar to amphotericin. It is not absorbed from the gastrointestinal tract and its use is restricted to infections of the skin and gastrointestinal tract.

9.9.3 Griseofulvin

Griseofulvin interacts with microtubules in fungal cells preventing the formation of the spindle during mitosis. It can be used topically and orally to treat skin and nail infections. Given orally the drug is sequestered into hair and nails. Recently the topical preparation has been added to the list of drugs that qualified, registered podiatrists can access and supply (see Chapter 8).

Adverse effects of griseofulvin are gastrointestinal upset, headache and photosensitivity. Allergy to griseofulvin can occur.

9.9.4 Flucytosine

Flucytosine is used to treat some systemic fungal infections. It is an antimetabolite that is converted to 5-fluorouracil in fungal cells but not in human cells, which then inhibits an enzyme necessary for DNA synthesis.

Adverse effects of flucytosine are gastrointestinal upset, anaemia, neutropenia, thrombocytopenia and alopecia, all mild and reversible.

9.9.5 Imidazoles

This group of drugs inhibits an enzyme necessary for ergosterol synthesis.

Some of these drugs (for example ketoconazole and fluconazole) are used orally to treat systemic fungal infections; others (for example econazole and tioconazole) are used topically to treat skin and nail infection.

Adverse effects of these drugs taken orally are generally mild such as gastrointestinal upset, headache and pruritis. In some patients, particularly those with HIV and on multiple therapy, more serious exfoliative dermatitis and Stevens–Johnson syndrome has occurred with fluconazole.

Some imidazoles (miconazole, fluconazole) are known to interact with oral hypoglycaemic drugs (Chapter 6) by enhancing their activity and they may increase the risk of myopathy with lipid-lowering drugs statins (Chapter 4). Some imidazoles enhance the anticoagulant properties of warfarin.

Adverse effects with topical preparations are irritation and itching of the skin.

Some of the imidazole antifungals that are on the list of drugs that qualified, registered podiatrist can access and supply are discussed in Chapter 8.

9.9.6 Terbinafine and amorolfine

These two drugs are used topically to treat skin and nail infections. They interfere with ergosterol synthesis. Their use is discussed in Chapter 8.

A summary of systemic antifungal drugs is given in Table 9.6.

Table 9.6 Systemic antifungal drugs

Drug	Action
Amphotericin	Binds to ergosterol in fungal cell membrane
Nystatin	Binds to ergosterol in fungal cell membrane
Griseofulvin	Inhibits spindle formation in fungal cell
Flucytosine	Inhibits enzyme necessary for DNA synthesis in fungal cell
Imidazoles	Inhibit ergosterol synthesis in fungal cell

9.10   Protozoa

Protozoa are microscopic single cell organisms, some of which can infect and cause disease in humans.

Pathogenic protozoa are able to evade the host's immune system by invading host cells. Examples of infections with protozoa include malaria, amoebiasis, leishmaniasis, trypanosomiasis and toxoplasmosis. Most of these infections tend to occur in countries of the developing world rather than in the United Kingdom. Nevertheless, occasional cases of them are seen in people arriving back from abroad.

9.10.1   Malaria

Malaria is a major killer worldwide and a major cause of chronic ill health. Malaria is not endemic in the United Kingdom, but it occurs in significant numbers of travellers coming back from countries where malaria is endemic. Malaria is caused by several species of protozoa called *plasmodia*. Plasmodia have complicated life cycles with a sexual cycle in the mosquito and an asexual cycle in man. Infection follows a bite from an infected mosquito and involves infection of liver cells and red blood cells. This makes drug treatment difficult because drugs effective against one stage in the life cycle are ineffective at other stages. Resistance to antimalarial drugs is now common.

The most severe form of malaria is caused by *Plasmodium falciparum* and can be fatal. Fever occurs every third day due to rupture of infected red blood cells. Other forms are less severe and rarely fatal.

9.10.2   Toxoplasmosis

Toxoplasmosis is common in the United Kingdom and in most cases causes a mild influenza-like illness, although infection during pregnancy can be a cause of abortion or neurological damage in the foetus. In patients with HIV, toxoplasma can be the cause of serious illness. Toxoplasma encephalitis is a common complication of HIV infection.

9.10.3   Pneumocystis pneumonia

Pneumocystis pneumonia is an opportunistic lung infection almost exclusively seen in patients with HIV. It is often the presenting symptom and a leading cause of death in patients with HIV. An organism called *Pneumocystis carinii*, which used to be classified as a protozoan, causes this type of pneumonia. However, the organism appears to have characteristics of both protozoa and fungi so its classification is currently uncertain. This organism has been renamed as *Pneumocystis jiroveci* and reclassified as a fungus by some authorities. However, pneumocystis pneumonia remains in the section on protozoan infections in the *BNF* and the drugs used to treat it are not typical antifungal drugs.

9.11 Antimalarial drugs

Drugs to treat different stages in the life cycle of the malaria parasite are available. Treatment of malaria can be complex. Guidelines are produced by the Health Protection Agency (Malaria Reference laboratory) in the United Kingdom and by the World Health Organization.

9.11.1 Drugs for clinical cure

These drugs treat the acute attack in the blood. They act on the form of plasmodium in the red blood cells. This group of antimalarial drugs includes quinine and chloroquine.

The mode of action of these two drugs is not fully understood. Chloroquine inhibits digestion of haemoglobin by the parasite (which it needs) and both drugs inhibit the disposal of haem, which is toxic to the parasite.

P. falciparum is resistant to chloroquine in most parts of the world. This appears to be due to decreased uptake and/or increased removal of the drug from the parasite. Quinine is now the main drug used against *P. falciparum* malaria. Chloroquine is still effective against other forms of malaria.

Both drugs have similar side effects of nausea, vomiting, dizziness, blurring of vision and headache, although those of quinine can be more severe.

Halofantrine is another antimalarial drug that is being used more now that resistance to chloroquine and quinine has developed. Its mode of action is not known.

Side effects are gastrointestinal disturbances and headache. Halofantrine can also cause cardiac arrhythmia and should be avoided in patients with arrhythmias.

9.11.2 Drugs for radical cure

These drugs aim to eliminate the resting stage of the parasite in the liver. Not all forms of malaria have a resting stage in the liver – *P. falciparum* does not. Those that do can be treated with primaquine, which prevents the disease from recurring. Primaquine is usually taken in combination with chloroquine.

Primaquine has few side effects limited to gastrointestinal disturbances, although it can cause haemolysis in susceptible individuals.

9.11.3 Drugs for prophylaxis

These drugs prevent infection of red blood cells. Treatment should begin one or two weeks before travelling to areas where malaria is endemic and be continued for at least four weeks after returning. No form of malaria prophylaxis is completely effective and other precautions should be taken. However, should an infection occur the symptoms are usually less severe.

Table 9.7   Drugs used to treat or prevent malaria

Drug group	Examples
Drugs for clinical cure	Quinine, halofantrine
Drugs for radical cure	Primaquine
Drugs for prophylaxis	Chloroquine, proguanil

Chloroquine or proguanil can be used for prophylaxis, although choice of drug will depend on whether the plasmodium is resistant in the area that is being visited.

A summary of drugs used to treat and prevent malaria is given in Table 9.7.

9.12 Drugs for toxoplasma and pneumocystis pneumonia

Most cases of infection with toxoplasma are self-limiting and do not need specific treatment. For treatment of toxoplasma infection of the eye (which can cause blindness) and general infection in the immunocompromized, the preferred treatment is with a combination of pyrimethamine (a folate antagonist) and sulfadiazine (a sulphonamide).

Pneumocystis pneumonia can be treated with co-trimoxazole, which is a mixture of trimethoprim and sulfamethoxazide (a sulphonamide). Co-trimoxazole is the drug of choice for pneumocystis prophylaxis in immunocompromized patients. Atovaquone is used to treat active infection in those who cannot tolerate co-trimoxazole and prophylactically in the immunocompromized.

9.13 Parasitic worms

Parasitic worms are multicellular organisms, which are not always microscopic and are difficult to eliminate by the immune system. They do elicit an immune reaction, mostly by production of immunoglobulin E (IgE) and attack by eosinophils, which can be harmful to the host. With few exceptions, the reproductive stage is either free in the environment or in another host, for example insects or snails. This means that in humans most infestations, in terms of numbers, are self-limiting.

Worms are known as *helminths* and the majority is not parasitic. Classes of worms that parasitize man are round worms (nematodes) and flat worms (platyhelminths). The flat worms are divided into tapeworms (cestodes) and flukes (trematodes). Only a few human parasitic worms are common in the United Kingdom, for example threadworms and round worms. In tropical and subtropical parts of the world, where abundant water and high temperatures provide an optimal environment for the larvae and intermediate hosts, parasitic worms are common and widespread. Table 9.8 lists some parasitic worms that can infect man, their effect and drugs used to treat them.

Table 9.8 Parasitic worms and treatment

Worm	Effects and mode of infection	Anthelmintics
Round worm	Intestinal worm	Mebendazole
Ascaris lumbricoides	Intestinal obstruction possible	Levamisole[a]
	Infection by food/soil contaminated with eggs	Piperazine
Hookworm	Intestinal worm	Mebendazole
Necator americanus	Draw blood, can cause anaemia	
Ancylostoma duodenale	Infection by larvae in soil through skin	
Whip worm	Intestinal worm	Mebendazole
Trichuris trichiura	May penetrate gut wall and cause peritonitis	
	Infection by food contaminated with eggs	
Strongyloidiasis	Intestinal worm	Ivermectin[a]
Strongyloides stercoralis	May penetrate gut wall and invade tissues	Tiabendazole[a]
	Infection through penetration of exposed skin by larvae	
Threadworm	Intestinal worm	Mebendazole
Enterobius vermicularis	Perianal itching and secondary infection	Piperazine
	Infection through contaminated food	
Trichina	Intestinal worm, but can infect muscle	Mebendazole
Trichinella spiralis	Infection from undercooked pork containing cysts	
Filaria	Insect vectors	
Wuchereria bancrofti	Elephantiasis	Diethyl carbamazine
Onchocerca volvulus	River blindness	(not in United Kingdom)
		Ivermectin
Tapeworm	Intestinal worm	Praziquantel[a]
Taenia saginata (beef)	Infection from beef or pork contaminated with cysts	Niclosamide[a]
Taenia solium (pork)	*T. solium* can invade tissues and cause serious disease	
Bilharzia	Intermediate host snail	Praziquantel[a]
Schistosoma haematobium	Larvae enter through skin from water	
	Infects liver, spleen, intestine	
Liver fluke	Infection from cattle via snails	Praziquantel[a]
Fasciola hepatica	Liver disease	
Hydatid disease	Primary host in dog	Albendazole[a]
Echinococcus granulosus	Larvae cause hydatid cysts	

[a]available on a named-patient basis from manufacturer.

The nervous system in helminths has important differences from that of humans and these differences form the basis of the selective toxicity of most drugs used to treat such infections. For example, nematode muscles have both excitatory and inhibitory neuromuscular junctions. The neurotransmitters are acetylcholine and gamma-aminobutyric acid (GABA), respectively.

9.14   Anthelmintics

Anthelmintics either act locally to expel worms from the gastrointestinal tract, or systemically to eradicate forms that have invaded tissues and organs. To be effective, the drug must be able to penetrate the worm.

Anthelmintics have the following modes of action:

- cause paralysis of the worm;

- damage the cuticle (outer covering) of the worm leading to partial digestion or attack by the immune system;

- interfere with the metabolism of the worm.

Many anthelmintics are available in the United Kingdom only on a 'named-patient' basis. These are unlicensed medicines that have not been given a Marketing Authorization by the Medicines and Healthcare Products Regulatory Agency and can only be obtained from the pharmaceutical company or manufacturer.

See Table 9.8 for drugs used to treat parasitic worm infections.

9.14.1   Benzimidazoles

This is a major group of anthelmintics. Mebendazole is a broad-spectrum anthelmintic effective against threadworms, hookworms and round worms. Its mode of action is to bind to β tubulin in the worm, which interferes with microtubular transport of glucose by the worm.

Mebendazole has few side effects; it may cause mild gastrointestinal disturbances.

9.14.2   Piperazine

Piperazine is effective against round worms and threadworms. Its mode of action is to inhibit neuromuscular transmission in the worm. It does this by acting on GABA receptors, which causes the opening of chloride channels and hyperpolarization of muscle cells in the worm. This causes paralysis of the worm.

Piperazine has few side effects, mainly gastrointestinal disturbances.

9.14.3 Niclosamide

Niclosamide is used against tapeworms. It works by damaging the head of the tapeworm so that it can no longer attach to the inside of the intestine.

Niclosamide has few side effects, but it has been superseded by praziquantel.

9.14.4 Praziquantel

Praziquantel (available on a named-patient basis) is a highly effective anthelmintic, which induces muscular contraction and spastic paralysis in trematodes and cestodes by increasing calcium ion fluxes.

Adverse effects are rarely of any clinical importance.

9.14.5 Ivermectin

Ivermectin is mostly used to treat filariasis and onchocerciasis, which are parasitic worm infections not seen in the United Kingdom. Filariasis causes elephantiasis (blockage of lymphatics and swelling of the limbs) and onchocerciasis causes river blindness in tropical and subtropical countries.

Ivermectin has occasional use against round worms, hookworms and whip worms and is available on a named-patient basis in the United Kingdom.

Its mode of action is by enhancing GABA mediated inhibition at the neuromuscular junction. This hyperpolarizes muscle cells so they cannot contract and causes flaccid paralysis of the worm.

Side effects of ivermectin are skin rashes, fever, dizziness, headache and muscle pain.

9.15 Summary

Infectious diseases can be caused by bacteria, viruses, fungi, protozoa or parasitic worms.

Bacterial infections are treated with antibiotics. There are many antibiotics available, but they fall into three major groups based on their mode of action: inhibitors of bacterial nucleic acid synthesis; inhibitors of cell wall synthesis; and inhibition of bacterial protein synthesis. Resistance of bacteria to commonly-used antibiotics has become a major problem necessitating the development of new antibiotics. Tuberculosis infection is difficult to treat and requires a combination of at least three different antibiotics.

Viral infections are normally overcome by the patient's immune system. However, the advent of HIV infections and AIDS has led to the development of several new antiviral drugs. Antiviral drugs work by inhibiting the synthesis of viral DNA, RNA or proteins; by reducing the release of viral genetic material inside host cells; by interfering with viral penetration of host cell membranes; and by interfering with attachment of virus to host cell membranes. Resistance to antiviral drugs can occur by a number of mechanisms. HIV infection is treated with a combination of antiviral drugs with different mechanisms of action.

Infection with fungi can be superficial affecting skin and nails, or systemic affecting internal organs, particularly the lungs. Systemic fungal infections have become more common due to overuse of antibiotics and increase in numbers of immunocompromized patients. Many antifungal drugs work by interfering with ergosterol production thus making the fungal cell membrane leaky. Others prevent fungal cells from dividing.

Infections with protozoa are not common in the United Kingdom, but are responsible for many deaths and ill health worldwide. Examples of protozoan diseases are malaria and toxoplasmosis.

The malaria parasite has a complicated life cycle and can be difficult to treat. Resistance to antimalarial drugs is common.

Toxoplasma infection is common in the United Kingdom and normally causes a mild illness that can go unnoticed. In pregnancy, it can however damage the developing foetus and may cause abortion. In patients with HIV infection, toxoplasma can cause encephalitis. Pneumocystis can cause serious pneumonia, almost exclusively in AIDS patients. Both toxoplasma and pneumocystis are treated with drugs containing sulphonamides.

Infection with parasitic worms in the United Kingdom is usually confined to round worms or thread worms. In tropical and subtropical parts of the world, infection with parasitic worms is common and widespread. There are many drugs available (in the United Kingdom on a named-patient basis) to treat parasitic worm infection. Most anthelmintic drugs work by interfering with neuromuscular transmission causing paralysis of the worm.

Useful websites

www.emc.medicines.org.uk Electronic Medicines Compendium.

http://www.hpa.org.uk/infections/topics_az/malaria/ Health Protection Agency. Malaria prevention and treatment guidelines.

http://www.merck.com/mmhe/sec17.html The Merck Manuals. Online Medical Library; Section on infections. Patient information leaflets and summaries of product characteristics.

http://www.nice.org.uk/TA58 National Institute for Health and Clinical Excellence. Anti-viral therapy guidelines 2003.

http://www.who.int/topics/malaria/en/ World Health Organization. Guidelines for the treatment of Malaria.

Case studies

The first two case studies are of hypothetical patients who might be seen by any health care professional. You should be able to offer professional advice to patients like these. The third case study is a diabetic patient who might be seen by podiatrist specializing in the care of diabetics, although other health care professionals might see him for other reasons.

Case study 1

Mrs Dexter is 77 and suffering from herpes zoster, commonly known as *shingles*. She is being treated systemically with acyclovir at a dose of 200 mg every 4 hours. She asks you about side effects of this drug. What should you tell her and what can you advise her if she should suffer from any of the side effects?

Systemic antiviral treatment reduces pain severity, duration and complications of shingles. It should be started within 72 hours of the first appearance of a rash.

Are there any cautions with use of this drug?

You see Mrs Dexter a few months later and she tells you that, although the rash has gone, she is still in considerable pain. She says that normal painkillers do not do any good. What advice can you give Mrs Dexter about what appears to be neuropathic pain? You may need to refer to Chapter 12 or look up the treatment of neuropathic pain in the *BNF*.

Case study 2

Mrs Bachelor is 80 years old. For her age, she is reasonably fit and active, but forgetful. She is currently suffering from a urinary tract infection and has been prescribed ampicillin by her doctor. Is there anything about taking antibiotics in general and ampicillin in particular that you might have concern with, especially in the light of Mrs Bachelor's forgetfulness?

Are there any adverse reactions and/or interactions with other drugs that should be taken into consideration?

Case study 3

Mr Pritchard is 61 and has had type 2 diabetes for several years; he has never been very good at controlling his blood glucose levels. Mr Pritchard is a heavy smoker and overweight. Mr Pritchard also has atrial arrhythmia and high lipid levels. The patient now has an infected diabetic ulcer on the sole of his left foot, which is showing no signs of healing. Clearly, the patient needs urgent antibiotic treatment.

Mr Pritchard is currently on the following drugs:

Warfarin 5 mg per day
Amiodarone 200 mg per day
Atorvastatin 20 mg per day
Gliclazide 80 mg with breakfast

Discuss the patient's treatment using the following questions as a guide.

- Mr Pritchard says he is allergic to penicillin; would it be safe to treat him with erythromycin?

- Is there anything about Mr Pritchard's current treatment that might be a cause for concern?

- What general advice can you give to this patient to improve his condition?

- Do you think Mr Pritchard would be a suitable patient for treatment under supplementary prescribing?

Chapter review questions

1. Describe the mode of action of the following groups of antibiotic drugs and give an example in each case:

   (a) sulphonamides

   (b) penicillins

   (c) tetracyclines

   (d) quinolones

   (e) macrolides

   (f) glycopeptides.

2. Discuss the problem of bacterial resistance and how it arises.

3. Describe how the following groups of antiviral drugs interfere with viral replication and give an example in each case:

   (a) nucleoside analogues

   (b) reverse transcriptase inhibitors

4. Describe three ways in which spread of viral particles can be halted by antiviral drugs.

5. What is it about fungal cells that is different to human cells and allows selective toxicity of antifungal drugs such as amphotericin, ketoconazole and terbinafine?

6. Why is malaria so difficult to treat?

7. Name three examples of anthelmintics that work by causing paralysis of the worm.

10
Cancer chemotherapy

10.1 Chapter overview

After cardiovascular disease, cancer is the second major cause of death in the developed world, being the cause of about one in five deaths. It is increasingly common over the age of 50 and rare in children under 10. Cancer covers a wide range of diseases including those commonly occurring such as lung cancer, breast cancer and colon cancer as well as much rarer conditions, some of which occur almost exclusively in children (for example retinoblastoma). Cancer is therefore not one disease but many, which may all be different and require different approaches to therapy.

Many advances in treatment and understanding of the molecular biology of cancer have been made since the mid-1970s. Some previously fatal cancers are now largely curable (for example cervical cancer, testicular cancer, lymphoma and leukaemia) and there have also been improvements in treatment of ongoing disease and in palliative care.

The three main approaches to treatment are surgery, radiotherapy and chemotherapy. The way in which they are combined depends on the type of cancer and its stage at diagnosis. Only a few cancers are treated with chemotherapy alone.

New approaches to treatment now include gene therapy, immunotherapy, stimulation of bone marrow, induction of differentiation in cancer cells and inhibition of angiogenesis in addition to the more standard chemotherapy.

This chapter will concentrate on the drugs used in cancer chemotherapy with consideration of some of the other forms of therapy.

10.2 Biology of cancer

Cancers are produced by the uncontrolled multiplication, growth and spread of abnormal body cells. This results from loss of mechanisms that normally regulate maturation and proliferation of cells.

Many approaches to therapy of cancer depend on interfering with cell division, and so it is necessary to consider the normal cell cycle of dividing cells. (See Figure 10.1.)

Pharmacology for the Health Care Professions   Christine M. Thorp
© 2008 John Wiley & Sons, Ltd

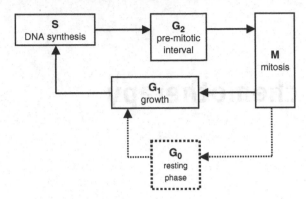

Figure 10.1 The cell cycle

Cell proliferation is regulated through the cell cycle, which is divided into five phases:

- G_1 – growth and preparation for DNA synthesis;

- S – synthesis of DNA;

- G_2 – growth and assembly of microtubules and centrioles;

- M – mitosis with separation of chromosomes and division of cell contents;

- G_0 – growth arrest.

If differentiation follows G_0, cells are no longer capable of division and they leave the cycle. Normally, mature cells have a limited life span and then die. Cells in tumours may remain in G_0 for prolonged periods and can be stimulated to divide again.

There are checkpoints between G_1 and S and between G_2 and M where, if a cell's DNA is damaged, cell division can be stopped by the action of tumour suppressor genes (also known as *suppressor oncogenes*). A damaged cell may be repaired and if that fails, the cell would normally self-destruct by a process known as *apoptosis*. Apoptosis is genetically programmed cell destruction that removes old, redundant and abnormal cells. This guards against mutations surviving and cells becoming malignant. Many cytotoxic anti-cancer drugs damage DNA and initiate apoptosis. Loss or mutation of tumour suppressor genes allows cells to undergo malignant transformations. For example, a tumour suppressor gene known as *p53* normally activates a key pathway in apoptosis. Loss or mutation of this gene allows continuous proliferation and survival of mutated cells and is seen in many types of human cancer. Mutated cells of this type may be resistant to drugs because they can survive DNA damage.

Cells are stimulated to divide by growth factors, which form part of the cell cycle control system. There are many growth factors and they are coded for by genes. Some of these genes are known as *proto-oncogenes*. They were given this name because they can be altered by mutation, carcinogens or viruses to become oncogenes. Oncogenes are capable of causing malignant transformation in cells.

An example of a proto-oncogene is the *ras* gene. This gene codes for a protein called Ras. Mutation of the *ras* gene leads to production of abnormal Ras, which activates cell division. Mutations of the *ras* gene are found in 20–30% of all human cancers.

It is believed that normal growth within tissues is also controlled by contact inhibition. That is, inhibition of abnormal cell division and growth by cell-to-cell contact factors called *integrins* and *adhesion molecules*. In addition, as cells are added during growth and repair, others die so overall size of a tissue or organ is maintained.

In summary, cell cycle control is lost in cancer cells possibly because of mutation of tumour suppressor genes, oncogene activity, abnormal growth factor function or abnormal cell cycle control function. Certainly, the genesis of cancer involves many factors working together.

Characteristics of cancer include uncontrolled cell division with excessive growth of undifferentiated cells. Cancer cells invade surrounding tissues, whereas normal cells do not survive outside their normal boundaries. Cancer cells may travel to distant sites in the body via the blood and lymph to form secondary tumours that are called metastases. As cells repeatedly divide a mass forms known as a *tumour*, which needs a blood supply. Angiogenesis is the process by which tumours develop their own blood supply.

By the time cancer has been diagnosed, cell growth and spread of malignant cells has probably occurred.

10.2.1 DNA structure, replication and protein synthesis

In order to understand how drugs used to treat cancer work, an overview of DNA structure and replication and the process of protein synthesis is necessary. There follows a brief outline of these processes.

DNA is a large molecule arranged in two long strands coiled around each other to form the double helix. Each strand is composed of millions of nucleotides. A nucleotide has three components: a base, which is an organic molecule containing nitrogen; a five-carbon sugar called *deoxyribose*; and a phosphate group. (See Figure 10.2.)

Nucleotides are joined to each other by bonds between the sugar of one nucleotide and the phosphate group of the next to form polynucleotides. There are four different bases in DNA: adenine, guanine, cytosine and thymine. The two polynucleotide strands of DNA coil around each other so that the bases project towards the inside of the helix. The structure is held in place by bonds that form between complementary bases. Adenine can only pair with thymine and guanine can only pair with cytosine. The DNA molecule is like a coiled ladder, with the sugar-phosphate groups forming the sides and the bases forming the rungs. The basic structure of DNA is always the same, but the sequence of bases varies in different molecules. The sequence of bases forms a set of instructions or a code for the assembly of amino acids into proteins. A stretch of DNA that codes for a particular protein is known as a *gene*. DNA is packaged into the nucleus of cells in the condensed form of chromosomes. In human cells there are 46 chromosomes (apart from sperm and egg cells where there are 23).

Replication of DNA takes place before a cell divides and necessitates the uncoiling and separation of the two strands. This requires the action of enzymes. New nucleotides pair

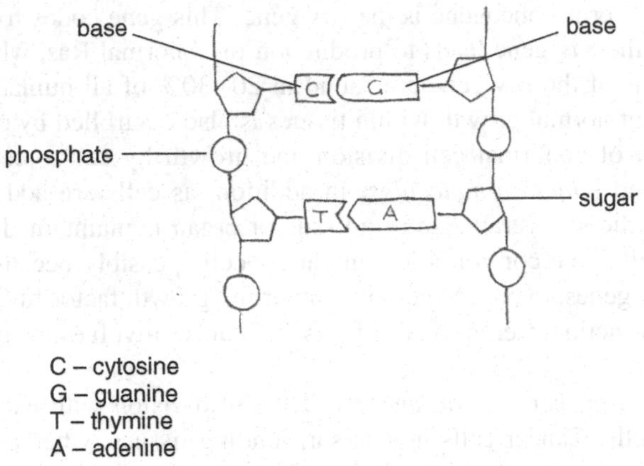

C – cytosine
G – guanine
T – thymine
A – adenine

Figure 10.2   Structure of a nucleotide

with the existing bases of each strand: adenine nucleotides pair with thymine; guanine nucleotides pair with cytosine. This process is catalysed by an enzyme called *DNA polymerase*. Bonds form between the sugar and phosphate groups of the new nucleotides and two new double strands of DNA are created, both identical to the original DNA molecule.

When not replicating, DNA serves as the code for synthesis of proteins. During protein synthesis, a small portion of DNA is copied by complementary base pairing of RNA nucleotides (adenine, guanine, cytosine and uracil instead of thymine) to form single stranded messenger ribonucleic acid (mRNA) in the nucleus of the cell. This process is known as *transcription* and is catalysed by an enzyme called *RNA polymerase*.

The mRNA leaves the nucleus and attaches itself to ribosomal ribonucleic acid (rRNA) in the cytoplasm of the cell. Without going into precise detail, rRNA reads the base-sequence code of mRNA and a third type of RNA, transfer ribonucleic acid (tRNA) brings the correct amino acid to the forming protein. This process is known as *translation*. Each amino acid is coded for by a sequence of three bases, known as a *triplet code*. For example, ACA codes for the amino acid cysteine. As amino acids are added in sequence a protein is built up.

10.3   Principles of chemotherapy

The aim of chemotherapy is either, to affect a cure by reducing the size of a tumour so that it can be surgically removed or it disappears altogether, or to prolong life and provide palliative care. For the best chance of success, drugs must be used early on in a patient's treatment along with irradiation or surgery when the cancer is most curable and the patient is most able to tolerate treatment. The general health of the patient, their nutritional status, their liver and kidney function and the viability of their bone marrow must all be taken into account. Above all the benefits of therapy must outweigh the risks.

Cells in tumours do not all grow and divide at the same rate. Initially, growth is exponential; that is each cell divides to produce two, which then divide to produce four and so on. As a tumour gets larger growth slows down, partly because it outgrows its blood supply and partly because not all cells in the tumour divide continuously.

Cells in a solid tumour can be in three different states:

- those that are dividing (rapidly), possibly in continuous cell cycle;
- those that are resting in the G_0 phase of the cell cycle;
- those that are no longer able to divide.

Cells that are dividing are the best targets for chemotherapy, but they may only represent 5% of the total of tumour mass. Cells that can no longer divide do not present a problem, but cells that are resting are often resistant to cytotoxic drugs and can re-enter the cell cycle after a course of chemotherapy.

Drugs used to treat cancer act on some stage in the cell cycle, either on DNA synthesis or spindle formation; or cause damage to preformed DNA. Phase-specific drugs act at a particular point in the cycle, usually S phase or M phase. Phase-non-specific drugs are cytotoxic at any point in the cell cycle and may be toxic to cancer cells in the resting phase.

However, most cancer drugs are effective in the S phase of dividing cells, preventing normal DNA synthesis so that the cells go into apoptosis. Drug therapy aims to prevent cell division or cause cell death in the tumour without damaging normal healthy cells. Cancer cells are similar to normal cells, so this is often impossible and explains many of the side effects of chemotherapy.

Cytotoxic drugs affect normal dividing cells producing unwanted adverse effects on bone marrow and the cells produced there; reduced healing; loss of hair due to damage to hair producing cells in hair follicles; damage to the gastrointestinal lining; reduced growth in children; sterility; and damage to the foetus. Nausea and vomiting are common with most anti-cancer drugs and are caused by toxic effects on the chemotrigger zone of the central nervous system. Another adverse effect of cancer chemotherapy is known as *tumour lysis syndrome*. This is a collection of symptoms caused by the death of many cells in the tumour. A significant symptom is the production of large amounts of uric acid as a result of the breakdown of nucleic acids. High plasma levels of uric acid can precipitate gout, or interfere with treatment for pre-existing gout, or uric acid can deposit in the kidneys and cause kidney damage. Concurrent use of allopurinol can reduce the production of uric acid (see Chapter 7).

Tumour cells can become resistant to chemotherapy. It seems that this is likely to happen as a result of low-dose single drug chemotherapy. Drugs are therefore generally more effective in combination and may act synergistically. This also means that severe toxicity effects can often be avoided. The drugs should be used at the maximum dose that can be tolerated and as frequently as possible to discourage tumour regrowth.

Mechanisms of acquired resistance to cancer chemotherapy include:

- decreased uptake of drug into cells, or increased expulsion from cells;

- increased production of substances that can inactivate the drug;

- increased activity of DNA repair enzymes;

- increased metabolism and inactivation of the drug;

- reduced conversion of drug into its active form.

10.4 Drugs used in cancer chemotherapy

All drugs used to treat cancer have a narrow therapeutic ratio and considerable potential for causing serious harm. An understanding of their pharmacology, drug interactions and pharmacokinetics is essential for safe and effective use, which should be under specialist guidance of an oncologist.

There are six groups of drugs used in cancer chemotherapy. The first four are all cytotoxic, which means they cause cell death. There are many drugs in each group. The descriptions that follow are not intended to give a comprehensive account of all anti-cancer drugs available, rather to explain the modes of action and to give examples from each group.

Toxic effects with all cytotoxic drugs are severe nausea and vomiting, hair loss and bone marrow depression.

10.4.1 Alkylating agents and related drugs

Alkylating agents act by forming bonds with guanine in DNA. This interferes with the normal base pairing of guanine with cytosine and allows cross-linking of guanine with other guanine molecules in the same strand and opposite strand of DNA. This denatures the DNA and as a result transcription of DNA for protein synthesis and replication of DNA for cell division are disrupted. The end result is cell death by apoptosis. Although alkylating agents can act on cells at any stage of the cell cycle, they are most effective in the S phase of the cell cycle and are therefore most cytotoxic to rapidly dividing cells.

10.4.1.1 Cyclophosphamide

Of the drugs in this group, cyclophosphamide is commonly used to treat chronic lymphocytic leukaemia, lymphoma and ovarian and testicular cancers.

Cyclophosphamide produces severe nausea and is often used with a 5HT$_3$ antagonist, ondansetron. Cyclophosphamide is metabolized to a toxic metabolite called *acrolein*, which can cause haemorrhagic cystitis, a rare but serious complication. This effect can be counteracted by a high intake of fluid and by using a drug called *mesna*. Otherwise, cyclophosphamide can cause bone marrow depression, sterility and increased risk of other cancers.

There are many drugs in this group of anti-cancer drugs. Some other examples are busulphan (for leukaemia), carmustine (for brain tumours), treosulphan (for ovarian cancer) and cisplatin (for lung, testicular, bladder, ovarian and cervical cancers).

10.4.2 Antimetabolites

Antimetabolites block or alter a metabolic pathway involved in DNA synthesis. They are analogues or antagonists of normal cell constituents.

10.4.2.1   Methotrexate

Methotrexate is a folate antagonist that competes for an enzyme (dihydrofolate reductase) necessary for the conversion of folate into its active form. The active form of folate is a cofactor needed for the synthesis of nucleotides adenine and guanine, which form part of DNA and RNA. In this way protein synthesis and cell division are interfered with.

(The observant reader may have noticed that this is similar to the mode of action of trimethoprim [Chapter 9, page 157]. Trimethoprim is selective for bacterial cells because bacterial dihydrofolate reductase is many times more sensitive to trimethoprim than is the human enzyme.)

Methotrexate is also used to treat rheumatoid arthritis (Chapter 7) and severe psoriasis (Chapter 8).

Resistance to methotrexate occurs possibly due to reduced uptake into cancer cells or altered enzyme activity.

Adverse effects of methotrexate include ulceration of the mouth and lower gastrointestinal tract and bone marrow suppression. With high doses of methotrexate synthetic folinates are used to prevent irreversible bone marrow damage.

10.4.2.2   Mercaptopurine

Mercaptopurine is a purine analogue that substitutes for purine bases (adenine and guanine) in the synthesis of DNA. It is converted to a false nucleotide that inhibits the formation of normal DNA.

Adverse effects of mercaptopurine are similar to methotrexate.

Mercaptopurine is metabolized by xanthine oxidase. It should be used with caution in combination with allopurinol because allopurinol inhibits xanthine oxidase.

Both methotrexate and mercaptopurine are used to treat acute leukaemia.

10.4.2.3   Fluorouracil

Fluorouracil is a pyrimidine analogue that substitutes for uracil in the synthesis pathway for thymidylate, a necessary step in the synthesis of DNA. Formation of a false nucleotide inhibits an enzyme necessary for DNA synthesis. This is similar to the mode of action of the antifungal drug, flucytosine (Chapter 9, page 167).

Fluorouracil is used to treat gastrointestinal and breast cancers, often together with synthetic folinates.

Adverse effects of fluorouracil are less likely, but similar to methotrexate.

10.4.2.4 *Cytarabine*

Cytarabine is a pyrimidine analogue, which is incorporated into both RNA and DNA and interferes with the action of DNA polymerase. Cytarabine is used to treat leukaemia.

Adverse effects of cytarabine are bone marrow suppression and gastrointestinal disturbances.

10.4.3 Cytotoxic antibiotics

Some cytotoxic antibiotics prevent cell division by direct action on DNA. Others form polynucleotides, which inhibit transcription of DNA to RNA and therefore suppress protein synthesis.

Examples of cytotoxic antibiotics are doxorubicin, dactinomycin and bleomycin.

10.4.3.1 *Doxorubicin*

Doxorubicin inhibits an enzyme called *topoisomerase* (also known as *DNA gyrase*). During replication of DNA, this enzyme brings about swivelling of individual strands of DNA so they can be copied without becoming entangled in each other. Doxorubicin prevents this.

Doxorubicin is used to treat leukaemia, lymphoma and many solid tumours.

Apart from nausea and vomiting and bone marrow suppression, adverse effects with doxorubicin are cardiotoxicity and possibility of cardiac failure.

10.4.3.2 *Dactinomycin*

Dactinomycin binds to guanine in DNA and prevents the movement of RNA polymerase, so interfering with protein synthesis.

Dactinomycin is mainly used to treat cancers in children.

Adverse effects are bone marrow suppression and nausea and vomiting.

10.4.3.3 *Bleomycin*

Bleomycin degrades DNA to cause fragmentation of the strands. Bleomycin is most effective in the G_2 phase and during mitosis of the cell cycle, but it is also effective against cells in the G_0 phase.

Bleomycin is used to treat squamous cell carcinoma.

Bleomycin is unusual among cytotoxic drugs in that it does not cause bone marrow suppression. However, it is associated with blistering skin rashes and serious progressive pulmonary fibrosis in 10% of patients. (See Chapter 8 for use of bleomycin in the treatment of viral warts.)

10.4.4　Vinca alkaloids and related drugs

This is a group of drugs that are derived from plants.

10.4.4.1　Vinca alkaloids

Vinca alkaloids are extracted from the periwinkle plant, *Vinca rosea*, although newer ones are semi-synthetic. They bind to tubulin and prevent its polymerization into microtubules. Microtubules form the spindle during mitosis, so vinca alkaloids halt mitosis at a certain stage.

Examples of vinca alkaloids are vinblastine, vincristine and vindesine.

These drugs are used to treat leukaemia, lymphoma, testicular cancer and lung cancer.

Adverse reactions of vinca alkaloids are relatively mild. Some are due to the inhibition of other functions of microtubules. Thus migration of white blood cells and axonal transport in neurons are inhibited leading increased susceptibility to infection and neurotoxicity respectively.

10.4.4.2　Taxanes

Taxanes are a group of drugs extracted from yew bark. The mode of action of taxanes is to promote microtubule formation and inhibit their breakdown, so preventing completion of mitosis.

The two taxanes currently in use are paclitaxel, which is used in the treatment of ovarian cancer and metastatic breast cancer and docetaxel, which is used to treat metastatic breast cancer, small cell lung cancer and prostate cancer.

Adverse effects of paclitaxel and docetaxel are similar, being hypersensitivity reactions, bone marrow suppression, peripheral neuropathy, hair loss and cardiac arrhythmias.

10.4.4.3　Podophyllotoxin

Podophyllotoxin is extracted from the mandrake plant, *Podophyllum peltatum*. A derivative of this chemical, etoposide, is used in small cell lung cancer, lymphoma and testicular cancer. The mode of action of etoposide is not completely known, but it may be due to an inhibitory effect on topoisomerase similar to that of doxorubicin (see above).

Adverse effects of etoposide are nausea and vomiting, hair loss and bone marrow suppression.

10.4.5　Hormones

Hormones are used specifically in cancers that are hormone dependent. Growth of such tumours can be suppressed by hormones with the opposite action, by hormone antagonists or by inhibition of hormone secretion.

Some examples are shown in Table 10.1.

Hormonal treatment plays an important role in the management of cancers of the breast, prostate and endometrium.

Table 10.1 Examples of hormones used to treat cancers

Hormone	Action	Used to treat
Glucocorticosteroids	Inhibit lymphocyte proliferation	Leukaemias
[a]GnRH analogues e.g. goserelin	Inhibit GnRH release	Advanced breast cancer and prostate cancer
Oestrogens e.g. ethinylestradiol	Block androgen action	Prostate cancer
Anti-oestrogens e.g. tamoxifen	Block oestrogen receptors	Breast cancer
Anti-androgens e.g. bicalutamide	Block androgen action	Prostate cancer
Radioactive iodine	Destroys cells in thyroid gland	Thyroid tumours

[a]GnRH – gonadotrophin releasing hormone

Glucocorticosteroids are used in leukaemia because they inhibit proliferation of lymphocytes and in palliative care of other cancers, because they can stimulate appetite and produce a sense of well-being.

Radioactive iodine, although not a hormone itself, is taken up by the thyroid gland and incorporated into thyroid hormones. During this process, it emits radiation that causes destruction of the thyroid gland. It is used specifically to treat thyroid tumours. Radioactive iodine is also used to treat thyrotoxicosis (Chapter 6).

10.4.6 Miscellaneous drugs

These anti-cancer drugs do not fit into the other groups. Most are cytotoxic in some way and cause the usual side effects. They are used for specific indications. Some examples are given below.

10.4.6.1 Procarbazine

Procarbazine inhibits RNA and DNA synthesis and interferes with mitosis. It is a drug that interacts with alcohol and generally increases the effects of central nervous system depressants and can produce hypertension.

10.4.6.2 Hydroxyurea and crisantaspase

Hydroxyurea inhibits ribonucleotide reductase, which is necessary for the conversion of ribonucleotides into deoxyribonucleotides.

Crisantaspase breaks down asparagine, which some tumour cells cannot synthesize. Normal cells can synthesize asparagine; therefore the drug is selective for certain tumours.

10.4.6.3 *Biological response modifiers*

Biological response modifiers are drugs that enhance the patients' natural response to cancer. They are used only in specific types of cancer, usually if more conventional chemotherapy has failed.

Interferons are produced by many cells of the immune system and some (interferon α) have anti-tumour effects.

Interleukin-2 regulates the proliferation of tumour-killing T lymphocytes and natural killer cells.

Tretinoin is the acid form of vitamin A. It induces differentiation in leukaemic cells.

Monoclonal antibodies (rituximab and alemtuzumab) have been developed recently which cause lysis of B lymphocytes. These drugs are intended to be used in particular types of leukaemia when other forms of treatment have failed to produce a long-lasting remission. The most serious adverse reaction with these drugs is the so-called cytokine release syndrome. This includes fever and chills, nausea and vomiting and allergic reactions such as rashes, itching, angioedema, bronchospasm and difficulty breathing. Fatalities have occurred and these drugs are still being monitored by the Medicines and Healthcare Products Regulatory Agency (MHRA).

(See Chapter 7 for use of rituximab in the treatment of rheumatoid arthritis.)

A monoclonal antibody against human epidermal growth factor receptors (HER2) has been developed for use in metastatic breast cancer. This drug is trastuzumab (Herceptin) and is also being monitored by the MHRA.

Table 10.2 gives a summary of drugs used in cancer chemotherapy.

10.4.7 New approaches to treatment

Drugs that target known oncogenes are under development. These drugs are known as *antisense* oligonucleotides (ON), which are small synthetic pieces of DNA that are complementary to a segment of mRNA. A hybrid of ON DNA and mRNA forms, which prevents translation of the mRNA. This means that the effect of the original oncogene is blocked.

Cancer cells produce enzymes that can breakdown the extracellular matrix between normal cells and allow cancer cells to invade and spread in normal tissue. New drugs are being developed that inhibit these enzymes.

Table 10.2 Drugs used in cancer chemotherapy

Drug group	Examples
Alkylating agents	Cyclophosphamide
Antimetabolites	Methotrexate, mercaptopurine, fluorouracil
Cytotoxic antibiotics	Doxorubicin, dactinomycin, bleomycin
Vinca alkaloids and related drugs	Vinblastine, paclitaxel, etoposide
Hormones	See Table 10.1
Miscellaneous drugs	Procarbazine, hydroxyurea, monoclonal antibodies (rituximab)

In order for tumours to grow they need to develop a blood supply. Drugs are being developed that inhibit angiogenesis, which is the growth of new blood vessels. A monoclonal antibody (bevacizumab) that inhibits vascular endothelial growth factor already exists for use in metastatic colorectal cancer. There are serious side effects, such as gastrointestinal perforation and impaired wound healing, and its use is being monitored by the MHRA.

Other areas of research include the development of drugs that block the capacity of oncogenes to turn cells malignant; antagonists of growth factors; inhibitors of cell cycle control factors; and gene therapy to restore tumour suppressor gene function.

10.5 Summary

Cancer occurs when mutated cells divide uncontrollably, taking up space and spreading to distant parts of the body. Treatment is usually with a combination of surgery, irradiation and chemotherapy. The aim of chemotherapy is to interfere with cell division in tumour cells without damaging normal healthy cells. However, because cancer cells are similar to normal cells, chemotherapy is toxic to rapidly dividing cells in bone marrow, hair follicles and the gastrointestinal tract. Resistance to cancer chemotherapy can occur and for this reason combinations of drugs with different mechanisms of action provide the best chance of a successful cure.

Antimetabolites are analogues or antagonists of normal cell constituents, which block or otherwise alter metabolic pathways in the synthesis of DNA. Cytotoxic antibiotics prevent cell division by direct action on DNA or by inhibition of transcription of DNA to RNA, thus preventing protein synthesis. Plant derived drugs work in various ways. Vinca alkaloids inhibit formation of microtubules so that cell division is halted because the spindle cannot form. Taxanes have a similar effect by inhibition of breakdown of microtubules once the spindle has formed. Hormones are used in some hormone-dependent cancers to suppress or antagonize the particular hormone involved. In addition there are miscellaneous anti-cancer drugs including biological response modifiers, which are intended to enhance the patients' natural immune response to cancer cells. They include interferons, interleukin-2, tretinoin and monoclonal antibodies.

The search for new more effective treatments for cancer goes on with the development of drugs to block the action of oncogenes, drugs to inhibit cancer cell enzymes and drugs to prevent growth of new blood vessels in tumours.

Useful websites

www.guidance.nice.org.uk National Institute for Health and Clinical Excellence guidelines. Useful for guidance on cancer chemotherapy and other topics.

www.cancerhelp.org.uk Cancer Research UK. Credible sources of information on cancer chemotherapy.

www.cancerresearch.org Cancer Research Institute USA. Useful links to information sources.

www.medicines.org.uk Datapharm Communications Ltd. Up-to-date information on UK medicines, electronic medicines compendium.

Case study

This is a case study that could be of relevance to any health care professional involved in the care of young people. You may need to look up the actions and side effects of chlorambucil, the drug mentioned, although it is typical of its group, alkylating agents.

John is a young lad of 10 who has leukaemia and is being treated with chlorambucil. You are seeing him with his parents. They are understandably anxious about their son's treatment and ask you about potential side effects they can expect with chlorambucil. What can you tell them about this drug and its side effects?

John's parents are keen to know how else they can help their son. They ask you for advice about making sure John eats well. How can you advise them about nutrition?

During your consultation with John and his family, you notice that John has said very little. How can you reassure a child who seems unwilling to talk?

Chapter review questions

You should be able to answer the following questions using the material in this chapter.

1. Describe the phases of a normal cell cycle.

2. What should happen if a cell's DNA is damaged?

3. Describe the function of oncogenes.

4. Discuss why cancer chemotherapy is not always successful.

5. Discuss the side effects of chemotherapy.

6. Discuss the mode of action of the following groups of cancer chemotherapy drugs and give one example in each case:

    (a) alkylating agents

    (b) antimetabolites

    (c) cytotoxic antibiotics

    (d) vinca alkaloids

7. Discuss new approaches to cancer treatment.

11

Disorders of the central nervous system

11.1 Chapter overview

This chapter describes some disorders of the central nervous system (CNS) and the drugs used to treat these conditions. It is an overview of a few disorders of the CNS, rather than an attempt to cover them all comprehensively. The emphasis is on conditions and drugs likely to be encountered in general practice.

Disorders of the CNS can be divided into psychiatric (or mental) disorders and neurological disorders, although the distinction is not always clear. Psychiatric disorders include types of depression, schizophrenia, anxiety and insomnia and attention deficit hyperactivity disorder (ADHD). Neurological disorders include epilepsy and Parkinson's disease, dementia and Alzheimer's disease.

In order to have some understanding of the nature of diseases of the CNS and the treatments available, it is necessary to have a basic knowledge of the brain and neurotransmitters that are involved in the transmission of information within the nervous system.

11.1.1 Brain physiology

The human brain is a complex organ made up of over 100 billion neurons and ten times as many glial cells. It is organized into systems, which sense, process and store, perceive and act on information received from outside and inside the body.

Structurally the brain can be divided into the central core, the deep brain nuclei and the cerebral cortex (see Figure 11.1).

11.1.1.1 Central core

The central core includes the brain stem (medulla, pons and midbrain), which controls automatic functions of the body like breathing and rate of heart beat.

Pharmacology for the Health Care Professions   Christine M. Thorp
© 2008 John Wiley & Sons, Ltd

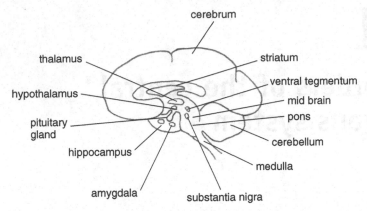

limbic system: striatum, hippocampus and amygdala
basal nuclei: substantia nigra and striatum

Figure 11.1 Structure of the brain

At the back of the brain stem is the cerebellum, which is concerned primarily with the fine control of movement.

Above the brain stem is the hypothalamus and thalamus, which together make the diencephalon. The hypothalamus is a small structure that plays an extremely important role in many aspects of emotion, motivation and response to stress. It has centres that control eating, drinking and sexual behaviour. It also regulates the autonomic nervous system and (through the pituitary gland) the endocrine system and maintains homeostasis.

The thalamus acts as a relay station for sensory information coming into the cerebrum and plays a role in the control of sleep and wakefulness.

The reticular activating system (RAS) is a network of neural pathways extending from the brain stem to the thalamus and other parts of the limbic system. It plays a role in controlling arousal and awareness. Sensory neurons from peripheral sensory receptors feed into the RAS, which appears to filter sensory messages going to the cerebral cortex, so that some sensory information reaches conscious awareness and some does not.

11.1.1.2 Deep brain nuclei

Around the central core of the brain are a number of structures that collectively make up the deep brain nuclei, which are organized into the limbic system, and the basal nuclei. These systems are closely connected to the thalamus, the hypothalamus and the sensory and motor areas of the cerebral cortex. The limbic system is concerned with emotions, behaviour and memory and the basal nuclei are concerned with control of movement, although there is overlap between the two. Malfunction of these areas is implicated in disorders such as Parkinson's disease and schizophrenia.

11.1.1.3   Cerebral cortex

The cerebrum is highly developed in humans. The cerebral cortex is a layer of neurons about 3-mm thick on the outer surface of the cerebrum. It is here that complex mental activities take place.

Sensory information coming into the cerebrum is interpreted by specific areas of the cortex. Movement of body parts is controlled by another area of the cortex. The rest and the largest part of the cortex, which is neither sensory nor motor consists of association areas. It is here that complex aspects of behaviour such as memory, thought and language reside.

11.1.2   Neurotransmitters

Within the nervous system neurons make connections with each other called synapses. At the synapse, the neurons do not touch each other but are separated by a microscopic gap, the synaptic cleft. When a nerve impulse arrives at a synapse, chemical substances are released. These are neurotransmitters, which are stored in pre-synaptic vesicles. On release they cross the synaptic cleft and bind to receptors on the post-synaptic neuron. Binding of neurotransmitter with its receptor is very brief, but it brings about stimulation or inhibition of the post-synaptic neuron. See Figure 11.2 for a diagram of synaptic transmission.

The events that take place at a synapse, that is synthesis of neurotransmitter, storage in vesicles, release, interaction with receptors and eventual inactivation provide targets for drug action (Chapter 3). Inactivation of a neurotransmitter can be by re-uptake into the neuron it was released from (as with noradrenaline and dopamine), or by the action of enzymes in the synaptic cleft (as with acetylcholine).

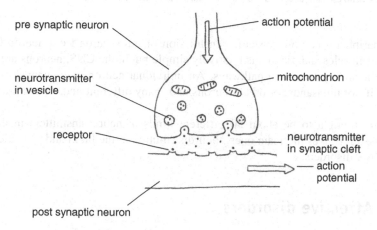

Figure 11.2   Synaptic transmission

Table 11.1 Neurotransmitters, distribution in brain and involvement in CNS disorders

Neurotransmitter	Distribution in brain	Involvement in CNS disorders
Noradrenaline	Most regions	Unipolar depression
		Mania?
		ADHD
Dopamine	Most regions	Schizophrenia
		Mania
		Parkinson's disease
		Bipolar depression
		ADHD
Serotonin (5-hydroxytryptamine)	Most regions	Unipolar depression
		Schizophrenia?
		Anxiety
Acetylcholine	Most regions	Depression?
		Dementias
		Mania?
Adrenaline	Midbrain and brain stem	Depression?
GABA (gamma-amino butyric acid)	Nigrostriatal system	Anxiety
	Cortex	Epilepsy
	Limbic system	Mania?
Glutamate and aspartate	Thalamus	Epilepsy
	Limbic system	Schizophrenia?
Glycine	Modulates glutamate NMDA receptors	Alzheimer's disease
		Epilepsy?

NB While the exact involvement of a particular neurotransmitter in a particular CNS disorder is unknown, a question mark denotes more uncertainty about the neurotransmitter involvement.

In the peripheral nervous system, transmission of information from neuron to neuron or effector (muscles and glands) is relatively simple but in the CNS, neurons are arranged into complicated systems and pathways. An individual neuron may synapse with many hundreds, if not thousands, of other neurons and many different neurotransmitters can be involved.

There are thought to be about 50 different kinds of neurotransmitters in the human brain. Table 11.1 gives the main ones, their distribution in the brain and possible involvement in CNS disorders.

11.2   Affective disorders

Affective disorders are a group of diseases in which there is an alteration of mood such that normal life is affected.

Table 11.2 Classification characteristics for depression

Common symptoms	Somatic symptoms	Psychotic symptoms
Decrease in concentration	Reduced activity/agitation	Hallucinations
Decrease in self-confidence	Reduced sexual activity	Delusions
Guilt	Weight loss	
Suicidal thoughts	Loss of interest/enjoyment	
Disturbed sleep	Loss of emotional reactivity	
Reduced appetite	General malaise	

Mood can be abnormally lowered, as in depression (also known as unipolar depression), abnormally elevated as in mania or an alternation of the two, as in manic-depression (also known as bipolar depression).

11.2.1 Unipolar depression

Depression is often divided into two sub-types – reactive depression and endogenous depression. Reactive depression is usually associated with stressful life events, whereas endogenous depression is a persistent alteration of mood with no obvious cause.

However, this classification has largely been abandoned because there is often overlap between the two. For example mild so-called reactive depression can occur without an obvious change in circumstances. Amongst psychiatrists one of two diagnostic classifications are now used. These are the *International Classification of Diseases 10th Edition* (*ICD-10*) and the *Diagnostic and Statistical Manual of Mental Disorders 4th Edition Revised* (*DSM-IV*). With these systems, a diagnosis of mental disorders is made from the presence of specific symptoms or a syndrome for a minimum of two weeks. This relies on the skills of recognizing patterns of symptoms. It is not based on pathology, diagnostic tests or the presence or absence of an immediate cause. Neither does it suggest a particular treatment. In diagnosing depression both systems allow for somatic symptoms, psychotic symptoms and other illness characteristics, some of which are given in Table 11.2.

ICD-10 gives guidelines for making a diagnosis. *DSM-IV* does not, but it includes specific criteria for a diagnosis. The advantage of using such systems is said to be that it has led to standardization of diagnostic practice across psychiatry and countries and allowed more accurate description of syndromes.

Possible diagnoses include a single depressive episode, several depressive episodes making up depressive disorder and alternations between depression and elevated mood (see manic-depression below). Table 11.3 gives a classification of mood disorders taken from the *DSM-IV*.

11.2.2 Mania and hypomania

Mania occurs less frequently than depression. It is characterized by overactivity and loss of social inhibitions and mood generally inappropriate to the circumstances. A manic

Table 11.3 Classification of affective disorders

Unipolar depressive disorders	Bipolar disorders

Major depressive episode
  Mild, moderate or severe
  Severe with psychosis
  Atypical forms
	Hypomanic episode
	
	Manic episode
	  Mild, moderate or severe
	  Severe with psychosis

Recurrent major depressive disorder	
	Bipolar I (defined by current episode)
Dysthymic disorder Low grade chronic depression	Mania predominates Manic, depressed, mixed

Depressive disorders otherwise not
  specified
  Recurrent brief episodes e.g. seasonal
  affective disorder (SAD)
	Bipolar II (defined by current episode)
	  Depression predominates
	  Hypomanic, depressed

	Cyclothymia Persistent mood instability with mild episodes

Adapted from American Psychiatric Association (2000) *Diagnostic and Statistical Manual of Mental Disorders*, 4th edn Revised, Washington, DC.

patient has an exaggerated sense of well-being and enthusiasm and talks non-stop. These signs may be combined with irritability, impatience and anger. Lack of judgement leads to a manic person becoming, for example overgenerous and reckless. They may have grandiose delusions.

Hypomania is a milder form of mania and may not require drug therapy.

11.2.3 Bipolar depression

Bipolar depression is depression that alternates with periods of mania. Patients with bipolar depression are generally younger at the outset than those with unipolar depression and are less likely to suffer from agitation and anxiety, suggesting a different biochemical abnormality to unipolar depression. Bipolar depression tends to run in families, so there may be a genetic element to its aetiology. First-degree relatives of patients with bipolar depression are more likely to develop mood disorders than relatives of those who do not have bipolar depression. Twin studies have shown concordance rates for bipolar depression of more than 50% in monozygotic twins compared with 20% in dizygotic twins. Bipolar depression appears to be a complex genetic illness, with susceptible genes interacting with environmental factors and adverse life events.

11.2.4 Biochemical theories of affective disorders

Both depression and mania are associated with changes in brain monoamines.

It was observed (in the 1960s) that the antihistamine imipramine and the anti-tuberculosis drug iproniazid had antidepressant activities. Imipramine was found to inhibit neuronal uptake of noradrenaline and serotonin by pre-synaptic neurons in the brain. Iproniazid was found to inhibit monoamine oxidase, the enzyme that breaks down monoamines. Both these actions result in raised levels of monoamines at the synapse. In addition, drugs that reduce the availability of monoamines cause symptoms of depression.

From such observations the monoamine theory of depression and mania was formed:

- Depression is due to an absolute or relative decrease in monoamines, or a decrease in receptor sensitivity, at certain receptor sites in the brain.

- Mania is due to an absolute or relative excess of monoamines or an increase in receptor sensitivity at these sites.

Even after years of research, it is still not clear which receptor sites or which monoamines are involved, although noradrenaline, serotonin and dopamine are all implicated. Neither does the theory explain how mania and depression can exist in the same patient. Much research has involved investigation of cerebrospinal fluid, blood and urine of patients to see if there are any abnormalities of monoamine metabolites. Although results may be complicated by diet, non-brain amines and the inevitable drug therapy, evidence indicates that bipolar depression is associated with a decrease in dopamine activity, mania with an increase in dopamine activity [or depletion of inhibitory gamma-amino butyric acid (GABA)] and unipolar depression with a decrease in noradrenaline or serotonin activity or both. Levodopa (a precursor of dopamine) has been shown to produce symptoms of mania in patients with bipolar depression but not in those with unipolar depression. This still does not explain the swings in mood and little is known about how or why alternation between depression and mania occurs.

Patients with unipolar depression are effectively treated with drugs, such as chlorimipramine and more recently fluoxetine, which are known to block neuronal uptake of serotonin. In addition, giving 5-hydroxytryptophan (a precursor of serotonin) either alone or in combination with chlorimipramine seems to benefit these patients.

Evidence against this serotonin theory includes the fact that low levels of 5HIAA (5-hydroxyindoleacetic acid, a metabolite of serotonin) are seen even in patients who have recovered and in some patients with mania.

It seems likely that low levels of serotonin or noradrenaline do not cause depression, but that more complicated mechanisms involving a change in balance between these monoamines are involved.

Despite evidence for and against, there remains one major problem with the monoamine theory of depression. All antidepressants take weeks to have an effect, which is far longer than it takes to alter brain amines. It has been suggested that effects of antidepressants are due to adaptive changes in the brain, which may involve down regulation of receptors or some other change in their sensitivity.

Some of the newer types of antidepressants do not facilitate monoamine transmission at all. Cholinergic systems may also be involved in depression since many antidepressants have strong anti-cholinergic activity.

High doses of adrenocorticotrophic hormone or cortisol can produce depression and depressed patients frequently have raised levels of cortisol in plasma. This could be due to increased pituitary adrenocorticotrophic hormone production, which in turn could be due to a disorder of hypothalamic control. Corticotrophin releasing hormone is known to be inhibited by noradrenaline and serotonin. However, a raised level of cortisol is also seen in mania and may just be a sign of a stressful condition.

Pathological changes have been deduced from neuroimaging studies with computed tomography (CT) and positron emission tomography (PET) scans. Mood disorders are associated with changes in regional brain function, which is different to those seen in non-depressed patients who feel sad. Newer techniques have allowed measurement of volume and size of discrete areas of the brain. Atrophy of the hippocampus (a part of the limbic system) is seen in depression possibly through neurodegeneration. Most antidepressants lead to the production of brain-derived neurotrophic factors that bring about neurogenesis and this could contribute to their effectiveness in depressed patients.

11.3   Drugs used to treat depression

There are a large number of drugs used to treat this condition, which with the exception of lithium all take approximately two to four weeks to achieve their optimum action.

All are given in courses that are of a number of months in duration rather than weeks. Classes of antidepressants are given in Table 11.4.

11.3.1   Tricyclic antidepressants

Tricyclic antidepressants were named after their characteristic chemical structure, although a number of the newer drugs in this group have a different chemical structure but still share similar pharmacological profiles. Tricyclic antidepressants inhibit the re-uptake of serotonin and noradrenaline into nerve endings. This leads to a build up of neurotransmitter at receptor sites in synapses. They also reduce the number of some receptors, which may contribute to the therapeutic action, but their complete mechanism of action is unknown. Examples of tricyclic antidepressants include imipramine and amitriptyline (for use of amitriptyline in neuropathic pain see Chapter 12).

Table 11.4   Classes of antidepressants

Class	Examples
Tricyclics	Amitriptyline
Selective serotonin re-uptake inhibitors	Fluoxetine, sertraline
Serotonin receptor blockers	Mirtazapine, trazodone
Monoamine oxidase inhibitors	Moclobemide

The main side effects of tricyclic antidepressants are sedation, anti-cholinergic effects (dry mouth, constipation and urinary retention) and postural hypotension. Some patients may experience mania, seizures or impotence. Tricyclic antidepressants should be used with caution in patients with liver disease.

Maprotiline is similar in structure to the tricyclics. It blocks only the re-uptake of noradrenaline and has a lower incidence of side effects.

11.3.2 Selective serotonin re-uptake inhibitors (SSRIs)

As the name suggests, these drugs selectively block the re-uptake of serotonin into nerve endings, thereby increasing the concentration of it at the synapse. Examples of SSRIs include fluoxetine and sertraline.

The principal side effects include nausea, diarrhoea, insomnia and agitation.

11.3.2.1 Other similar drugs

There are other drugs not classified as SSRIs but which work in similar ways.

Venlafaxine selectively inhibits the re-uptake of both serotonin and noradrenaline. It has benefits in patients with symptoms of melancholia, anxiety or agitation.

Reboxetine selectively inhibits the re-uptake of noradrenaline and is particularly useful in patients with negative symptoms such as withdrawal and flattening of emotional response.

11.3.3 Serotonin receptor blockers

Drugs have been developed that block serotonin receptor sub-types while still enhancing serotonin transmission. This reduces the effects of insomnia, agitation and sexual dysfunction. Their mode of action is complicated by the fact that currently seven serotonin receptor sub-types are recognized, $5HT_1$–$5HT_7$, and some of them are subdivided further into $5HT_{1A}$ receptors, for example. (An alternative name for serotonin is 5-hydroxytryptamine, 5HT). Although they are widely distributed in the body, the significance of the 5HT receptor sub-types is not fully understood. All are found in the brain and most is known about $5HT_1$, $5HT_2$ and $5HT_3$ sub-types. $5HT_1$ receptors are inhibitory autoreceptors. Such receptors are found pre-synaptically and their stimulation regulates release of neurotransmitter. $5HT_2$ and $5HT_3$ sub-types are excitatory post-synaptic receptors.

Mirtazapine enhances both noradrenergic and serotonergic transmission through block of pre-synaptic α-adrenergic receptors and $5HT_1$ receptors and post-synaptic $5HT_2$ and $5HT_3$ receptors.

Due to the blockade of $5HT_3$ receptors in the chemotrigger zone of the medulla, this drug does not produce the nausea associated with other serotonergic drugs. Mirtazapine does cause sedation and weight gain.

Trazodone antagonizes both pre-synaptic α-adrenergic and post-synaptic $5HT_2$ receptors. It has sedative properties and causes fewer anti-cholinergic side effects than the tricyclics.

11.3.4 Monoamine oxidase inhibitors (MAOIs)

Monoamine oxidase inhibitors (MAOIs) are drugs that irreversibly inhibit the action of the enzyme monoamine oxidase. Monoamine oxidase exists in two forms: MAO-A and MAO-B. MAO-A is the enzyme preferentially responsible for degrading serotonin and noradrenaline. Inhibition of this enzyme allows the concentration of these monoamines to increase in the brain, thereby producing an antidepressant effect.

However, a major problem with MAOI drugs is that they produce a number of potentially fatal interactions with other drugs and common foods. Many foods contain tyramine, for example red wines, some pickled foods, yeast extract, particular cheeses and some vegetables, including soya and broad beans. Tyramine and indirectly-acting sympathomimetic drugs (such as ephedrine found in decongestants and cold remedies) are normally broken down by monoamine oxidase in the tissues. If the enzyme is inhibited these substances cause rapid elevation in blood pressure leading to a hypertensive crisis, which can be fatal. Newer selective drugs inhibit MAO-A reversibly, allowing displacement by exogenous amines and so are less likely to interact with drugs and food substances and are safer. An example is moclobemide.

Nevertheless, MAOIs should be regarded as second line drugs and are indicated when the patient is resistant to other therapy, or is phobic, hysterical or has other atypical symptoms.

All patients taking MAOIs should carry a warning card highlighting the potential dangers of interaction with other drugs and food substances.

11.4 Drugs used to treat bipolar depression and mania

11.4.1 Lithium

Lithium is used for the prophylactic control of mania and hypomania and bipolar depression. It also has a use in unipolar depression that is unresponsive to other antidepressants.

Lithium is similar to sodium in that it forms positive ions (Li^+) and can pass through sodium ion channels in neuronal cell membranes. It tends to accumulate inside neurons and may interfere with nerve action potentials or the activation of second messenger systems within the neuron (see Chapter 3, page 63). In addition, lithium may inhibit the release of monoamines from nerve endings and increase their uptake. However, the exact mode of action of lithium in affective disorders is unknown.

Lithium has a narrow therapeutic ratio and blood concentration must be carefully monitored to avoid toxicity. It is important that the patient uses the same brand of lithium, as bioavailability may differ with different brands.

Toxic effects of lithium can be fatal. Early signs of lithium toxicity are vomiting and severe diarrhoea followed by tremor, ataxia, renal impairment and convulsions. Many non-steroidal anti-inflammatory drugs interact with lithium to increase its plasma concentration and risk of toxicity. Aspirin is an exception and safe to use with lithium. Angiotensin-converting enzyme inhibitors similarly can lead to increased lithium levels and the two drugs should only be used together with caution and frequent monitoring of lithium blood levels. Patients should carry a lithium treatment card.

11.4.2   Other drugs

Benzodiazepines or anti-psychotics, for example flupentixol, can be used in mania initially until lithium takes effect.

Carbamazepine and valproate can be effective in bipolar depression that is unresponsive to lithium and appear to exert their action by depressing the limbic system. Both these drugs are normally used to treat epilepsy, see page 218. In bipolar depression, this is an unlicensed use.

11.5   Psychoses

Psychoses are a group of disorders in which patients have a distorted perception of reality and they include reactive psychosis, paranoid/delusional psychosis, some types of mania and schizophrenia.

Schizophrenia is the most common psychosis, affecting about 1% of the general population. It often affects people in late adolescence and can be chronic and disabling. Schizophrenia has a number of sub-types and there have been attempts at classifying them, but this has proved difficult. Modern diagnostic classifications depend on the presence or absence of positive or negative symptoms as well as the duration of symptoms according to either the *ICD-10* or the *DSM-IV*.

Negative symptoms are social withdrawal, lack of emotional responsiveness and apathy; positive symptoms are hallucinations (usually auditory), thought disturbances, delusions, restlessness and aggression.

Some psychoses can be secondary to other diseases such as infection and peripheral vascular disease or intoxication with drugs or alcohol, but most often the aetiology is unknown. Schizophrenia appears to have a genetic basis. The risk for the disease is greater in relatives of schizophrenics than that in the general population. The concordance rate for monozygotic twins is 50% compared with 10% in other siblings. However, inheritance is complex and likely to be an interaction between several genes and environmental factors.

11.5.1   Biochemical theories of schizophrenia

In schizophrenia, there appears to be overactivity of dopamine systems in areas of the brain associated with complex mental and emotional functions. This is the dopamine hypothesis of schizophrenia.

11.5.1.1 *Dopamine systems in the brain*

There are four main neuronal systems in the brain in which dopamine is a transmitter: the basal nuclei; the hypothalamic-pituitary pathway and the mesolimbic and mesocortical pathways (see Figure 11.3).

A basic knowledge of these systems helps to understand how anti-psychotic drugs work and why they produce certain side effects.

The basal nuclei are a collection of paired structures that modulate motor pathways from the cerebral cortex and play a role in the fine control of movement. Part of this is a pathway from the substantia nigra in the midbrain to an area called the striatum in the basal nuclei. The neurotransmitter of this pathway is dopamine and loss of dopamine-containing neurons here causes Parkinson's disease (see page 212).

The hypothalamic-pituitary dopamine pathway inhibits the release of prolactin from the anterior pituitary gland.

The mesolimbic and mesocortical pathways are dopamine-containing neurones that run from an area in the midbrain called the ventral tegmentum to the limbic system and the prefrontal area of the cerebral cortex respectively. These are areas of the brain that are normally involved in behavioural and emotional functions.

In addition, the chemotrigger zone of the medulla contains dopamine-containing neurones that initiate vomiting in response to noxious and possibly toxic substances.

It is thought that there is an abnormality of dopamine receptors or increased release of dopamine in the mesolimbic and mesocortical pathways in schizophrenics. However, no reproducible changes in dopaminergic systems have been found in schizophrenia and the abnormality may be in another system that is somehow linked to dopaminergic neurones. More recently, it has been suggested that schizophrenia may be a developmental disorder of the prefrontal cortex where there is actually a deficiency of dopamine, which leaves dopamine activity in the mesolimbic pathway unbalanced.

There is some evidence for the involvement of serotonin (5HT) and possibly other transmitters interacting with dopamine pathways.

Most anti-psychotic drugs seem to work by blocking dopamine receptors in the brain, although some of the new atypical anti-psychotics also block serotonin receptors.

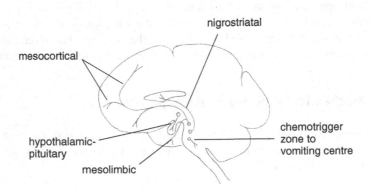

Figure 11.3 Dopamine pathways in the brain

11.5.1.2 Dopamine receptors

There are two main types of dopamine receptor: the D_1 type, which includes D_1 and D_5, and is excitatory; and the D_2 type, which includes D_2, D_3 and D_4, and is inhibitory. The significance of different types of dopamine receptors is still unclear, but drugs that are effective in schizophrenia appear to have an affinity for D_2 type receptors.

11.6 Drugs used to treat schizophrenia

Anti-psychotic drugs are used in the treatment of schizophrenia and there are many in use. The choice of drug depends on the severity of the condition and the diagnosis of positive or negative symptoms. Other factors to take into consideration are whether sedation is required and how susceptible the individual is to Parkinsonian (extra-pyramidal) side effects. Most drugs are more effective at relieving the positive symptoms and all provide better results in acute schizophrenia. Long-term treatment may be necessary to prevent relapse and chronic illness. Withdrawal of treatment must be cautious to prevent relapse.

There can be difficulty maintaining treatment if the patient is uncooperative, forgetful or does not take their drugs as prescribed. Depot preparations are sometimes used for such patients.

Older anti-psychotic drugs are classified according to their chemical structure, while the newer ones are known as atypical anti-psychotics. The main distinction is that the newer drugs produce fewer motor side effects. (See Table 11.5).

11.6.1 Phenothiazines

Phenothiazines were developed after it was noticed that the antihistamine, promethazine had a sedative action.

Chlorpromazine is the most well known and is the standard against which all other anti-psychotics are compared. It is still widely used.

Anti-psychotic drugs block dopamine receptors in all four dopamine systems of the brain. Blocking of dopamine D_2 receptors in the mesolimbic system is assumed to be responsible for the efficacy in therapy of schizophrenia.

Table 11.5 Classes of anti-psychotic drugs

Class	Examples	Motor side effects
Phenothiazines	Chlorpromazine, pericyazine	+++
Thioxanthines	Flupentixol	+++
Butyrophenones	Haloperidol	+++
Benzamides	Amisulpiride	++
Atypical anti-psychotics	Risperidone, olanzapine, quetiapine	+

Positive symptoms of delusions, hallucinations and thought disorder are improved.

Many anti-psychotic drugs have what are called Parkinsonian or extra-pyramidal side effects. These are abnormalities of movement that resemble Parkinson's disease (see page 211) and are due to the blocking of dopamine D_2 receptors in the basal nuclei. Anti-psychotics that also have anti-cholinergic activity (for example pericyazine) cause fewer Parkinsonian side effects.

A more serious and irreversible side effect related to Parkinsonism is tardive dyskinesia. This occurs after many years of anti-psychotic drug use and is characterized by abnormal and disabling involuntary movements of the face, tongue, trunk and limbs.

Phenothiazines have pronounced sedative effects. They are also antiemetic (due to the blocking of dopamine receptors in the chemotrigger zone) and cause increased levels of prolactin (by inhibition of the effects of dopamine in the hypothalamic-pituitary pathway, see Chapter 6, Table 6.1) which leads to breast swelling, pain and lactation in both men and women.

11.6.2   Thioxanthines

This group of drugs is similar to phenothiazines with mode of action by block of dopamine D_2 receptors. Thioxanthines produce less sedation than phenothiazines, but otherwise they are very similar. An example from this group is flupenthixol.

Side effects of thioxanthines are as expected, being antiemetic, Parkinsonism, tardive dyskinesia and raised prolactin levels.

11.6.3   Butyrophenones

This group of anti-psychotic drugs has a similar mode of action and side effects to the phenothiazines. Butyrophenones are of particular use for the rapid control of hyperactive states and in older people because they are less likely to cause a drop in blood pressure. An example is haloperidol.

There is a high incidence of Parkinsonian side effects with this group as well as the expected antiemetic and raised prolactin level effects.

11.6.4   Benzamide derivatives

These drugs have been developed to be more specific for D_2 receptors in the mesolimbic pathway and therefore avoid the Parkinsonian side effects and particularly tardive dyskinesia. They control the positive symptoms in high doses and negative symptoms in low doses.

An example of a benzamide derivative is amisulpiride.

11.6.5 Atypical anti-psychotics

These are the newer drugs used to treat schizophrenia. Atypical anti-psychotic drugs are recommended for newly diagnosed schizophrenics and those who cannot tolerate other drugs or who are not getting adequate control from other drugs.

All are thought to improve negative as well as positive symptoms of schizophrenia in addition to causing fewer motor side effects and are less likely to affect prolactin secretion.

Examples of atypical anti-psychotic drugs are clozapine, risperidone, olanzapine and quetiapine.

Clozapine can cause agranulocytosis and should only be used if other drugs are not effective.

Other adverse effects of most anti-psychotic drugs are anti-cholinergic effects such as dry mouth, constipation and urinary retention and α-adrenergic blocking effects, primarily vasodilation leading to a fall in blood pressure especially in the elderly.

Neuroleptic malignant syndrome is a rare but potentially fatal adverse reaction to some anti-psychotic drugs and requires immediate withdrawal of the drug. Symptoms are hyperthermia, fluctuating level of consciousness, muscular rigidity and autonomic dysfunction and can last five to seven days after withdrawal of the drug.

Actions on D_2 receptors in the chemotrigger zone make anti-psychotic drugs useful in the prevention of nausea and vomiting, particularly in patients being treated with cancer chemotherapy or radiotherapy.

11.7 Anxiety and insomnia

Anxiety and insomnia are discussed together because the drugs used to treat them are often the same.

11.7.1 Anxiety

Anxiety may be described as an unpleasant experience or fear of the unknown. Symptoms are those of increased sympathetic activity including increased heart rate and blood pressure, sweating and tremor and a dry mouth. Anxiety can be associated with pain and breathing difficulties.

Anxiety can be a reaction to major life events, such as moving house, changing jobs or bereavement. As such, it is a normal, but short-lived response to a potentially threatening event. Anxiety becomes a medical problem when it is excessive, long lasting or inappropriate and incapacitating. Then it can be called 'anxiety neurosis', or an anxiety syndrome.

According to the *DSM-IV* , anxiety syndromes are obsessive-compulsive disorder, panic attacks for no apparent reason, phobias and post-traumatic stress disorder. Anxiety can be associated with depression or psychosis.

The term generalized anxiety disorder is used to describe persistent anxiety with three or more of the following symptoms:

- increased motor activity (tremor);

- autonomic hyperactivity (sweating, increased heart rate, difficulty breathing, dizziness and gastrointestinal disturbances);

- apprehension (fear and worry);

- insomnia;

- fatigue;

- loss of appetite;

- lack of concentration.

Symptoms like these are familiar to most people and do not constitute an anxiety syndrome providing they are mild, short-lived and for a reason. In such cases, drug therapy is not usually necessary.

For anxiety due to fear of flying or after dinner speaking for example a one-off treatment with a benzodiazepine or a β-blocker may be justified.

It is generally acknowledged that treatment of bereavement with antianxiety drugs is not appropriate. Grieving is a natural process that serves a purpose in recovery from loss.

It is important to realize that some medical conditions can mimic the symptoms of generalized anxiety disorder, for example hyper- or hypothyroidism, hypoglycaemia, drug or alcohol withdrawal and cardiac arrhythmias.

Anxiety and fear are both associated with general arousal of the CNS and may be due to over activity in the RAS. Neurotransmitters in the RAS are serotonin and GABA. It is thought that GABA normally moderates the activity of serotonin-containing neurons and that in anxiety syndromes some of this moderation is lost due to changes in receptor sensitivity, or due to overactivity of a natural inhibitor of GABA. An inhibitor of GABA has not been identified, but it has been tentatively called GABA modulin. It is possible that the RAS can be activated through pathways from the cerebral cortex if a situation is perceived as being threatening.

11.7.2 Insomnia

Sleep is complicated and involves the brain stem RAS. When the RAS is switched off sleep is possible.

Acetylcholine, noradrenaline and serotonin all play a role in inducing sleep. GABA, dopamine and some neuropeptides influence sleep-inducing activity in cholinergic neurons.

Insomnia is sleep disturbance and can be manifest as difficulty falling asleep, poor quality of sleep or premature awakening. Insomnia can be secondary to other conditions, for example anxiety or depression. Other causes are factors such as stress and excessive noise, jet lag, shift work, physical illness and pain, stimulants (coffee and tea), sleep apnoea or poor habits at bedtime.

Primary insomnia is insomnia of no obvious cause.

11.8   Anxiolytics and hypnotics

An anxiolytic is a drug that reduces feelings of anxiety (also known as a sedative), whereas a hypnotic is a drug that induces sleep. The two terms are interchangeable in the sense that most anxiolytics will induce sleep when given at night and most hypnotics will cause sedation when given during the day. Drugs used to treat anxiety and insomnia are often the same and because these conditions are common, the drugs are widely prescribed.

11.9   Treatment of anxiety

Drug therapy is not the only possibility. Psychotherapy therapy techniques such as relaxation, cognitive behavioural therapy and counselling can benefit some patients.

Drug therapy of anxiety is mainly with a group of drugs called benzodiazepines.

11.9.1   Benzodiazepines

Benzodiazepines are commonly used anxiolytics and hypnotics. They act at benzodiazepine receptors that are found close to GABA receptors. GABA receptors are found on chloride ion channels at inhibitory synapses in the CNS, including the RAS (see Figure 11.4).

When chloride channels open, chloride ions flow inwards making the neuronal membrane hyperpolarized. This means that there is a reduction in impulses passing on to other parts of the brain and this effect normally produces a reduction in anxiety and wakefulness. Benzodiazepines enhance the effects of GABA at its receptor, but they do not bring about the opening of chloride ion channels in the absence of GABA. Because of the existence of benzodiazepine receptors, it seems likely that there must be either an endogenous benzodiazepine or an endogenous anxiolytic substance (GABA modulin), although none has been conclusively identified yet.

For treatment of anxiety, benzodiazepines such as diazepam and oxazepam are used because they have a slower onset of action with a longer duration than others.

Benzodiazepine anxiolytics have been prescribed to almost anyone with stress-related symptoms, unhappiness or minor physical trauma. This kind of use is now considered unsuitable. Neither are they considered appropriate for treating depression, phobic or obsessional states or chronic psychosis.

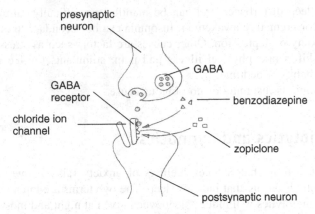

Figure 11.4 GABA receptor and Chloride ion channel

On the advice of the Committee on the Safety of Medicines (now the Commission on Human Medicines), benzodiazepines are indicated for the short-term relief of anxiety that is severe, disabling or subjecting the individual to unacceptable distress, occurring alone or in association with insomnia or short-term psychosomatic, organic or psychotic illness.

Benzodiazepines should be used for as short a term as possible (two to four weeks is recommended) because both physical and psychological dependence develop to these drugs and withdrawal can be difficult after only a few weeks. Abstinence symptoms are worse and more common with short-acting benzodiazepines and can appear within a few hours. With the longer-acting drugs, it may take up to three weeks for the withdrawal syndrome to develop and some symptoms may continue for weeks or months after stopping benzodiazepines entirely. Symptoms of withdrawal are similar to the original complaint and include anxiety, psychosis, restlessness, insomnia, confusion, irritability and possibly convulsions. Withdrawal must be gradual to avoid symptoms and the need to return to drug use.

Tolerance develops to the sleep effects of benzodiazepines but not the antianxiety effects.

Benzodiazepines can sometimes cause paradoxical effects such as an increase in hostility and aggression, excessive talkativeness and excitement or increased anxiety and perceptual disorders. These effects can usually be reduced by altering the dose.

11.9.2 Buspirone

Buspirone is used as an anxiolytic and response to treatment may take two to four weeks. This drug is thought to work as an agonist on $5HT_{1A}$ receptors. These are pre-synaptic receptors that limit the release of serotonin. Buspirone does not have the dependence potential of the benzodiazepines.

Table 11.6 Drugs used to treat anxiety and insomnia

Drug group	Action	Uses
Benzodiazepines, for example diazepam	Enhance GABA activity	Anxiety, insomnia
Buspirone	Stimulates $5HT_{1A}$ receptors	Anxiety
Beta blockers, for example oxprenolol	Reduce somatic symptoms of palpitations and tremor	Anxiety
Zolpidem and zopiclone	Enhance GABA activity	Insomnia
Clomethiazole	Enhance GABA activity?	Insomnia

11.9.3 Beta blockers

Beta blockers (for example propranolol and oxprenolol) have a place in the treatment of some anxiety states. Although they do not affect psychological symptoms, such as worry, tension and fear, they are useful for patients with predominantly somatic symptoms such as palpitations and tremor; treatment of these symptoms may prevent the onset of worry and fear.

Beta blockers are used to treat arrhythmia, angina and hypertension and are discussed more fully in Chapter 4.

11.9.4 Anti-psychotics

Anti-psychotics, in low doses, are also sometimes used in severe anxiety for their sedative action, but long-term use should be avoided in view of a possible risk of tardive dyskinesia (see page 333).

See Table 11.6 for a summary of drugs used to treat anxiety and insomnia.

11.10 Treatment of insomnia

Drug therapy is not always necessary or desirable for insomnia. The cause of the insomnia should be established and, where possible, underlying factors should be treated. Alternative approaches include counselling and relaxation training and avoiding stimulants, alcohol and exercise late in the evening.

Psychiatric disorders such as anxiety, depression and abuse of drugs and alcohol should be treated rather than the insomnia.

Use of hypnotics is justified to treat insomnia in the short-term if no obvious reasons for it can be found. However, the use of hypnotics rarely benefits chronic insomnia.

11.10.1   Benzodiazepines

Long-acting benzodiazepines such as nitrazepam are no longer used as hypnotics because of the hangover effects, which may affect the ability to drive or operate machinery. Loprazolam, lormetazepam and temazepam have a shorter duration of action and little or no hangover effect.

As a rule, benzodiazepines should be used to treat insomnia only when it is severe, disabling or subjecting the individual to extreme distress.

11.10.2   Zolpidem and zopiclone

Although neither zolpidem nor zopiclone is a benzodiazepine, they both act on the benzodiazepine receptor (see Figure 11.4). Both of these drugs are used as hypnotics. They have a short duration of action with little or no hangover effect and a lower dependence potential than benzodiazepines.

11.10.3   Clomethiazole

Clomethiazole is used as a hypnotic for older patients because of its lack of hangover effect but it can still produce dependence. It should be used as a hypnotic only in the short-term. Hypnotics generally should be avoided in older people because of the risk of confusion, ataxia and falls as a consequence.

11.10.4   Antihistamines

Some antihistamines, for example diphenhydramine and promethazine, are available for occasional insomnia; their prolonged duration of action may often lead to drowsiness the following day.

See Table 11.6 for a summary of drugs used to treat anxiety and insomnia.

11.11   Attention deficit hyperactivity disorder (ADHD)

ADHD occurs in childhood and adolescence and may continue into adulthood. It is more common in boys than girls and may affect between 3 and 5% of all children. It is a behavioural syndrome (or syndromes) with variable symptoms including hyperactivity, impulsiveness and difficulty concentrating and being attentive. Affected individuals may have learning difficulties, but are usually of average or above average intelligence.

The cause of ADHD is unknown. It is possible that there is loss of inhibitory control in the limbic system because of a disorder in the right frontal cerebral cortex. Dopamine D_4 receptors are implicated in ADHD and there is a catecholamine hypothesis of a deficit of noradrenaline and/or dopamine release in the prefrontal cerebral cortex, which is based on the effects of drugs used to treat the condition (see below).

There may be involvement of defective nicotinic receptors because the use of transdermal nicotine patches has been shown to improve some symptoms of the disorder.

11.12 Treatment of ADHD

Behavioural therapy or counselling may be helpful in some individuals with ADHD.

Psychostimulant drugs are used to treat ADHD, where paradoxically they have a calming effect and improve concentration and attentiveness in about 70% of cases. Treatment should be under the guidance of a specialist in ADHD.

Examples of drugs used to treat ADHD are dexamfetamine, methylphenidate and atomoxetine. These drugs inhibit the re-uptake of noradrenaline and dopamine. Increase in levels of these two transmitters in the prefrontal cortex is thought to increase inhibitory control in the limbic system.

Side effects of psychostimulant drugs include insomnia, restlessness, irritability, nervousness, dizziness, tremor and sometimes psychosis.

11.13 Parkinson's disease

Dr James Parkinson first described this disease in 1817 as *paralysis agitans* or the shaking palsy. Parkinson's disease is the second most common neurodegenerative disease after Alzheimer's disease, affecting 2% of the population over the age of 65.

The disease is progressive and usually begins with a fine tremor of the extremities. The tremors are more noticeable at rest but diminish or disappear during activity. Movement of the limbs assumes a typical pattern, often being described as a cogwheel or ratchet movement. The muscles become progressively more rigid and there is decrease in the frequency of voluntary movement. Rigidity of the voluntary respiratory muscles increases the effort required for ventilation. Table 11.7 shows characteristics seen in fully developed Parkinson's disease.

There is a typical stance with the head flexed on the chest, body bowed, arms, wrists and knees bent. The centre of gravity is forward and the patient walks on the front of the

Table 11.7 Characteristics of Parkinson's disease

Characteristic	Effects
Hypokinesia	Slowness and poverty of movement due to muscle weakness and fatigue
Rigidity	Uniform increase in muscle tone
Tremor	Involuntary in resting muscles, particularly in the extremities, typically 'pill-rolling' with index finger and thumb

feet with quick short steps. There is difficulty in initiating movements and stopping once movement has begun. Fine movements of the hand become increasingly difficult, for example writing becomes progressively smaller, until eventually illegible. Patients often have a mask-like facial expression with staring eyes, which may give the impression of decreased mental ability, but they are probably as mentally alert and agile as others of the same age. However, in about a third of patients with Parkinson's disease dementia eventually develops.

Parkinson's disease is a disorder of the basal nuclei in the brain. Post-mortem studies of brains from patients with Parkinson's disease have shown greatly decreased levels of dopamine in the basal nuclei. The main pathology is degeneration of the dopaminergic neurons of the pathway from the substantia nigra in the brain stem to the corpus striatum (the nigrostriatal tract).

Parkinson's disease was the first disease to be associated with deficiency of a specific neurotransmitter in the brain.

It can be difficult to determine onset of disease. Because compensation occurs, there must be 80% depletion of dopamine in the basal nuclei before symptoms appear. Compensation takes the form of increased activity in the remaining dopaminergic neurones and increase in numbers of receptors.

Dopamine pathways normally balance cholinergic pathways in the basal nuclei and many of the symptoms of Parkinson's disease are thought to be due to over activity of cholinergic pathways. Therapy therefore aims to improve dopaminergic transmission or inhibit cholinergic pathways to restore the balance.

11.13.1 Aetiology of Parkinson's disease

Idiopathic Parkinson's disease occurs due to degeneration of dopamine neurons with no obvious cause. There may be a genetic susceptibility, but no causal gene has been found, although mutation of the so-called Parkin gene is associated with early-onset Parkinsonism (under 50 years).

Post-encephalitic Parkinsonism followed an epidemic of viral encephalitis between 1918 and 1925.

Drug-induced Parkinsonism can occur as an adverse effect to drugs used to treat psychotic disorders (see page 203).

In the early 1980s a group of Californian drug addicts suddenly developed a Parkinson-like paralysis. The cause was found to be a contaminant called methylphenyltetrahydropyridine (MPTP), which causes severe degeneration of the nigrostriatal tract. This led to the suggestion that idiopathic Parkinson's disease may be caused by some, as yet, unidentified neurotoxin. It is possible that overactivity of glutamate could cause neuronal damage [as may also be the case in motor neuron disease (Chapter 7, page 200) and Alzheimer's disease (page 366)].

Other known toxic causes include Wilson's disease, where there is deposition of copper in the basal nuclei, and mercury and manganese poisoning.

Environmental factors that may play a role in the aetiology of Parkinson's disease include: living in a rural environment; using well water; exposure to pesticides; a diet high in animal fats or carbohydrates; or low consumption of foods rich in antioxidants such as vitamin C and E.

Rarely, Parkinsonism can also follow head injuries, cerebral tumours and cerebral ischaemia.

At post-mortem, brains of patients with Parkinson's disease are found to contain Lewy bodies. Lewy bodies are intra-neuronal inclusions and can be found throughout the brain and cerebral cortex. There appear to be three types of Lewy body disease: brain stem predominant (Parkinson's disease); limbic (associated with dementia) and neocortical (associated with dementia).

Cigarette smokers appear to be at a reduced risk of developing Parkinson's disease. The reasons for this are not know, but suggestions include nicotine induction of detoxifying enzymes, inhibition of harmful enzymes or stimulation of factors that protect neurons.

11.14 Drugs used to treat Parkinson's disease

Treatment of Parkinson's disease is not usually started until symptoms are interfering with daily life and should be initiated under supervision of a specialist.

The aim of drug therapy is to slow progression of the disease, but there is no cure. There are four groups of drugs, which augment the action of the dopaminergic system, plus anti-cholinergic drugs. (See Table 11.8.)

11.14.1 Dopaminergic drugs

Levodopa, (laevo-dihydroxyphenylalanine) is the principle drug in this group. Replacement therapy with dopamine itself is not possible because it does not pass the blood-brain barrier. Levodopa is the precursor of dopamine and it does penetrate the brain where it is converted in the neurons to dopamine by decarboxylation.

Levodopa is given orally and in order to maximize the amount that actually reaches the brain, it is given in combination with a decarboxylase inhibitor, carbidopa, which does not enter the brain. Because the drug is not converted into dopamine outside of the brain, this greatly increases the effective dose of levodopa and reduces peripheral side

Table 11.8 Drugs used to treat Parkinson's disease

Drug group	Example	Action
Dopaminergic drugs	Levodopa plus carbidopa	Converted into dopamine
Dopamine receptor agonists	Bromocriptine	Stimulates dopamine receptors
Dopamine release stimulants	Amantadine	Stimulates release of dopamine
Monoamine oxidase-B inhibitors	Selegiline	Reduces metabolism of dopamine
Anti-cholinergic drugs	Benzatropine	Inhibits overactive cholinergic neurones

effects. This combination seems to be the best treatment for most Parkinson's patients, at least in the early stages.

The site of decarboxylation of levodopa in patients with Parkinson's disease is uncertain, but it may be that there is sufficient enzyme in the remaining dopamine nerve terminals or that conversion takes place in other neurons (noradrenergic or serotoninergic) as they also contain decarboxylase enzymes. Release of dopamine replaced in this way must be very abnormal but most patients seem to benefit.

Improvement takes about two to three weeks. Hypokinesia and rigidity are both reduced but tremor is not. Overall, there is improvement in facial expression, manual dexterity, gait and speech. There is also elevation of mood, although this may be due to alleviation of distressing symptoms.

Side effects of levodopa can be severe. Initially, side effects result from widespread stimulation of dopamine receptors. Peripheral actions include nausea, vomiting and postural hypotension, although these are lessened by the use of carbidopa.

Central side effects with long-term therapy of levodopa can be serious enough for treatment to be stopped. These include dyskinesias, restlessness, anxiety, confusion, disorientation, insomnia or a schizophrenia-like syndrome. Dyskinesias can be particularly severe, involving unusual writhing movements of the limbs and grimacing and chewing movements of the face.

After five years of treatment about half of patients will experience the drug becoming less effective and a gradual recurrence of symptoms, especially hypokinesia, occurs. Another type of deterioration is the shortening of action of each dose with time ('end of dose deterioration') and unpredictable fluctuations ('on-off effect') in response to treatment, which can happen quite abruptly. It is not known why these effects occur, but they may be due to advance of the disease process. End of dose deterioration can be alleviated to a certain extent by the use of modified release preparations of levodopa or by the concurrent use of catechol-o-methyl transferase (COMT) inhibitors, for example entacapone. COMT inhibitors prevent the peripheral breakdown of levodopa by an enzyme, COMT.

Vitamin B_6 interferes with the therapeutic effects of levodopa. Multivitamin supplements should not be taken together with levodopa and foods high in vitamin B_6 (for example liver, walnuts and bananas) should be limited during use of levodopa.

11.14.2 Dopamine receptor agonists

These act directly on dopamine receptors in the brain and are selective for the D_2 subtype found in the basal nuclei. An example is bromocriptine.

Dopamine receptor agonists can be used alone in newly diagnosed patients, but are most often used together with levodopa in the later stages of Parkinsonism. This combination reduces some of the problems seen in long-term therapy with levodopa.

Apomorphine is a potent dopamine receptor agonist used in advanced disease. It has to be administered by subcutaneous injection and causes such severe nausea and vomiting that an antiemetic (domperidone) has to be administered for two days prior to treatment.

Dopamine receptor agonists have similar, but less severe, central side effects to levodopa.

11.14.3 Dopamine release stimulators

Amantadine stimulates release of dopamine from nerve terminals and reduces the re-uptake of dopamine. Rigidity, hypokinesia and tremor are all improved.

Amantadine has few side effects, mainly gastrointestinal disturbances and ankle oedema, but becomes less effective over time as the dopaminergic neurones are progressively lost.

11.14.4 Monoamine oxidase-B (MAO-B) inhibitors

Monoamine oxidase is an enzyme that metabolizes monoamines. MAO-B is more specific for dopamine, than for noradrenaline or serotonin. Inhibition of MAO-B slows the breakdown of dopamine thereby increasing its availability at synapses and prolonging its effects.

An example of a MAO-B inhibitor is selegiline, which is used in combination with levodopa. The dose of levodopa can be reduced and there is evidence that this combination has a neuroprotective effect and slows the progression of the disease. This may be because in Parkinson's disease, MAO-B is involved in the production of an endogenous neurotoxin. Combination therapy with levodopa also reduces end of dose deterioration.

Side effects of MAO-B inhibitors include nausea, dry mouth, dizziness, postural hypotension, insomnia, confusion and hallucinations.

11.14.5 Anti-cholinergic drugs

Anti-cholinergic drugs were originally used prior to the development of levodopa because many signs of excessive parasympathetic activity are seen in Parkinson's patients (salivation, sweating and incontinence).

The basal nuclei contain many excitatory cholinergic neurons. As the nigrostriatal tract progressively degenerates, the release of inhibitory dopamine declines and the cholinergic system becomes relatively overactive. The site of action of anti-cholinergic drugs appears to be at muscarinic receptors in the basal nuclei. However, as the disease progresses, the loss of dopaminergic activity becomes so great that anti-cholinergic drugs cannot compensate for it. An example of an anti-cholinergic drug used in Parkinson's disease is benzatropine, but this drug has to be given by injection, either intramuscularly or intravenously. Trihexyphenidyl is an anti-cholinergic drug that can be given orally.

Anti-cholinergic drugs are most effective against tremor, but less so at improving muscle rigidity and hypokinesia. Because of this and unacceptable side effects, such as dry mouth, blurred vision, constipation and urinary retention, the use of anti-cholinergic drugs is limited.

11.14.6 Foetal tissue transplants

Recently, transplantation of foetal brain tissue has been carried out in a number of patients in the hope of reversing the symptoms of Parkinson's disease by providing neurons that produce dopamine.

Initially the results looked promising, with some patients with severe disease showing some degree of improvement after surgery. However, it now appears that deterioration with time is likely, although trials are still being evaluated.

Alternative approaches have been considered using genetically modified cells, for example fibroblasts from the skin, or retinal cells as sources of dopamine production. These approaches are still very much in the experimental stages.

11.15 Epilepsy

Epilepsy affects approximately 0.5% of the general population, of which between 70 and 80% can be controlled by drug therapy. There is no cure for epilepsy.

The main symptoms are seizures, often together with convulsions, but there are many types of epilepsy.

Seizures are caused by abnormal high-frequency firing of a group of neurons. The extent to which this abnormal activity spreads determines whether the epilepsy is generalized or partial.

Partial (focal) epilepsy is often associated with damage to a part of the brain, whereas generalized epilepsy does not seem to have an obvious cause. An attack of either type may have motor, sensory and/or behavioural effects, with loss of consciousness if the RAS is involved in generalized seizures.

11.15.1 Generalized epilepsy

The commonest forms of generalized epilepsy are tonic–clonic (grand mal) and absence seizures (petit mal). Tonic–clonic seizures are major convulsions involving spasm of all muscles. There is loss of consciousness and possibly incontinence. The abnormal brain activity involves the whole of the cerebral cortex. Seizures last for several minutes and may be preceded by an 'aura', which is a kind of warning that an attack is about to happen.

Status epilepticus is a series of seizures without recovery of consciousness in between.

Absence seizures are brief periods of loss of consciousness (15 seconds or so) with little motor activity, maybe just eyelid blinking or muscle jerking. Sometimes there is complete relaxation of muscles. This type of epilepsy is common in children.

11.15.2 Partial epilepsy

Common forms of partial epilepsy are Jacksonian and psychomotor epilepsy. In Jacksonian epilepsy, the abnormal brain activity is confined to the motor area of the cerebral cortex. An attack consists of jerking of particular muscle groups for a few minutes with no loss of consciousness. With psychomotor epilepsy, the abnormal activity is confined to the temporal lobe of the cerebral cortex resulting in brief periods of confused behaviour.

It is not known what causes epileptic seizures, but suggestions include increased excitatory nerve transmission, decreased inhibitory transmission or some abnormality of the nerve cells themselves. Such abnormalities may be the result of head trauma, stroke or tumours, but in most cases, the cause is unknown.

However, it seems that abnormalities in partial epilepsy are different to those in generalized epilepsy.

11.16   Drugs used to treat epilepsy

The choice of drug depends on the type of epilepsy. Combination therapy with two or more drugs may be necessary. In this case, careful monitoring is needed because unpredictable interactions between antiepileptic drugs are possible. Withdrawal from antiepileptic drugs should be gradual to avoid the risk of precipitating rebound seizures.

It is not known exactly how antiepileptic drugs work, but two mechanisms seem to be important for drugs in current use, while newer drugs may work by other means yet to be elucidated.

11.16.1   Enhancement of GABA action

Many well-established antiepileptic drugs (for example phenobarbital and benzodiazepines) are known to facilitate the opening of GABA chloride ion channels at synapses. However, this is not the complete explanation as, for example phenobarbital is far more effective as an anticonvulsant than other barbiturates with the same enhancing effects on GABA.

Newer drugs have been shown to interfere with GABA in other ways: by inhibiting GABA-transaminase, the enzyme responsible for inactivating GABA (vigabatrin and valproate) or by directly stimulating GABA receptors.

11.16.2   Blocking of sodium ion channels

Sodium ion channels in the neuronal membrane open during the depolarization phase of an action potential. Phenytoin and carbamazepine have been shown to block sodium ion channels, thus reducing the excitability of the neuronal membrane. Sodium ion channels can be in three different states – open, closed or inactive. It seems that this type of antiepileptic drug binds to the channels in the inactive state and prevents them from opening again. These drugs also show 'use-dependence'. That is, they preferentially block neurons that are transmitting action potentials at a high frequency (as occurs during an epileptic seizure) because more channels will be in the inactive state under such circumstances.

11.16.3 Phenytoin

Phenytoin is effective in many forms of epilepsy, but not absence seizures. Its mode of action appears to be by blocking sodium ion channels. Phenytoin can be a difficult drug to use for the following reasons.

Phenytoin is extensively bound to plasma albumin (80–90%) and other drugs such as salicylates and valproate displace it from its binding site. However, this does not necessarily lead to increased drug activity because metabolism of increased amounts of free drug is also enhanced. Thus, phenytoin may show increased or decreased activity. Other drugs, for example phenobarbital, compete for the same liver enzymes.

Phenytoin induces liver enzymes and can affect the metabolism and activity of other drugs, anticoagulants for example. Liver enzymes can also be saturated which means that plasma levels of phenytoin can rise unpredictably as its inactivation slows down.

The therapeutic ratio for phenytoin is very small and can vary from patient to patient.

Adverse effects of phenytoin range from mild vertigo, ataxia, headache and nystagmus, to the more serious but reversible confusion and intellectual deterioration. Hyperplasia of the gums and hirsutism can occur. Hypersensitivity reactions, usually skin rashes, are quite common while hepatitis, which can be severe and lymph node enlargement are less common. Increased incidence of fetal malformation, particularly cleft palate and cardiac malformation is suspected in epileptic mothers who take phenytoin.

11.16.4 Carbamazepine

Carbamazepine appears to have similar pharmacological action to phenytoin. It is particularly useful in treating psychomotor epilepsy.

It induces liver enzymes, which reduces its effectiveness and the effectiveness of other drugs such as phenytoin, oral contraceptives, warfarin and corticosteroids.

Adverse effects of carbamezepine are similar to those seen with phenytoin. They include the relatively mild drowsiness, dizziness and ataxia, to the more severe mental and motor disturbances. The drug can also cause gastrointestinal disturbances and arrhythmia. More rarely, severe hypersensitivity reactions including fatal bone marrow depression have been reported.

(For use of carbamazepine in neuropathic pain see Chapter 12.)

11.16.5 Valproate

Valproate is effective in all types of epilepsy and is especially useful in children and older people because it does not cause sedation and generally has low toxicity. It is considered to be first-line treatment for generalized seizures.

The exact mode of action of valproate is unclear. It inhibits enzymes that inactivate GABA, which causes an increase in GABA content of the brain and it possibly enhances the action of GABA.

Side effects of valproate are few. It causes thinning and curling of the hair in about 10% of patients. It is potentially hepatotoxic but this is rare. Valproate has been linked with increased risk of neural tube defects in the unborn child. Folic acid supplements are recommended as a precaution.

11.16.6   Ethosuximide

Ethosuximide is effective in absence seizures, but not other types of epilepsy. It may even precipitate tonic–clonic seizures in susceptible patients.

The mechanism of action of ethosuximide is not understood.

Adverse effects of ethosuximide are nausea, anorexia and sometimes lethargy. Very rarely, it can cause severe hypersensitivity reactions.

11.16.7   Phenobarbital

Phenobarbital is very similar to phenytoin in use and not effective in absence seizures. It appears to work, at least in part, by enhancing GABA transmission.

Phenobarbital is a potent enzyme inducer and as such can reduce the effects of other drugs, for example oral contraceptives, warfarin and corticosteroids.

Adverse effects of phenobarbital include sedation, which can occur at therapeutic doses and is a serious disadvantage in a drug that would have to be taken for many years. Other adverse effects include megaloblastic anaemia, mild hypersensitivity reactions and osteomalacia.

Phenobarbital is dangerous in overdose, producing coma and respiratory and circulatory failure and consequently is not much used nowadays.

11.16.8   Benzodiazepines

Most benzodiazepines are not useful in treating epilepsy because of their sedative effects and exacerbation of seizures on withdrawal. However, intravenous diazepam is used to treat *status epilepticus*.

11.16.9   New antiepileptic drugs

Not all epileptic seizures can be controlled by conventional therapy, and so new drugs have and are being developed. Some of them already in use are discussed below.

11.16.9.1   Vigabatrin

Designed to inhibit GABA-transaminase, vigabatrin is an anticonvulsant that can be effective in patients resistant to other drugs.

Adverse effects of vigabatrin include sedation, dizziness and behavioural changes similar to those seen with phenytoin. Vigabatrin may cause irreversible visual field defects, which limits its use.

11.16.9.2 Lamotrigine

Lamotrigine is similar in action to phenytoin. It appears to work on sodium ion channels to cause inhibition of the release of excitatory amino acids. Lamotrigine is considered to be first-line treatment for generalized and partial seizures and is considered more appropriate than valproate for women of childbearing age.

Adverse effects of lamotrigine are similar to phenytoin.

11.16.9.3 Gabapentin

Gabapentin was designed as a GABA analogue and is an effective anticonvulsant but it does not work on GABA receptors.

The mode of action remains unknown.

Gabapentin is used in partial epilepsy together with other drugs that have failed to provide control alone.

Adverse effects of gabapentin are gastrointestinal disturbances, mild sedation and ataxia.

11.16.9.4 Topiramate

Topiramate is a recently developed anticonvulsant with similar actions to phenytoin. It appears to block sodium ion channels and enhances the action of GABA. Topiramate is used, either alone or in addition, when other drugs do not provide adequate control. However, topiramate is teratogenic and should not be used in women of childbearing age.

Adverse effects of topiramate are gastrointestinal disturbances, dizziness and drowsiness.

Antiepileptic drugs are summarized in Table 11.9.

Table 11.9 Antiepileptic drugs

Drug	Action
Phenytoin	Blocks sodium ion channels
Carbamazepine	Blocks sodium ion channels
Valproate	Enhances action of GABA?
Ethosuximide	*Not understood*
Phenobarbital	Enhances action of GABA
Vigabatrin	Inhibits GABA-transaminase
Lamotrigine	Blocks sodium ion channels
Gabapentin	*Not understood*
Topiramate	Blocks sodium ion channels and enhances action of GABA

11.17   Alzheimer's disease

Alzheimer's disease is the term given to dementia that cannot be attributed to an obvious cause, for example brain trauma, a stroke or alcohol abuse. Alzheimer's disease is a gradual and irreversible decline in intellectual ability that usually appears after the age of 60. The disease has been estimated to affect 5% of over 65 year olds. Patients experience a progressive loss of cognitive function, usually beginning with loss of short-term memory followed by loss of other functions such as ability to calculate and ability to use everyday objects.

Alzheimer's disease is characterized by atrophy of the cerebral cortex, with loss of neurons particularly in the hippocampus and forebrain. Pathological features of Alzheimer's disease are extracellular plaques of β-amyloid protein and neurofibrillary tangles. β-amyloid plaques are found particularly in the hippocampus and association areas of the cerebral cortex. Similar plaques are seen in unaffected individuals, but there are many more in patients with Alzheimer's disease. Their formation may be due to abnormal metabolism of a precursor protein. Neurofibrillary tangles are found in neurons and may be collections of abnormal microtubule proteins, which form as a result of cell destruction. These appear to be a form of Lewy body (see page 213). There is a general deficiency of acetylcholine and loss of cholinergic neurons in the brains of people with Alzheimer's disease, although other neurotransmitters may also be involved. It is possible that an increase in glutamate activity leads to neuronal damage.

11.18   Treatment of Alzheimer's disease

There are no effective therapies for Alzheimer's disease and no cure. Treatment aims to enhance cholinergic transmission. The most useful drugs are central acetylcholinesterase inhibitors, for example donepezil. Acetylcholinesterase is the enzyme that normally breaks down acetylcholine after it has interacted with its receptors at the synapse. Inhibition of this enzyme in the brain increases the amount of acetylcholine available and prolongs its action. These drugs produce a modest improvement in memory or slow progression of symptoms in some patients. The response to anti-cholinesterase drugs may take several weeks. Their use is limited by side effects, which can be severe.

Side effects of acetylcholinesterase inhibitors include abdominal cramps, nausea, vomiting and diarrhoea and liver damage.

Acetylcholine precursors such as lecithin have been tried, but are of limited effectiveness.

Glutamate acts on N-methyl-D-aspartate (NMDA) receptors in the brain. Drugs have been developed to block these receptors. An example is memantine.

11.19   Summary

There are many disorders of the CNS. Psychiatric disorders include types of depression, psychosis and anxiety. Neurological disorders include Parkinson's disease, epilepsy and

Alzheimer's disease. There are many neurotransmitters that carry information from neu-rone to neurone in the nervous system. Although the cause of most CNS disorders is unknown, alteration in the function of neurotransmitters or their receptors is implicated in many of them. Drugs used to treat these disorders often interact with neurotransmitters or receptors. However, the full explanation of how the drugs work is still unknown and it can take many weeks for them to have an effect.

Unipolar depression is associated with a decrease in noradrenaline and/or serotonin activity and drugs effective in its treatment increase the concentration of one or both these transmitters at their synapses.

Bipolar depression is associated with a decrease in dopamine activity and mania with an increase in dopamine activity. Both these conditions can be treated prophylactically with lithium, although the mode of action of lithium is unknown.

Schizophrenia is the most common psychotic disorder. It may be due to an abnormality of dopamine receptors or increased release of dopamine in particular regions of the brain, the mesolimbic and mesocortical pathways. In addition, there may be an abnormality of serotonin pathways that interact with dopamine. Most drugs effective in the treatment of schizophrenia block dopamine receptors and some of the newer ones block serotonin receptors. Adverse effects of antipsychotics can be severe and are largely due to the blocking of dopamine receptors in other parts of the brain. For example Parkinsonism and tardive dyskinesia are the result of dopamine receptor blocking in the basal nuclei.

Anxiety and insomnia are relatively common and sometimes can be managed by psychotherapy techniques rather than with drugs. Antianxiety drugs are mainly the ben-zodiazepines, which work by enhancing the effects of GABA in the brain. This relieves the effects of anxiety. In higher doses, the hypnotic effect of benzodiazepines can be useful in short-term treatment of insomnia.

ADHD is a behavioural disorder seen in children and adolescents. It may be due to a deficit of noradrenaline and/or dopamine in the prefrontal cerebral cortex. Treatment with stimulant drugs, for example dexamphetamine, seems to have a calming effect in some patients.

Parkinson's disease was the first neurological disorder to be associated with deficiency of a specific neurotransmitter. In Parkinson's disease, there is a deficiency of dopamine in the basal nuclei, which leads to overactivity of cholinergic pathways and the characteristic hypokinesia, rigidity and tremor. Drugs used to treat Parkinson's disease aim to replace dopamine; stimulate dopamine receptors or the release of remaining dopamine; reduce breakdown of dopamine or reduce excessive parasympathetic activity. More recently, attempts have been made to replace dopamine-secreting cells by transplantation of fetal brain tssue.

Epilepsy is caused by abnormal high-frequency firing of neurons, either in the whole of the cerebral cortex (generalized epilepsy) or in discrete areas of the cerebral cortex (partial epilepsy). The abnormal activity may be due to increased excitatory neuronal transmission, possibly involving abnormal sodium ion channels or decreased inhibitory transmission, possibly due to abnormality at GABA receptors. Certainly, most antiepilep-tic drugs either block sodium ion channels or enhance the action of GABA or both.

Alzheimer's disease is characterized by atrophy of the cerebral cortex and loss of neu-rons, leading to gradual decline in intellectual ability and short-tem memory. There is no cure for Alzheimer's disease, but treatment with central acetylcholinesterase inhibitors aims to enhance cholinergic transmission and glutamate blocking drugs reduce the possibly damaging effects of glutamate.

Useful websites

www.nice.org.uk/ National Institute for Health and Clinical Excellence. Guidelines on ADHD expected in 2008.

http://www.nice.org.uk/CG020 National Institute for Health and Clinical Excellence. Guidelines on Epilepsy 2004.

http://www.nice.org.uk/CG022 National Institute for Health and Clinical Excellence. Guidelines on Anxiety 2007.

http://www.nice.org.uk/CG023 National Institute for Health and Clinical Excellence. Guidelines on Depression 2007.

http://www.nice.org.uk/CG0235 National Institute for Health and Clinical Excellence. Guidelines on Parkinson's disease 2006.

http://www.nice.org.uk/TA43 National Institute for Health and Clinical Excellence. Guidelines on Schizophrenia 2002.

http://www.nice.org.uk/TA111 National Institute for Health and Clinical Excellence. Guidelines on Alzheimer's disease 2007.

www.parkinsons.org.uk Parkinson's Disease Society. Section for Healthcare Professionals.

www.rcpsych.ac.uk/ Royal College of Psychiatrists. Mental health information.

Case studies

This is a long chapter and as such warrants several case studies.

The following case studies are of patients with Alzheimer's disease, Parkinson's disease, depression, bipolar disorder and epilepsy.

Case study 1

The following case study is of a patient that any health care professional might see on a regular basis for whatever reason.

Mr Jones, 72, has recently been diagnosed with Alzheimer's disease, which so far is relatively mild, being mainly increased forgetfulness. His consultant at the hospital has prescribed him donepezil.

After two weeks, Mr Jones's family is concerned that he does not seem to be any better. They also tell you that since starting on the donepezil, Mr Jones has been suffering from nausea and they wonder if this can be a side effect of the drug.

What can you tell the family about Mr Jones's apparent lack of improvement?

Would you expect nausea to be a side effect of this drug? What other side effects could be expected?

Case study 2

The following case study is of a patient that any health care professional might see on a regular basis for whatever reason.

Mrs Cooper is an elderly patient who has recently needed treatment.

Although she has Parkinson's disease Mrs Cooper still lives at home, being visited regularly by her daughter and has the help of a home care assistant.

Mrs Cooper is being treated with amantadine for her mild tremor and slowness of movement. While Mrs Cooper is managing quite well at home, her daughter is keen to know if there is anything she should be aware of regarding the treatment for Parkinson's disease.

Are there any side effects of amantadine that she should look out for?

Mrs Cooper's daughter asks you if her mother's condition is likely to get worse, how quickly, and if there is anything else that can be done if it does. What can you tell her about this and is there anything else you could advise the patient and her family about?

Case study 3

You are treating Pete, a young man of 28, for an injury sustained after an accident. In conversation, he tells you that he has depression and is currently going through a particularly bad patch that has lasted several months. When taking his history, you noted that he was taking amitriptyline, 30 mg four times a day. Pete does not think his medication is improving his mood and tells you that his family thinks his medication should be changed. Could you offer any advice?

Assuming you have advised Pete to make an appointment with his doctor to review his medication, when you next see him he tells you the doctor has increased the dose and told Pete to give the drug more time to have an effect. Given the increase in dose, what are the side effects Pete might experience?

Several months later Pete is still taking amitriptyline, which seems to be having a positive effect but he is found to have signs of liver disease. Would you expect this to lead to a change in Pete's drug therapy?

Case study 4

You are seeing the following patient, Mr Davies who is 54 years old, on a regular basis for treatment following an injury.

Apart from the recent injury, Mr Davies has bipolar disorder, which was diagnosed when he was 26. In addition, he has a two-year history of hypertension, which is being treated, but seems to be inadequately controlled at the moment. Mr Davies is quite knowledgeable about his bipolar disorder and the combination of drugs he uses to control the symptoms, but he is less sure about what having hypertension means. He tells you that he has his blood tested every three months and that last time he went to the clinic they prescribed him a new drug that he had not taken before and told him to come back in a week. Mr Davies says he has not been feeling quite right lately and wonders if it could be the new drug.

Mr Davies current medication is as follows:

- Lithium carbonate, 800 mg at night

- Sodium valproate, 500 mg twice a day

- Atenolol, 50 mg in the morning

- Amlodipine 5 mg in the morning

- Lisinopril, 5 mg in the morning, recently prescribed

- Chlorpromazine, 50 mg to be taken as required

Discuss this patient using the questions as a guide.

- Identify which of Mr Davies drugs are for hypertension and which are for bipolar disorder.

- Consider if there are likely to be any interactions between the many drugs that Mr Davies is taking.

- What would the regular three-monthly blood test be for and what is the significance of the next blood test being in a week's time?

- What are the usual signs of lithium toxicity?

- One of the drugs, valproate, is normally prescribed for epilepsy. Mr Davies is aware of this and asks you to explain why he has been prescribed it because he is pretty sure he does not have epilepsy.

- Why would chlorpromazine be prescribed 'as required'?

- Would it be safe to treat Mr Davies injury with a non-steroidal anti-inflammatory drug?

Case study 5

This patient, Mrs Phillips, is 26 years old and has been epileptic since childhood. However, her epilepsy is well controlled and she does not let it interfere with her life. You are to perform a minor procedure on Mrs Phillips, which may require follow-up appointments. During the treatment you learn that Mrs Phillips is a primary school teacher, has not been married very long and is planning to have a family of her own in a few years time. You may need to look up the use of oral contraceptives.

Mrs Phillips current medication is quite straightforward:

Carbamazepine 600 mg twice a day

Norimin combined contraceptive pill daily for 21 days each cycle

Discuss this patient's situation using the questions as a guide.

- Norimin is a relatively high-strength combined contraceptive pill. Is there any reason why this is preferable to a low-strength combined contraceptive pill or a progesterone-only contraceptive pill for this patient?

- What are the potential disadvantages to using a high-strength combined contraceptive pill?

- If your patient said she was thinking of stopping with the contraceptive pill in order to become pregnant, what should your advice be?

- Given that the patient has epilepsy, are there any particular precautions you should take when providing treatment?

- Assuming that Mrs Phillips does get pregnant later on and has to continue with antiepileptic medication, what special precautions could she take to help ensure she has a healthy baby?

- What is the advice on breast-feeding for mothers who take anticonvulsant medication?

Case study 6

Mr Warchowski is a 59-year-old patient you have not seen before. On taking a medical history, you discover that he is being treated for a number of conditions. The patient was diagnosed with hypertension five years ago. He also has a history of depressive disorder, but it has been seven years since his last episode. Mr Warchowski has recently been diagnosed with benign enlargement of the prostate gland and is due to have suspected ischaemic heart disease investigated at the cardiology clinic.

Mr Warchowski's current medication is as follows:

- Aspirin 75 mg daily

- Atenolol 50 mg daily

- Doxazosin 4 mg daily

- Amlodipine 5 mg daily

Discuss this patient's medical conditions and treatment using the questions as a guide.

- Determine which medication is for which condition.

- Are there likely to be any interactions between the drugs Mr Warchowski is currently taking?

- Mr Warchowski tells you he has been feeling a bit depressed again lately. Consider whether any of his medication could have contributed to this feeling.

- In the past, Mr Warchowski's depressive episodes have been successfully treated with amitriptyline. Would it be appropriate to treat Mr Warchowski with amitriptyline this time, given his other medication?

- What alternatives to amitriptyline are there for a patient like Mr Warchowski?

- What alternatives to doxazosin are there for a patient like Mr Warchowski?

Chapter review questions

You should be able to answer these review questions from the information in this chapter.

1. Briefly describe the monoamine theory of depression and mania. What is the major problem with this theory?

2. Which group of drugs, used to treat depression, can produce serious interactions with many drugs and common food substances and why?

3. How are tricyclic antidepressants thought to work?

4. Lithium can be used to treat bipolar depression and mania. Why is it important that a patient always uses the same brand of lithium?

5. What are the four main dopamine pathways in the brain and which ones are implicated in schizophrenia?

6. Most anti-psychotic drugs block dopamine receptors. How does this explain Parkinsonian side effects?

7. Benzodiazepines are commonly prescribed for anxiety. How do benzodiazepines exert their action?

8. What is the Committee on the Safety of Medicines (now Commission on Human Medicines) recommendation for the use of benzodiazepines?

9. Give examples of four types of drugs available for the treatment of insomnia. What are their advantages and disadvantages?

10. Levodopa is used to treat Parkinson's disease. Explain why it is used together with drugs that inhibit dopa decarboxylase and COMT.

11. What are the alternatives to treatment of Parkinson's disease with levodopa?

12. Describe the different forms of epilepsy.

13. Explain the two proposed mechanisms of action for antiepileptic drugs.

14. Discuss the difficulties of using phenytoin to treat epilepsy.

15. Review alternatives to phenytoin for treating epilepsy.

12

Anaesthesia and analgesia

12.1   Chapter overview

Anaesthesia and analgesia are closely related; both terms can mean loss of sensation.

General anaesthesia aims to achieve loss of consciousness in the patient together with loss of all sensation. Usually this is to allow surgery or manipulation of a body part.

Local anaesthesia is the loss of sensation in a particular part of the body while the patient remains conscious. This can allow minor surgery or other treatment of a body part.

Analgesia is the absence or relief of pain. There are many reasons why analgesia is desirable. Pain can be due to chronic illness, acute injury, or because of surgery or manipulation.

Although it is unlikely, but not impossible, that health care professionals would be administering general anaesthetics, they are included here for the sake of completeness and because it is possible that overdose of local anaesthetics can cause general anaesthesia.

Registered podiatrists who hold a certificate of competence are allowed to administer local anaesthetics in the course of their professional practice from a list specified in the Medicines Act. Physiotherapists and radiographers may administer local anaesthetics in some circumstances under patient group directions.

Certain analgesics can be accessed and supplied by podiatrists in the course of their professional practice, some of which are prescription-only medicines. Any health care professional who trains to be a supplementary prescriber could prescribe prescription-only analgesics if it was appropriate to the individual clinical management plan.

(See Chapter 14 for details of how the law on the use and administration of medicines applies to members of the health care professions.)

12.2   General anaesthesia

General anaesthesia refers to a loss of sensation accompanied by a change or loss of consciousness. The effects of all general anaesthetics should be reversible.

Pharmacology for the Health Care Professions   Christine M. Thorp
© 2008 John Wiley & Sons, Ltd

The development of general anaesthesia has quite a long history, which is worth relating here.

Humphrey Davey first suggested the use of nitrous oxide as an inhalation anaesthetic in 1800. This gas was used for many years as 'laughing gas' for entertainment. Around this time, an American dentist, Horace Wells, used it medicinally during the extraction of one of his own teeth.

The use of ether had a similar reputation for public abuse (so-called ether frolics) until it was used for dental extraction in 1846 by another American, William Morton. This dentist persuaded a surgeon to allow him to administer ether to a patient during a surgical procedure. This was conducted on 16 October 1846 in Massachusetts General Hospital by the chief surgeon Dr Warren in front of an audience of other surgeons. The demonstration was a resounding success and surgeons rapidly adopted the process of anaesthesia.

Also in 1846, Professor Simpson of Glasgow used chloroform to relieve the pain of labour. This practice was widely condemned by the clergy as the work of the devil until 1853 when Queen Victoria gave birth to her seventh child under its influence.

Today, either inhalation or intravenous anaesthetics or a combination of the two is used to produce general anaesthesia.

12.2.1 Mechanism of action of general anaesthetics

Chemically, general anaesthetics are a diverse group of drugs; they are all small lipid soluble molecules, but their mechanism of action is unknown. However, there are two main theories to explain the mechanism of action of general anaesthetics. These are the lipid theory and the protein theory.

As anaesthetic potency is closely related to lipid solubility rather than to chemical structure, the lipid theory proposes that general anaesthetics interact with the lipid bilayer of the cell membrane. Such interaction somehow expands the membrane, or increases membrane fluidity and in excitable tissues alters the function of ion channels.

According to the protein theory, the site of action is a hydrophobic region of cellular proteins, possibly receptors or proteins involved in transmitter release. Interaction with general anaesthetic would result in reduction in the excitability of the cell and inhibition of nerve transmission. This could be through potentiation of neurotransmitter release at inhibitory synapses or by inhibition at excitatory synapses. Gamma-aminobutyric acid (GABA) is an inhibitory neurotransmitter and glutamate is an excitatory neurotransmitter in the brain. Intravenous general anaesthetics at least seem to work by enhancing the effects of GABA or blocking the effects of glutamate.

The protein theory is currently considered more likely to explain the process of anaesthesia, because general anaesthetics affect synaptic transmission rather than axonal conduction. Both the release of transmitter and the response of the postsynaptic receptors are affected.

General anaesthesia involves three main responses of the nervous system: loss of consciousness, loss of motor reflexes and loss of sensory reflexes. Because areas of the brain responsible for consciousness are the most sensitive it is possible to administer anaesthetics at concentrations that produce unconsciousness without seriously depressing the cardiovascular and respiratory centres or the heart muscle, but the margin of safety is narrow. At doses in excess of those used for anaesthetic purposes, all general anaesthetics can cause death from cardiorespiratory failure. Most general anaesthetics depress the cardiovascular system due to direct effects on the myocardium and blood vessels combined with inhibition of neuronal activity.

General anaesthesia usually involves administration of several drugs with different actions for premedication, induction and maintenance of anaesthesia.

Premedication is intended to prevent the parasympathetic effects of anaesthesia (excessive salivation and bronchial secretion and a reduction in heart rate) and to reduce anxiety or pain. Premedication and adjuncts to general anaesthesia are discussed further on page 234.

Induction is most likely to be achieved using an intravenous anaesthetic, which produces unconsciousness within seconds.

Maintenance of anaesthesia is then usually by administration of an inhalation anaesthetic in a mixture of air or oxygen.

12.3   Inhalation anaesthetics

Inhalation anaesthetics are either gases or volatile liquids. Apart from nitrous oxide, which is still widely used, earlier inhalation anaesthetics are no longer used. Ether is not suitable because it is explosive and irritant to the respiratory tract. Chloroform cannot be used because it is toxic to the liver. Inhalation anaesthetics currently in use are the volatile liquids halothane (since 1956) and more recently isoflurane, desflurane and sevoflurane and nitrous oxide gas.

12.3.1   Halothane

Halothane is a potent general anaesthetic, which is non-explosive and non-irritant and it provides a pleasant smooth induction.

Side effects of halothane include hypotension, cardiac arrhythmias and hangover. The high incidence of severe liver damage on repeated exposure to halothane has limited its use. The Committee on the Safety of Medicines (now the Commission on Human Medicines) recommends that repeated use of halothane should be avoided.

12.3.2   Isoflurane, desflurane and sevoflurane

Isoflurane, desflurane and sevoflurane have superseded halothane. None of them is as potent as halothane, but they are less likely to cause liver toxicity. Isoflurane is widely used. Its main adverse effects are that it can cause myocardial ischaemia in patients with heart disease and it depresses respiration.

All of these anaesthetics (and suxamethonium, see page 235) have the potential to cause malignant hyperthermia. This is a rare but potentially lethal complication of anaesthesia. It is characterized by a rapid rise in body temperature, due to excessive muscle contractions, together with increased heart rate and acidosis. It is treated as an emergency with dantrolene, which causes muscle relaxation.

12.3.3   Nitrous oxide

Nitrous oxide has a low potency and must be used in combination with other inhalation anaesthetics for general anaesthesia. Nitrous oxide provides rapid induction and recovery. It also has an analgesic action and is used as a 50% mixture with oxygen to provide analgesia without loss of consciousness during labour and manipulations of injured body parts.

Adverse effects of nitrous oxide are the risk of bone marrow depression with repeated or prolonged use and megaloblastic anaemia due to interference with the actions of vitamin B_{12}. Used in labour, nitrous oxide can depress respiration of the newborn.

12.4   Intravenous anaesthetics

The mechanism of action of intravenous anaesthetics is thought to be by enhancement of inhibitory pathways or inhibition of excitatory pathways in the brain. Enhancement of the actions of GABA or inhibition at glutamate receptors seems most important, although modulation of other receptors may play a role as well.

Intravenous anaesthetics can be used for short surgical procedures of 10–20 minutes. Their elimination from the body is too slow to allow rapid control of the depth of anaesthesia needed for long procedures. They are most often used for induction because, even the fastest acting inhalation anaesthetics take a few minutes to act and cause excitement before anaesthesia is produced, which would be unpleasant for the patient. It also means that the amount of inhalation anaesthetic can be reduced.

See Table 12.1 for a comparison of individual intravenous anaesthetics.

12.4.1   Thiopental

Thiopental is a barbiturate, which is highly soluble in lipid, with a rapid onset and shorter duration of action than many other barbiturates. Its mode of action is by enhancement of inhibitory actions of GABA in the brain.

Thiopental is an anaesthetic but has no analgesic action; indeed, it may increase perception of pain.

Because thiopental causes myocardial and respiratory depression and has a very narrow margin of safety, it has been largely superseded by non-barbiturates.

Table 12.1 Individual intravenous anaesthetics

Drug	Actions	Adverse effects	Comments
Thiopental	Rapid onset, slow recovery due to accumulation	Cardiorespiratory depression	Narrow margin of safety
Etomidate	Rapid onset, fast recovery	Cardiorespiratory depression Muscle twitching	Greater margin of safety
Propofol	Rapid onset, very rapid recovery	Cardiorespiratory depression	Painful on injection
Ketamine	Relatively slow onset and recovery	Hallucinations, nightmares	Limited usefulness
			Minor use in children

12.4.2 Etomidate

Etomidate has similar effects on the brain to thiopental, producing anaesthesia but no analgesia.

The site of action of etomidate is assumed to be activation of GABA receptors. Recovery from etomidate is rapid with no hangover.

Compared to thiopental, etomidate has a greater margin of safety between anaesthetic dose and the dose that produces cardiorespiratory depression.

The main adverse effects with etomidate are pain on injection and muscle twitching during induction, both of which can be reduced by using an opioid analgesic. It also causes suppression of the adrenal cortex.

12.4.3 Propofol

Propofol is the most recently developed intravenous anaesthetic.

The site of action of propofol is probably the GABA receptor. Propofol has a fast onset of action and very fast recovery time. It is metabolized rapidly and has an antiemetic effect.

Propofol can be used for both induction and maintenance of anaesthesia. Because of this, it is used alone for short procedures, for example podiatric surgery.

A disadvantage of propofol is that it is painful on injection.

12.4.4 Ketamine

Ketamine produces a different effect to other intravenous anaesthetics; analgesia, sensory loss, amnesia and muscle paralysis are produced without loss of consciousness, so-called dissociative anaesthesia, and with minimal respiratory depression.

The site of action of ketamine is thought to be through inhibition at glutamate receptors.

Ketamine is a phencyclidine derivative and as such has abuse potential. Phencyclidine was developed in the 1950s as an anaesthetic, but its use was abandoned because it caused hallucinations and delirium. It became popular as PCP (phenylcyclohexylpiperidine), a drug of abuse, in the 1970s.

Ketamine has limited use for induction of anaesthesia in children.

Adverse effects of ketamine are that it causes muscle twitching and excessive salivation together with nightmares and hallucinations (apparently less so in children).

12.5 Premedication and adjuncts to general anaesthesia

Premedication is used to prepare the patient for general anaesthesia. The objective is to reduce feelings of anxiety and lightly sedate the patient. In addition, drugs may be used to prevent parasympathetic effects of some general anaesthetics.

Four main groups of drugs are used as premedication and as adjuncts to general anaesthesia: sedatives; antimuscarinic drugs; muscle relaxants; and analgesics. Table 12.2 shows actions and examples of drugs used to supplement general anaesthesia.

12.5.1 Sedatives

Often, short-acting sedatives may be used to allay anxiety and produce some sedation and amnesia.

Oral benzodiazepines, such as diazepam or midazolam are commonly used. These drugs also reduce muscle tone, which is useful during surgery. They have no analgesic effect, so may be used together with opioid analgesics.

As a side effect benzodiazepine sedatives produce a degree of respiratory depression.

Table 12.2 Premedication and adjuncts to general anaesthesia

Drug group	Actions and uses	Examples
Sedatives	Reduce anxiety and produce sedation and amnesia	Diazepam, midazolam
Antimuscarinics	Prevent salivation and bronchial secretions Protect heart from arrhythmia	Hyoscine
Muscle relaxants	Non-depolarizing neuromuscular blockers Short-lasting muscle paralysis	Pancuronium, atracurium
	Depolarizing neuromuscular blockers Short-lasting muscle paralysis	Suxamethonium
Analgesics	Opioids At induction and during operation to control pain	Fentanyl, morphine
	[a]NSAIDs Prevention and treatment of post-operative pain	Diclofenac, ibuprofen

[a]NSAIDs, non-steroidal anti-inflammatory drugs.

12.5.2 Antimuscarinic drugs

Antimuscarinic drugs are used as premedication to prevent salivation and bronchial secretions during an operation and to protect the heart from arrhythmias caused by inhalation anaesthetics such as halothane and propofol. They are less commonly used nowadays because modern anaesthetics are less irritant.

Hyoscine is antiemetic, sedative and produces some amnesia.

12.5.3 Muscle relaxants

Muscle relaxants are also known as *neuromuscular blocking drugs*. They are used as adjuncts to general anaesthesia, to cause muscle relaxation where this would aid the surgical process. Artificial ventilation is required because respiratory muscles are also paralysed.

There are two types of neuromuscular blocking drugs used during surgery: non-depolarizing neuromuscular blockers and depolarizing neuromuscular blockers.

12.5.3.1 Non-depolarizing neuromuscular blockers

Non-depolarizing neuromuscular blockers are competitive antagonists of acetylcholine at the neuromuscular junction. Normally, binding of acetylcholine to its receptor triggers muscle contraction. These drugs produce a short-lasting paralysis that can be reversed by anticholinesterase drugs. They have no sedative action or analgesic effects.

Examples of non-depolarizing neuromuscular blockers are pancuronium and atracurium.

Side effects of non-depolarizing neuromuscular blockers are mainly caused by histamine release and include flushing, increased heart rate, hypotension and bronchospasm.

12.5.3.2 Depolarizing neuromuscular blockers

Suxamethonium is the only example of a depolarizing neuromuscular blocker in current clinical use. This drug interacts with the acetylcholine receptor at the neuromuscular junction, initially stimulating it and then preventing further stimulation by endogenous acetylcholine, resulting in paralysis. Suxamethonium is metabolized at its site of action by the enzyme pseudocholinesterase. The effect of suxamethonium is not reversed by the use of anticholinesterase drugs, but would be prolonged.

Side effects of suxamethonium are bradycardia and, rarely, malignant hyperthermia.

Individuals with genetically determined less active plasma pseudocholinesterase levels are likely to experience prolonged paralysis with suxamethonium (see Chapter 2, page 23).

12.5.4 Analgesics

Opioid analgesics can be used at induction to relieve anxiety and during an operation to control pain and to reduce the need for post-operative analgesia.

Examples of opioid analgesics are fentanyl (most often used) and morphine.

However, opioid analgesics cause respiratory and cardiovascular depression, which is additive to that of general anaesthetics.

Some non-steroidal anti-inflammatory drugs (NSAIDs) are licensed for the prevention and treatment of post-operative pain. They may be inadequate for relief of very severe pain. Examples in use are diclofenac and ibuprofen.

Side effects of NSAIDs are discussed in Chapter 7 and later in this chapter, but are unlikely to be a problem with short-term use as adjuncts to general anaesthesia.

12.6   Local anaesthesia

Local anaesthesia is the reversible loss of sensation in an area of the body. Anaesthesia means loss of sensation and analgesia means loss of pain. Since local anaesthetics are often used to prevent pain, they are also known as *local analgesics*.

12.6.1   Local anaesthetic techniques

There are several techniques used to administer local anaesthetics, which are briefly described below. The first three are techniques that can be used by podiatrists in the course of their professional practice providing they are registered with the Health Professions Council and hold a certificate of competence in the use and administration of local anaesthetics.

Other techniques for local anaesthetic administration are given for the sake of completeness.

12.6.1.1   *Topical anaesthesia*

EMLA® (eutectic mixture of local anaesthetics; a combination of two local anaesthetics, lidocaine and prilocaine in a cream) is capable of producing skin thickness analgesia from topical administration. The cream is applied for a period of 1–2 hours under an occlusive dressing. EMLA is indicated prior to venepuncture (particularly in children) and whenever pre-injection analgesia of the skin is required.

This technique may be used by physiotherapists and radiographers under patient group directions (see Chapter 14).

12.6.1.2   *Infiltration*

The anaesthetic is injected in or around the area requiring anaesthesia. The drug diffuses through the tissues and directly affects the nerve endings in the skin preventing conduction of nerve impulses. This method is commonly used when stitching wounds or during minor surgery limited to the skin thickness.

12.6.1.3   Nerve or conduction block

The anaesthetic is injected near to a nerve trunk where it reversibly prevents conduction along many of the nerve fibres in the trunk. This method requires good anatomical knowledge so that the anaesthetic solution can be injected accurately around the required nerves.

12.6.1.4   Spinal anaesthesia

This technique is used in obstetrics and genito-urinary surgery and is also known as *intrathecal block*.

The anaesthetic is injected into the subarachnoid space via a lumbar puncture between vertebrae L2 and L3 or L3 or L4. This places the anaesthetic in the cerebro-spinal fluid. Efferent nerves (motor and sympathetic) and afferent fibres (sensory) are affected by the local anaesthetic. This method carries a risk of damage to the spinal cord.

12.6.1.5   Epidural anaesthesia

This technique is similar to spinal anaesthesia. It is technically more difficult but is considered to be safer and has largely replaced spinal anaesthesia.

The local anaesthetic is introduced into the extradural space between the bone of the spine and the dura mater, which is filled with fatty tissue and blood vessels. The drug then diffuses into nervous tissue.

12.6.1.6   Intravenous regional anaesthesia (bier's block)

This method is used for the complete anaesthesia of a single limb, usually the arm. It allows surgery without the risks accompanying general anaesthesia. Blood flow in the limb is prevented by application of a tourniquet before the anaesthetic is injected. When the tourniquet is removed, there is the risk of systemic side effects due to large amounts of anaesthetic reaching the general circulation.

12.6.2   Mechanism of action of local anaesthetics

All local anaesthetics block sodium ion channels so that sodium ions cannot flow into neurons. This inhibits transmission of action potentials (nerve impulses) along individual neurons. (See Figure 12.1.)

Sodium ion channels can be in three different states – open, closed or inactivated. Local anaesthetics preferentially bind to channels in the inactive state and prevent them from opening again. This is similar to the mechanism of action of some antiepileptic drugs (see page 217) and it explains the use of lidocaine in arrhythmia (page 66). Moreover, like antiepileptic drugs, local anaesthetics show 'use dependence' in that they preferentially block neurons that are transmitting impulses at high frequency (as pain sensory neurons do).

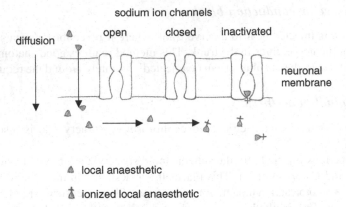

▲ local anaesthetic

🔺 ionized local anaesthetic

Figure 12.1 Site of action of local anaesthetics

Local anaesthetics are weak bases and at the slightly alkaline pH of body fluids, they exist in both ionized and unionized forms in equilibrium.

Only the unionized form can reach its site of action, which is inside the neuron. This is because the drug has to be in a lipid soluble form to diffuse through the myelin sheath and neuronal membrane.

Once inside the neuron dissociation is necessary, because it is the ionized form binds to the sodium ion channel. Local anaesthetic activity is dependent on pH, because pH determines the degree of dissociation into ions. This becomes of clinical importance in inflamed and infected tissue, which often has a more acid pH. Acid conditions result in increased degree of ionization and reduced diffusion of local anaesthetic into neurons. This makes them less effective anaesthetics.

However, it is recommended that local anaesthetics should not be injected into inflamed tissue as this increases absorption into blood and consequently increases the risk of side effects.

Neurons are differentially sensitive to local anaesthetics. Small diameter neurons are more susceptible than larger diameter neurons and unmyelinated neurons more than myelinated neurons.

Fortunately, neurons that carry the sensation of pain are of the smallest diameter unmyelinated and myelinated type and as such are affected by low doses of local anaesthetic, leaving larger motor neurons unaffected.

12.6.3 Complications of local anaesthesia

Complications can arise because of technical errors such as broken needles or practitioner error, for example injection of the wrong solution or accidental injection into a blood vessel.

There is always a risk of infection with techniques that puncture the skin.

Neuralgic pain, after the anaesthetic has worn off, and nerve palsy due to direct nerve damage are rare but possible complications.

12.6.4 Overdose

Overdose occurs when plasma levels of local anaesthetic are such that systemic effects appear. Overdose can be due to the patient receiving a dose larger than is appropriate (absolute overdose), or because the patient is intolerant of the local anaesthetic and has increased sensitivity to its effects (relative overdose).

The symptoms of overdose are graduated according to plasma levels of the drug; they range from mild, where the patient becomes anxious and restless to severe, where there may be convulsions and cardiac and respiratory failure requiring emergency treatment.

12.6.5 Intolerance

Intolerance reactions are often unpredictable with no family history of similar events. Generally, the sick, frail, aged and malnourished are most likely to experience adverse effects in response to usual doses.

12.6.6 Allergy

Allergy to local anaesthetics is possible, although this is not common. It is most likely in people who have regular contact with local anaesthetics such as dentists and anaesthetists.

See Chapter 3 for details of allergic reactions to drugs.

12.6.7 Adverse effects of local anaesthetics

Local anaesthetics used in podiatry are all readily absorbed from their sites of administration into the bloodstream, so potentially they can have systemic side effects. Metabolism is by enzymes in the liver, and to a lesser extent in the kidneys. Excretion occurs via the kidneys but only a small percentage of the drug appears in the urine in its original form. Therefore, the risk of adverse effects is increased in patients with hepatic or renal disorders.

In addition to their action as local anaesthetics, the following actions on other parts of the body are possible. These effects are not clinically significant except in intolerant individuals, those with idiosyncratic reactions, in cases where absorption into the blood stream is unexpectedly rapid or in those with impaired metabolism and/or excretion.

Local anaesthetics can pass into the central nervous system through the blood brain barrier. They have the same effect on central nervous tissue as they have on peripheral nervous tissue; that is, they inhibit conduction of nerve impulses. The first effects are tremor and restlessness because inhibitory pathways are affected first. As the dose increases, convulsions may occur.

Initially, due to central nervous system effects, the rate of respiration increases. As the drug concentration increases, depression of the medulla results in breathing becoming rapid and shallow. Larger doses depress respiration and lower the level of consciousness of the patient.

Local anaesthetics can affect both autonomic nerves and the neuromuscular junction. Apart from their local anaesthetic action, local anaesthetics have anticholinergic activity. This is due partly to prevention of acetylcholine release and partly to non-depolarizing competitive block of the acetylcholine receptors on the postsynaptic membrane. The importance of this clinically is that local anaesthetics should be used with great care in patients suffering from myasthenia gravis because of the danger exacerbating the condition and precipitating paralysis, particularly of the respiratory muscles.

Such patients should be treated as high risk.

Cardiac muscle cells are excitable, and hence the diffusion of sodium ions is affected as it is in nervous tissue. The result of this is a slowing of the heart rate. Indeed lidocaine is used for this purpose in the treatment of ventricular arrhythmias.

Local anaesthetics cause varying degrees of vasodilation. The combination of a fall in heart rate and peripheral vasodilatation can result in a fall in blood pressure and fainting.

12.6.8   Drug interactions with local anaesthetics

Interactions of local anaesthetics with other drugs are relatively rare and are most likely to occur if both drugs are given intravenously. With the doses and routes of administration used in podiatry, physiotherapy and radiography drug–drug interactions can be expected to be a rare event. Nevertheless, interactions are possible. Table 12.3 shows some possible drug–drug interactions with lidocaine.

12.6.9   Cautions

Local anaesthetics must be used with extra caution in the medical conditions shown in Table 12.4. Patients with these conditions should be considered as high risk.

Table 12.3   Drug interactions with local anaesthetics

Drug group	Effect of interaction
Barbiturates	Additive respiratory depression
β adrenergic blockers	Prolonged action of lidocaine
Phenytoin	Increased risk of central side effects and cardiovascular collapse
Procainamide	Potentiation of central effects of both drugs
Quinidine	Increased risk of cardiotoxicity

Table 12.4 Medical conditions requiring cautious use of local anaesthetics

Medical condition	Effect/caution
Myasthenia gravis	Danger of respiratory paralysis
Epilepsy	May provoke seizures
Cardiac arrhythmia	Exacerbated by local anaesthetics
Liver disease	Reduced metabolism increases risk of side effects
Kidney disease	Reduced excretion increases risk of side effects

12.7 Local anaesthetics

The drugs described below, and summarized in Table 12.5, are on the list of local anaesthetics that registered podiatrists who hold a certificate of competence are allowed to administer in the course of their professional practice.

Some local anaesthetics, lidocaine and bupivacaine, can be used in combination with adrenaline. Adrenaline is a vasoconstrictor and its use increases the speed of onset and prolongs the duration of action of the local anaesthetic. Vasoconstrictors should never be used with local anaesthetics in digits or appendages, because of the risk of vasoconstriction leading to ischaemic necrosis. See page 277 for a list of local anaesthetics and other injectable drugs that can be administered by podiatrists.

Table 12.5 Local anaesthetics

Drug	Comments
Lidocaine	Most commonly used
	Rapid onset of action
	Duration of action 90 min
	Can be used with adrenaline
Bupivacaine	Two to four times potency of lidocaine
	Slow onset of action, up to 30 min
	Long duration of action, up to 8 h
	Can be used with adrenaline
Prilocaine	Similar to lidocaine
	Longer duration of action
	Can be used with felypressin
Mepivacaine	Short onset of action
	Short duration of action
	Mostly used in dentistry
Levobupivacaine	Isomer of bupivacaine
	Similar actions but fewer adverse effects
Ropivacaine	Less potent than bupivacaine
	Less cardiotoxic
	Longer duration of action.

12.7.1 Lidocaine

Lidocaine is the most commonly used local anaesthetic. It acts rapidly and is more stable than most others are. It has a duration of action of about 90 minutes.

Repeated injections have been shown to produce tachyphylaxis; that is successive administration produces a reduced effect.

Lidocaine may be used with adrenaline to increase its speed of onset and prolong its duration of action.

The most common side effect is transient drowsiness. Allergy can occur in susceptible individuals.

Lidocaine should not be used in patients with hypovolaemia, as this would result in higher plasma levels and consequent increase in risk of toxic effects, or in bradycardia, because there may be further slowing of the heart.

12.7.2 Bupivacaine

Bupivacaine has two to four times the potency of lidocaine and shows similar toxic effects with high doses.

It has a slow onset of action, up to 30 minutes, and a long duration of action, up to eight hours when used for nerve block. Bupivacaine can also be used with adrenaline.

Bupivacaine is often used to produce continuous epidural anaesthesia during labour.

12.7.3 Prilocaine

Prilocaine is similar to lidocaine but has a longer duration of action and is less toxic. It can be used in combination with felypressin (a vasoconstrictor), but not by podiatrists.

High doses of prilocaine can cause methaemoglobinaemia and cyanosis. For this reason, it is used with caution in patients with anaemia, cardiac failure, congenital or acquired methaemoglobinaemia or impaired respiratory function.

12.7.4 Mepivacaine

Mepivacaine has a short onset of action and a short duration of action. It is mostly used in dentistry.

12.7.5 Levobupivacaine and ropivacaine

Levobupivacaine and ropivacaine were recently (November 2006) added to the list of local anaesthetics that registered podiatrists are allowed to administer.

Levobupivacaine is an isomer of bupivacaine with similar actions but fewer adverse effects.

Ropivacaine is similar, less potent than bupivacaine but less cardiotoxic and with a longer duration of action.

12.8 Analgesia

Analgesia means without pain.

Analgesics are used to relieve pain irrespective of its cause.

They are commonly used to remove a patient's discomfort, while the body's natural repair mechanisms take place, or, until measures can be taken to resolve the underlying cause. Indiscriminate use of analgesics can be dangerous if the cause of pain is ignored.

The type of analgesic used depends on the source and severity of the pain.

Pain-relieving drugs can be divided into peripherally acting analgesics and centrally acting analgesics.

12.8.1 Pathophysiology of pain

Pain is an unpleasant sensory and emotional experience in response to tissue damage, threat of tissue damage or perceived tissue damage. Experience and interpretation of pain is subjective and influenced by previous experience and an individual's physical and mental condition at the time.

Sensory nerve endings are found throughout the body in the skin, muscles, joints, blood vessels and internal organs. These nociceptors are sensitive to the effects of potentially damaging mechanical, thermal and chemical stimuli.

When cells are damaged they release a variety of chemical mediators, which can activate or sensitize nociceptors to other chemicals. This explains acute pain. Chronic pain is more difficult to explain, especially if it goes on beyond the initial tissue damage. Chronic pain is thought to be associated with changes to the normal physiological pain pathway.

The pathway for pain perception is a chain of three neurons (see Figure 12.2).

Stimulation of nociceptors results in transfer of pain signals by first order sensory neurons to the dorsal horn of the spinal cord. The neurotransmitters released here are glutamate and a protein known as *substance P*. From the dorsal horn, second-order neurons cross over the spinal cord and pass the signal on to the thalamus in the brain. In the thalamus the general feeling of pain is interpreted and the information is then transmitted to a third order neuron, which terminates in the somatosensory area of the cerebral cortex. It is here that the details about localization, intensity and qualities of pain are perceived. The pain pathways also go to other parts of the brain, such as the hypothalamus, the amygdala and reticular activating system (RAS). In these regions, somatosensory information is processed and associated with previous experiences that are related to pain.

Inhibitory neurons in the spinal cord are activated by descending pathways from the brain. Here the transmitters are GABA and enkephalin. Enkephalin is one of the endogenous endorphins, which are natural opioids. Enkephalin reduces the release of substance

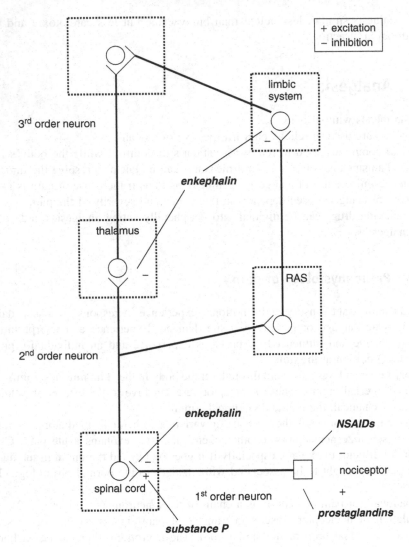

Figure 12.2 Pain pathways

P at the sensory neuron terminals in the spinal cord. It is also released in areas of the midbrain and medulla where it enhances the descending inhibitory pathways, which terminate in the spinal cord.

12.9 Peripherally acting analgesics

This group of analgesics contains aspirin-like drugs and paracetamol.

Table 12.6 lists commonly used peripherally acting analgesics.

Table 12.6 Peripherally acting analgesics

Drug	Comments
Aspirin	First choice for mild to moderate pain
	Not in children under 16
Paracetamol	First choice for mild to moderate pain, including children; little anti-inflammatory action
Ibuprofen	Mild to moderate pain; first choice for inflammatory conditions
Fenbrufen	An alternative to ibuprofen
Naproxen	Similar to ibuprofen
Diclofenac	Similar to ibuprofen
Mefenamic acid	Mild to moderate pain; less anti-inflammatory action
Piroxicam	Similar to mefenamic acid
Tenoxicam	Similar to mefenamic acid
Indometacin	Anti-inflammatory; less analgesic action
	Serious side effects

12.9.1 Non-steroidal anti-inflammatory analgesics

Aspirin belongs to a group of drugs that all possess analgesic, anti-inflammatory and antipyretic properties to different degrees. Because of the anti-inflammatory effects they are known as *non-steroidal anti-inflammatory drugs*, or NSAIDs (see Chapter 7 and Figure 7.1)

NSAIDs exert their effects by inhibition of the enzyme cyclo-oxygenase (COX). This results in reduced production of prostaglandins and thromboxanes because COX normally catalyses their formation from arachidonic acid. Prostaglandins play a role in pain by sensitising sensory nerve endings to the effects of other mediators such as bradykinin.

Thromboxanes promote platelet aggregation.

There are many different drugs of this type, none of which is a more potent analgesic than aspirin. Some of them, however, have a more potent anti-inflammatory action.

The major difference between individual NSAIDs is the range of side effects experienced by patients. Individuals have different reactions to different NSAIDs. Some tolerate a particular drug while others do not. In the treatment of chronic pain, a number of drugs may be tried until one is found that produces the most acceptable result to the patient.

12.9.2 Aspirin

Aspirin, is often the first choice of analgesia for mild to moderate pain in adults.

In large doses, aspirin has anti-inflammatory effects and has a role in the treatment of chronic inflammatory disease (Chapter 7).

Aspirin also reduces the adhesive and aggregative properties of platelets thereby decreasing the tendency of thrombi formation. Use is made of this in the treatment of thromboembolic disease (Chapter 4).

Adverse effects of aspirin are dyspepsia or gastritis and it can cause gastric ulceration. This is because normally prostaglandins inhibit the secretion of stomach acid.

Aspirin is not recommended for children or adolescents under 16 due to the risk of Reye's syndrome. Reye's syndrome is a rare disorder of the liver and brain with a mortality rate of 20–40% that can follow acute viral infection. It is not known for certain what role aspirin plays in its development.

12.9.3   Other NSAIDs

Ibuprofen, fenbrufen, naproxen and diclofenac all have similar actions and are generally well tolerated. They have a lower incidence of side effects than aspirin, but are more expensive and probably no more effective.

Mefenamic acid, and piroxicam and tenoxicam have a spectrum of side effects between that of aspirin and the ibuprofen type. The oxicams have a markedly longer duration of action than the majority of other NSAIDs and are therefore useful in chronic pain.

Indometacin is a much more potent anti-inflammatory drug but poorer analgesic than aspirin. It is more likely to produce serious side effects and is only recommended for chronic pain when other analgesics do not work.

12.9.4   Paracetamol

Paracetamol relieves pain and fever in adults and children, and is the most widely accepted medicine in the United Kingdom for this purpose. It is used mainly for its pain relief properties either as prescribed by a doctor, or as an over-the-counter medicine. Paracetamol is suitable for most situations of mild to moderate pain and for all age groups including the very young. It may be used following immunization procedures, and it is available in special liquid formulations for children.

The analgesic and antipyretic effects of paracetamol are similar to those of aspirin and it works in a similar, though not identical, way. The exact mechanism of action of paracetamol is unknown.

Unlike aspirin, increasing the dose of paracetamol does not result in anti-inflammatory activity. Paracetamol is therefore not of value in chronic rheumatic diseases, or in situations where the pain is due to inflammation.

When taken at the recommended dosage, there are few side effects with paracetamol and no major interactions with other drugs.

In substantial overdose, severe, life-threatening liver damage is likely to occur due to excessive production of a toxic metabolite. There is an antidote to paracetamol overdose (acetylcysteine) and provided it is given within eight hours a complete recovery can be made. In the liver, acetylcysteine is converted to glutathione, which forms a harmless conjugate with paracetamol (see Table 2.2, page 22).

12.10   Centrally acting analgesics

Centrally acting analgesics are also known as *narcotics* or *opioids*. Strictly speaking, narcotic refers to the ability of a drug to induce sleep, which drugs in this group do in high dosage, but the term can be misleading and is best avoided.

Opioids are drugs that have morphine-like actions and the name comes from the source of morphine and opium, the opium poppy, *Papaver somniferum*. Opium itself is a mixture of substances that occur in the sap of the opium poppy.

12.10.1   Morphine

The most common extract of opium is morphine, which is currently used for the treatment of moderate to severe pain. The effects of morphine are characteristic of many of the opioid drugs used in analgesia. The actions of opioids are mediated through opioid receptors. These are the receptors for the natural endorphins. Endorphins are endogenous analgesics, which act by inhibiting sensory neurons in the spinal cord and thalamus, thereby preventing or altering perception of pain. See Figure 12.2 and explanation on page 243. The inhibition comes about by the opening of potassium ion channels. Efflux of potassium ions leads to hyperpolarization of the sensory neuron membrane. Thus, transmission of pain signals to the brain is inhibited.

The actions of all opioids can be blocked by naloxone, which is used in cases of overdose.

The effects of opioids on the central nervous system are to produce analgesia (particularly effective in chronic or acute pain of a constant nature), elevation of mood (euphoria), respiratory depression (occurs at therapeutic doses), cough suppression, nausea and vomiting and miosis (pin-point pupils).

Other effects include reduction in tone and motility of the gastrointestinal tract, producing constipation, and the release of histamine from mast cells, causing local pain and itching at the injection site. Histamine released systemically may induce bronchoconstriction, bradycardia and hypotension. Tolerance and dependence are not usually a problem in clinical use.

Other opioid analgesics are compared to morphine in Table 12.7.

12.11   Neuropathic pain

Neuropathic pain is severe chronic pain due to damage to sensory nerves. This is not because of tissue injury and can occur as a consequence of central nervous system disorders, such as stroke or multiple sclerosis, or because of malignancy, amputation (phantom limb pain) diabetic neuropathy or following infection with *Herpes zoster* (as shingles). Mechanisms underlying neuropathic pain are poorly understood. The pain is described as

Table 12.7 Opioid analgesics

Drug	Comments
Morphine	Prototype
	Analgesia, euphoria, respiratory depression
	Produces dependence
Codeine and	Less potent than morphine
Dihydrocodeine	Little or no euphoria
	Dependence rarely occurs
Pethidine	Short-acting
	Used in labour
	Causes restlessness
Diamorphine (heroin)	More potent than morphine
	Short-acting
Fentanyl	Highly potent
	Used as adjunct to general anaesthesia
Methadone	Less sedative than morphine
	Produces fewer withdrawal symptoms

burning, shooting or scalding pain and it can be difficult to treat with conventional analgesics. Often a combination of physiotherapy and psychological treatment together with one or more of the following approaches is needed. Opioids, for example methadone, sometimes work and techniques like nerve block with local anaesthetic or transcutaneous electrical nerve stimulation (TENS) or the topical use of capsaicin as a counter-irritant may help. Unlicensed use of amitriptyline, gabapentin, lamotrigine or carbamazepine is indicated in severe cases, as is intravenous infusion of either lidocaine or ketamine under specialist supervision.

12.12 Summary

Anaesthesia and analgesia both mean loss of sensation; general anaesthesia leads to loss of consciousness as well, local anaesthesia means loss of sensation in part of the body and analgesia means absence of pain specifically.

The mechanism of action of general anaesthetics is unknown, but there are two theories to explain their action: the lipid theory and the protein theory. The lipid theory states that general anaesthetics interact with lipids in the neuronal cell membrane and disrupt neurotransmission and the protein theory states that general anaesthetics interact with membrane proteins to alter release of neurotransmitters. The protein theory is thought most likely.

All general anaesthetics cause reversible loss of consciousness, but they are also capable of depressing respiratory and cardiovascular centres in the brain and of depressing the cardiovascular system directly. This makes them potentially dangerous in use.

Inhalation anaesthetics are gases or volatile liquids. Nitrous oxide and isoflurane are commonly used.

Intravenous anaesthetics are used to induce general anaesthesia, followed by an inhalation anaesthetic for maintenance, and for short surgical procedures. Propofol is commonly used for both induction and maintenance anaesthesia.

Premedication is the use of drugs to reduce a patient's anxiety and to prevent parasympathetic effects of the anaesthetic. Muscle relaxants and analgesics are used as adjuncts to anaesthesia, respectively to cause muscle relaxation and aid surgery and to reduce patient discomfort post-operatively.

Local anaesthetics are also known as *local analgesics* because they cause the loss of the sensation of pain in a part of the body. There are several techniques used to administer local anaesthetics: topical, infiltration and nerve block are techniques that can be used by qualified, registered podiatrists in the course of their clinical practice.

Local anaesthetics prevent nerve transmission by blocking sodium ion channels from the inside of the neuron. Sensory nerves that carry the stimulus of pain are the most sensitive to the actions of local anaesthetics. Accidental overdose can cause systemic effects including cardiorespiratory depression and loss of consciousness. Lidocaine is probably the most commonly used local anaesthetic. Qualified, registered podiatrists are allowed to access and administer lidocaine, bupivacaine, prilocaine, mepivacaine, levobupivacaine and ropivacaine.

Analgesia can be achieved by peripherally acting and centrally acting analgesics. The most commonly used peripherally acting analgesics are paracetamol and aspirin for mild to moderate pain. For moderate to severe pain morphine is commonly used. Peripherally acting analgesics work by inhibiting the enzyme necessary for prostaglandin synthesis. Centrally acting analgesics work by acting on opioid receptors to inhibit pain pathways in the spinal cord and brain stem.

Neuropathic pain can be difficult to treat and may need the use of drugs not conventionally used as analgesics.

Useful websites

www.frca.co.uk/ Anaesthesia UK. Local anaesthetic pharmacology.

http://www.frca.co.uk/SectionContents.aspx?sectionid=66 Anaesthesia UK. Section on pharmacology.

www.painrelieffoundation.org.uk The Pain Relief Foundation. Research into chronic pain.

www.thepainweb.com The Pain Web. Website for health professionals dealing with pain.

Case studies

Case study 1

One of your patients, with a learning disability, is due to have a minor operation. She has never had an operation before, is feeling quite nervous about the prospect and has many concerns she would like to discuss with you.

She has been told that she will have a premedication and asks you what this is for. What can you tell her?

Can you give some examples of drugs used for premedication?

Apart from the anaesthetic itself, what other types of drugs might be used during the operation?

Case study 2

The following case study shows a situation where a physiotherapist working in a pain clinic is involved in the management of a patient.

A 50-year-old male patient has been referred to a chronic pain clinic with a six-month history of low back pain and leg pain for pain management. The patient is otherwise fit and well, but has not worked for six months.

The patient has been diagnosed as having nociceptive low back pain and neuropathic leg pain, which was confirmed by very mild nerve root compression as seen on the magnetic resonance imaging (MRI) scan. In conjunction with the consultant pain physician, the patient's pain management plan was devised to include listing for epidural therapy, rationalization of his medication and referral to a concurrent pain management programme.

The patient's current medication is as follows:

Co-codamol 8/500, up to a maximum of eight tablets per day

Diclofenac, maximum 150 mg per day

This regime did not appear to be providing a good therapeutic effect and the patient complained of gastric irritation.

Discuss this patient's treatment, using the following questions as a guide.

- What could the physiotherapist advise the patient to do about the gastric irritation?

- Would you recommend any changes to the patient's treatment regime?

- Are there any alternative medicines that could be tried for the neuropathic pain?

- Do you think this patient would be suitable for treatment under patient group directions or supplementary prescribing? (Refer to Chapter 14).

Chapter review questions

You should be able to answer the following review questions from the material presented in this chapter.

1. What are the advantages of the inhalation anaesthetics isoflurane, desflurane and sevoflurane over the older ones ether, chloroform and halothane?

2. What are the advantages of using intravenous anaesthetics for induction of general anaesthesia and why are they generally unsuitable for sole use in lengthy operations?

3. What are the two main types of muscle relaxant used in conjunction with general anaesthesia and how do they work?

4. Describe three methods of producing local anaesthesia that can be used by podiatrists.

5. List three methods of local anaesthesia that would not be used by podiatrists.

6. Where is the site of action of local anaesthetics and what process within the neuronal membrane is inhibited by local anaesthetics?

7. Explain why injection of local anaesthetic into an inflamed area is likely to be less effective than expected.

8. Explain how it is possible for local anaesthetics to affect the sensation of pain without inhibiting motor function.

9. Summarize the possible side effects of local anaesthetics.

10. With which groups of drugs do local anaesthetics produce interactions?

11. Describe three medical conditions in which local anaesthetics should be used with caution.

12. List five peripherally acting analgesics.

13. How do NSAIDs exert their action and what are their potential side effects?

14. Why is aspirin not recommended in children under 16?

15. Why should a medical practitioner be consulted if analgesia is needed for more than a few days?

16. List five centrally acting analgesics.

17. Under what circumstances is pain best treated with opiates?

13

Contrast agents and adjuncts to radiography

13.1 Chapter overview

Radio-opaque substances are used in various ways to facilitate radiological examination of parts of the body. They are known as contrast agents or contrast media and are usually based on iodine, barium or gadolinium containing compounds. These substances are injected, swallowed or introduced rectally. Like all drugs, once they have entered the systemic circulation, contrast agents have the potential to be distributed to all parts of the body, although they do not cross cell membranes easily. They may be metabolized, but most are excreted unchanged via the kidneys. Contrast agents used to visualize the bowel do not usually enter the circulation and are excreted rectally. The ideal contrast agent should be non-toxic, should not be absorbed or metabolized and should be excreted rapidly. However, all contrast agents have the potential to cause adverse drug reactions and some interact with other drugs. Radiographers must be trained to administer drugs and manage adverse reactions to contrast agents.

Adjuncts to therapeutic radiography, used under patient group directions, include analgesics, laxatives, anti-diarrhoeals, antiemetics and drugs for wound care and skin reactions.

Adjuncts used in addition to contrast agents in diagnostic radiography, under patient group directions, include laxatives, drugs to relax the bowel, normal saline and topical anaesthesia.

13.2 Contrast agents

Contrast agents are substances that alter, or attenuate, X-rays. The density of a substance determines the degree of attenuation of radiation. Air and gas (oxygen, carbon dioxide) can be used as contrast agents. They provide negative contrast because they absorb X-rays

Pharmacology for the Health Care Professions Christine M. Thorp
© 2008 John Wiley & Sons, Ltd

less than the surrounding tissue. They appear darker on X-ray film and are used to show the outline of surrounding soft tissue. For example, gas is introduced into the stomach or colon during double contrast barium examination.

Other contrast agents are said to be radio-opaque and as such they absorb X-rays. That is, they attenuate X-rays positively and appear white or lighter on X-ray film. They are used to enhance radiological examination of parts of the body not normally visualized by X-rays, by providing a greater contrast of structures such as blood vessels, ducts and hollow organs of the body. Such contrast agents are based on iodine or barium because these particular atoms absorb X-rays well. The more radio-opaque atoms per molecule in a contrast agent, the greater the X-ray absorption is.

Ideally, contrast agents should be non-toxic, should not be absorbed or metabolized and should be excreted rapidly after imaging.

13.2.1 Development of contrast agents

Iodine contrast agents are the most convenient, most effective and least toxic for general use. The background of their development is presented here for non-specialists.

Historically oily contrast agents were used, such as the ethyl esters of iodinated fatty acids of poppy seeds. They are no longer routinely used, having been superseded by newer, safer agents.

Water-soluble iodine contrast agents form the largest group of contrast agents in use today. They were originally developed from mono-iodinated pyridine derivatives (in the 1920s as by products of research for drugs to treat syphilis). Further development to reduce toxicity led to tri-iodinated benzene derivatives.

Nowadays all such contrast media are derivatives of tri-iodinated benzoic acid. Monomeric contrast agents contain one benzene ring and have been used orally for examination of the gall bladder. Dimeric contrast agents contain two benzene rings and are used intravenously. Dimers are less likely to diffuse out of the blood circulation into tissues and are therefore less likely to cause adverse reactions. All are either sodium salts or meglumine salts or mixture of the two. (Meglumine is a contraction of methyl glucamine.) Meglumine salts are said to be less toxic, but more viscous than sodium salts. A mixture of the two is a compromise of these two important parameters. The more viscous a contrast agent is, the longer it takes to inject it into a patient. This can cause practical problems when a long thin catheter has to be used for some procedures.

Contrast agents can be ionic or non-ionic. Ionic contrast agents dissociate into ions in solution to form an anion containing iodine and a cation containing the rest of the molecule, sodium or meglumine. This means they can have an osmolality of seven or eight times that of plasma. Osmolality is a measure of the effect a substance can have on the movement of water and depends on the number of molecules per kilogram of water. It is measured in milliosmoles per kilogram (mOsm/kg) of water. Cells and body fluids are normally in osmotic equilibrium and have an osmolality of about $300 \, \text{mOsm} \, \text{kg}^{-1}$. Addition of ions to them in the form of ionic contrast agents can alter osmolality (the

Table 13.1 Monomeric and dimeric non-ionic contrast agents

Monomeric	Uses
Iohexol	All used for urography and angiography
Iopamidol	
Iopentol	
Iopromide	
Ioversol	
Dimeric	Uses
Iotrolan	Myelography
Iodixanol	Urography and angiography

osmolality of ionic contrast agents ranges between 600 and 1000 mOsm kg^{-1}) and cause adverse effects, particularly the loss of water from cells, especially blood cells. Other adverse effects (see page 257) are pain on injection, endothelial damage, thrombosis, thrombophlebitis, disturbance of the blood–brain barrier, bradycardia (in cardioangiography) and increased pulmonary blood pressure.

Ionic monomeric contrast agents are known as high osmolar contrast agents (HOCAs) and are not much used nowadays because of the risk of adverse effects.

Since the 1980s, low osmolar contrast agents (LOCAs) have been available. They are non-ionic contrast agents and therefore do not dissociate in solution. Because of this, they are less toxic and produce fewer side effects. However, they contain fewer radio-opaque iodine atoms and therefore are less effective as contrast agents. The higher the concentration of iodine in a contrast agent the greater is the positive radiographic contrast that can be achieved. The latest non-ionic dimeric contrast agents have six iodine atoms per molecule. Some are nearly isosmolar with plasma (iotrolan), others are hypo-osmolar (iodixanol) and they have very low toxicity. Because of this low toxicity, they are used most often nowadays, although monomeric, ionic meglumine ioxaglate is said to cause less pain and heat on injection.

Table 13.1 lists some monomeric and dimeric non-ionic contrast agents and their uses.

13.2.2 Use of contrast agents

Intravenous iodine contrast agents are used together with X-rays to image the gall bladder (cholecystography/cholangiography), the bladder and kidneys (urography), blood vessels (angiography), lymphatics (lymphography), nerves (myelography) and joints (arthrography). Arthrography can be used to aid assessment prior to podiatric surgery. Figure 13.1 shows the use of iopamidol to visualize the capsular lining of a lesser metatarsophalangeal joint of the foot during an investigation of a possible capsular tear.

Contrast agents can also be used to enhance computed tomography (CT) scans and magnetic resonance imaging (MRI). Contrast agent use with CT scans is particularly

(a)

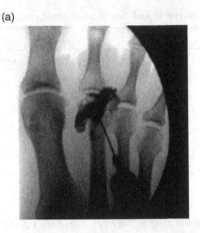

(b)

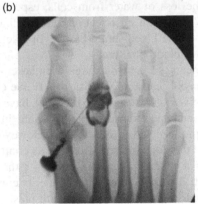

Investigation of a capsular tear

Iopamidol used to visualize the capsular lining of a lesser metatarso-phalangeal joint of the foot

(a) injection of contrast agent

(b) proximal and distal margins of the joint capsule highlighted by the contrast agent

Pictures provided by A. Waddington

Figure 13.1   Arthrogram using Iopamidol. Reproduced with permission of Anthony Waddington

useful for the diagnosis, staging and follow-up of malignant disease and with MRI for imaging the spinal cord.

The digestive tract can be imaged following ingestion of barium sulphate suspended in a gel. The preparation must be taken with plenty of water to avoid it causing intestinal obstruction. Barium sulphate is insoluble and not absorbed from the gastrointestinal tract and therefore it is not normally toxic. Severe and potentially fatal inflammation can occur if the gastrointestinal tract is perforated and the barium contrast agent escapes into the peritoneal cavity.

13.3 Cautions in use of contrast agents

Iodine contrast agents should be used with caution in patients with thyroid disorders. Both Graves's disease (hypersecretion of thyroid hormone), and Hashimoto's thyroiditis (hyposecretion of thyroid hormone), can be adversely affected by contrast agents. Iodine-containing drugs in general, including contrast agents can cause hypothyroidism in susceptible patients.

(Radiological contrast agents are among the most commonly used iodine-containing drugs. Others are iodine-containing antiseptics and expectorants.)

Use of iodine contrast agents is contraindicated in pregnancy because they can cross the placenta and interfere with foetal thyroid development.

Special precautions should be taken in patients with known hypersensitivity to iodine and in those with other allergies, for example to foods or other drugs. Prophylactic corticosteroids, for example prednisolone, are recommended as premedication 12–18 hours prior to imaging in such high-risk patients.

Contrast agents should also be used with caution in patients with epilepsy because they can provoke seizures; in hepato-renal syndrome (renal failure following liver cirrhosis and jaundice); in severe respiratory disease, especially asthma; and in diabetes.

Iodine contrast agents interact with metformin, an oral hypoglycaemic drug used to treat type 2 diabetes (see Chapter 6). Iodine contrast agents increase the risk of a patient developing lactic acidosis with metformin, particularly if their kidney function is impaired. Metformin has to be stopped prior to radiological examination and should not be restarted until normal renal function has resumed.

13.4 Complications of intravenous administration of contrast agents

With intravenous administration, there is always a risk of damage to veins through mechanical or chemical means. A good cannulation technique is essential to avoid extravasation, tissue damage or the introduction of air into the circulation. Air embolism is a rare but life-threatening complication. Techniques that puncture the skin can introduce infection through poor aseptic technique or contaminated equipment. Antihistamines and corticosteroids, if needed for adverse reactions, should never be mixed in the same syringe because precipitation would occur.

13.5 Adverse reactions to contrast agents

Contrast agents are among the safest medicinal products in use. They are not intended to have pharmacological activity; nevertheless adverse reactions are possible. Adverse reactions to iodine contrast agents are related to iodine concentration, osmolality, whether the agent is ionized or not, the rate and frequency of injection and the dose given and the chemical composition of the agent. Adverse reactions to contrast agents can range from trivial skin rashes to life-threatening anaphylaxis. The commonest side effects are nausea and vomiting and hypersensitivity with intravenous use.

High osmolality causes dehydration of red blood cells and increases the risk of thrombosis and causes vasodilation and sensation of heat on injection.

Ionization can adversely affect the heart and central nervous system and may cause allergic reactions. Entry of contrast agent into the central nervous system is unlikely unless the integrity of the blood–brain barrier has been compromised by the osmotic effects of the contrast agent. Dehydration of endothelial cells of brain capillaries may allow diffusion of contrast agent into the brain. If this happens, the contrast agent may provoke seizures. LOCAs are the safer contrast agents in this respect.

There is a link between osmolality of iodine contrast media and the risk of renal toxicity. LOCAs (600–800 mOsm kg^{-1}) have lower renal toxicity than HOCAs ($>$1500 mOsm kg^{-1}). Not surprisingly, isosmolar (300 mOsm kg^{-1}) contrast agents are safer and recommended as first choice.

The frequency of contrast agent injection and dose are known risk factors for renal toxicity. Ideally, in patients at high risk of renal toxicity, there should be five days between administrations and the lowest possible dose of either non-ionic monomer or non-ionic dimer should be used.

Adverse reactions include the risk of acute renal failure due to blockage of tubules because contrast agents are relatively insoluble.

Adequate water intake by the patient is essential to reduce the risk of renal toxicity. In some high-risk patients or in emergency situations, renal toxicity can be prevented by the use of acetylcysteine. This is possibly because of its antioxidant properties. However, it is probably better to use alternative methods of investigation, for example ultrasound in high-risk patients.

The common conditions predisposing to high risk of renal toxicity with contrast agents are listed below:

- Pre-existing renal failure
- Serum creatinine above 1.2 mg dl^{-1}
- Hypovolaemia (heart failure, nephritic syndrome, cirrhosis)
- Ventricular dysfunction
- Hypertension
- Dehydration
- Nephrotoxic drugs (NSAIDs, diuretics, ACE inhibitors)
- Diabetes mellitus with renal impairment
- Advanced age ($>$70)
- Multiple myeloma
- Any patient requiring high dose

Table 13.2 Adverse reactions to contrast agents

Mild	Moderate	Severe
Nausea	Erythema	Paralysis
Warm feeling	Chest pain	Seizures
Sneezing	Abdominal pain	Pulmonary oedema
Runny nose	Vasovagal syncope	Bronchospasm
Metallic taste	Facial oedema	Anaphylactic shock
Headache		Cardiac arrest
Pruritis		Respiratory arrest
Sweating		

Toxicity is also related to chemical composition of a contrast agent. This is not completely understood, but appears to be related to protein binding capacity of the molecule. Binding of contrast agent to enzymes in particular can inhibit their activity. For example, inhibition of acetylcholinesterase (the enzyme that normally breaks down acetylcholine) leads to increased parasympathetic effects. Consequences of this are bronchospasm, a fall in heart rate and hypotension. Protein binding is usually due to the electrical charge of ions, but can also be due to the hydrophobic parts of the molecule (the benzene ring). Newer non-ionic contrast agents have structures whereby a hydrophilic side chain shields the benzene ring and this makes the molecule less toxic.

Adverse reactions to iodine contrast agents, rated mild, moderate and severe are listed in Table 13.2. Conventionally, mild adverse reactions need reassurance of the patient but no treatment; moderate adverse reactions may interfere with radiological examination but usually require no treatment; and severe adverse reactions require the examination to be stopped and emergency treatment.

13.5.1   Other contrast agents

Nephrogenic systemic fibrosis (nephrogenic fibrosing dermopathy) has been linked with the use of gadolinium contrast agents. This is a condition where there is fibrosis of the skin, connective tissues and muscle causing restriction of flexion and contraction of the limbs. Internal organs including lungs, liver and heart can all be affected and the condition can prove fatal.

Use of gadolinium contrast agents is therefore contraindicated in patients with renal impairment.

Gadolinium contrast agents are used in MRI.

13.6   Management of acute adverse reactions to contrast agents

Serious adverse reactions to contrast agents are rare occurrences and the vast majority are of a minor nature. The management of serious adverse reactions to contrast agents

Table 13.3 Emergency treatment of adverse reactions to contrast agents

Drug	Use
Adrenaline	Restores blood pressure
Antihistamine, e.g. chlorphenamine	Reduces effects of histamine (vasodilation and oedema)
Hydrocortisone	Reduces oedema and inflammation
Salbutamol	Bronchodilation
Atropine	Restores heart rate in bradycardia
Aminophylline	Bronchodilation

(less than 0.05% incidence rate) can involve the emergency use of adrenaline, oxygen, antihistamine, hydrocortisone and salbutamol. Radiographers who inject contrast agents must be trained in emergency treatment and basic life support.

The most serious adverse reaction to contrast agents is anaphylactic shock, which must be treated as an emergency. See Chapter 3 for details of allergic reactions to drugs. In this type of hypersensitivity reaction, the contrast agent causes the release of histamine from mast cells and basophils.

Anaphylactic shock includes laryngeal oedema, bronchospasm and hypotension. Immediate treatment means making sure the patient can breathe, laying them flat with the feet raised and injection of adrenaline (500 μg intramuscularly, repeated every five minutes until blood pressure, pulse and breathing are restored). Intravenous atropine (0.3–1 mg) may be necessary if the hypotension is accompanied by bradycardia. Oxygen must be administered. An antihistamine, for example chlorphenamine, can be given by slow intravenous infusion after adrenaline and for 24–48 hours to prevent relapse (maximum 40 mg in 24 hours). If recovery is not apparent, intravenous fluids to maintain blood pressure and circulating blood volume (normal saline or gelatine infusion) and intravenous aminophylline or nebulized salbutamol in addition to oxygen are indicated. Intravenous corticosteroid can also be given to prevent further deterioration. See Chapter 5 for details of salbutamol, aminophylline and antihistamines. Corticosteroids suppress all phases of allergic reactions. Their actions and uses in long-term inflammatory disease are discussed in Chapter 7.

Less serious reactions to contrast agents are nausea and vomiting, mild skin reactions (hives) and more serious generalized skin reactions with urticaria. Nausea and vomiting rarely require treatment with antiemetics, for example domperidone. Skin reactions can be treated with oral or intravenous antihistamines, which also have an antiemetic effect. Intravenous corticosteroids may be required for serious urticaria.

Table 13.3 summarizes drugs used in emergency treatment of adverse reactions to contrast agents.

13.7 Adjuncts to radiography

A number of drugs are used as adjuncts to radiography. They can be administered by qualified, registered and trained radiographers under patient group directions.

See Chapter 14 for the legislation surrounding administration of drugs by members of the health care professions.

13.7.1 Analgesia

Analgesia may be necessary for the after-effects of venepuncture or following therapeutic radiographic procedures. Peripherally acting analgesics, for example paracetamol, aspirin or ibuprofen should be sufficient. For pain that is more severe codeine may be necessary. See Chapter 12.

Topical anaesthesia prior to venepuncture may be necessary. This technique with EMLA® (eutectic mixture of local anaesthetics) cream may be used by radiographers under patient group directions. See Chapter 12.

13.7.2 Laxatives and bowel evacuants

Laxatives and evacuants are used to empty the bowel prior to the administration of barium contrast agents and abdominal radiological examination.

There are four types of laxatives depending how they work: bulk-forming laxatives; softeners/lubricants; osmotic laxative; and stimulant laxatives.

13.7.2.1 Bulk-forming laxatives

Bulk-forming laxatives work by increasing the volume of non-absorbable solid residue in the bowel. They act by absorbing water, swelling and stimulating peristalsis.

Bulk-forming laxatives are long chain polysaccharides, which are not broken down by the normal process of digestion. An example of a bulk-forming laxative is methyl cellulose.

Because they take several days to have an effect, bulk-forming laxatives are not suitable for use before radiological examination.

Bulk-forming laxatives are used to control diarrhoea associated with therapeutic radiography.

13.7.2.2 Faecal softeners and lubricants

Faecal softeners act as detergents on the surface of faeces causing them to absorb water and soften. An example is docusate, which can be used prior to imaging. However, other types of laxatives (see below) are better for this purpose.

Lubricants work by coating the faeces and rectal wall, thereby aiding expulsion of faeces. A traditional example of a lubricant laxative is liquid paraffin. Lubricants do not work fast enough to be useful prior to radiological imaging.

13.7.2.3 Osmotic laxatives

Osmotic laxatives are salts that act osmotically by retaining the water they are given with, or by drawing water into the bowel from the body. This accelerates transfer of gut contents through the small intestine with an abnormally large volume entering the colon. This causes distension and reflex peristalsis. Osmotic laxatives must be given with plenty of water. They may cause abdominal cramps. Examples of osmotic laxatives used prior to imaging are magnesium citrate and magnesium sulphate. Lactulose is a disaccharide with a similar action. Polyethylene glycol is an osmotic laxative used for bowel cleansing (see below).

Although only small amounts of these drugs are absorbed, osmotic laxatives should not be used in patients with poor renal function because of the risk of magnesium accumulation. High concentrations of magnesium in the body can interfere with the function of calcium ions in the heart, skeletal muscle and the central nervous system. Effects of this include neuromuscular block or central nervous system depression.

13.7.2.4   Stimulant laxatives

Stimulant laxatives work by stimulating enteric nerves (in the wall of the intestine), which results in smooth muscle contraction and an increase in peristalsis. At the same time they increase fluid secretion from the intestinal mucosa. Stimulant laxatives can cause abdominal cramps.

Stimulant laxatives are unsuitable for prolonged use because this can lead to excessive loss of potassium ions and a colon that no longer responds to stimulation. They are, however, useful for emptying the bowel before surgery and radiological imaging. They have an effect within 6–12 hours of being given orally. Examples of stimulant laxatives used in this way are bisacodyl and sodium picosulfate.

Some stimulant laxatives, for example bisacodyl, can be given in suppository form for a rapid action within 15 minutes of administration.

In some preparations, a stimulant laxative is combined with a softener or osmotic laxative, for example Picolax. Picolax is commonly used to empty the bowel before imaging.

Preparations recommended for use prior to diagnostic imaging of the abdomen are shown in Table 13.4.

Table 13.4   Preparations used prior to diagnostic imaging of the abdomen

Single preparation	Type of laxative
Magnesium sulphate	Osmotic
Polyethylene glycol (macrogol)	Osmotic
Docusate sodium	Softener and stimulant
Bisacodyl	Stimulant
Sodium picosulfate	Stimulant
Combination preparation	Type of laxative
Sodium picosulfate and magnesium citrate (Picolax)	Stimulant and osmotic

13.7.2.5 Bowel evacuants

Bowel evacuants are known as bowel cleansing solutions. They are used before surgery or radiological procedures to empty the bowel. They can cause bloating and nausea. Examples are magnesium salts and Picolax and polyethylene glycol.

13.7.3 Miscellaneous drugs

13.7.3.1 Buscopan

Buscopan contains hyoscine, which is an anticholinergic drug used to inhibit gastric secretion and intestinal motility during gastrointestinal imaging. Buscopan is given intravenously.

Buscopan produces typical anticholinergic side effects of blurred vision, dry mouth, tachycardia and urinary retention. It should not be used in patients who have glaucoma or an enlarged prostate as it will make both of these conditions worse.

13.7.3.2 Glucagon

Glucagon is a hormone normally produced by α-cells of the pancreas in response to low blood glucose levels. Glucagon normally has a hyperglycaemic effect. However, it also produces smooth muscle relaxation and is used as an adjunct to barium imaging of the gastrointestinal tract. Glucagon is given intravenously. It is more effective than Buscopan and has a shorter duration of action. A disadvantage is that, because it is a protein, glucagon can provoke hypersensitivity reactions.

13.7.3.3 Metoclopramide

Metoclopramide is a dopamine antagonist that stimulates gastric emptying and small intestine transit and is used to speed up transit time of barium follow-through examinations. Metoclopramide is also an antiemetic, which can be useful in radiological examinations and to counteract the effects of therapeutic radiography. The antiemetic effect is due to the blocking of dopamine (D_2) receptors in the chemotrigger zone of the medulla. The chemotrigger zone is sensitive to potentially harmful chemicals in blood.

Adverse effects of metoclopramide include rashes and pruritis.

In high dose, injection of metoclopramide can cause sedation and facial muscle spasms due to effects on dopamine receptors in the brain. This is similar to the adverse reactions seen with antipsychotic drugs (see Chapter 11).

13.7.3.4 Other antiemetics

Domperidone is an alternative to metoclopramide. It too acts on dopamine (D_2) receptors in the chemotrigger zone of the medulla. It does not penetrate other areas of the brain, therefore it is less likely to cause sedation and muscle spasms.

Table 13.5 Miscellaneous adjuncts to radiography

Drug	Action/use
Buscopan (hyoscine)	Inhibits gastric secretion and intestinal motility
Glucagon	Smooth muscle relaxation in intestine
Metoclopramide	Stimulates gastric emptying and small intestine transit
Loperamide	Decreases motility of the intestine

Granisetron is a serotonin antagonist. Serotonin ($5HT_3$) receptors are also found in the chemotrigger zone of the medulla and in the gastrointestinal tract. Stimulation of them plays a role in emesis.

Both domperidone and granisetron are used to counteract emetic effects of therapeutic radiography.

13.7.3.5 *Loperamide*

Loperamide is an opiate drug that decreases motility of the intestine. It does not cross the blood–brain barrier easily and therefore does not have the central nervous system effects (euphoria, respiratory depression, nausea and vomiting and dependence) of other opiates. Loperamide is prescribed to counteract gastrointestinal side effects (diarrhoea) of therapeutic radiography.

Adverse effects of loperamide are constipation, abdominal cramps, drowsiness and dizziness.

Miscellaneous drugs used as adjuncts to radiography are summarized in Table 13.5.

13.7.4 Wound care

Wounds may become infected and need topical antibacterial treatment. Silver sulfadiazine 1% cream is commonly used for this purpose (see Chapter 8). Severe infection with cellulitis requires systemic treatment with an oral antibiotic, for example flucloxacillin or erythromycin.

13.8 Summary

Contrast agents are used to enhance radiological examination of the body. Although air and gas can be used to provide negative contrast, positive contrast agents are most commonly used. These are based on iodine or barium molecules which absorb X-rays well.

Development of intravenous iodine contrast agents has a relatively long history, going back to the 1920s. Modern, water-soluble iodine contrast agents are based on iodinated benzene rings. They have been developed to be as safe as possible. Non-ionic dimers are considered to be less toxic and less likely to produce adverse effects when compared to monomers and ionic contrast agents. This is because they do not easily diffuse out of the blood and they have an osmolality similar to that of body fluids.

Nevertheless, iodine contrast agents must be used with caution in patients with thyroid disease and in those with known hypersensitivity to iodine or allergies to food and other drugs. Premedication with corticosteroids is recommended for such patients. Cautious use of contrast agents is indicated in patients with epilepsy, hepato-renal syndrome, severe respiratory disease and diabetes. They should not be administered to patients who are taking metformin for Type 2 diabetes. Interaction between metformin and iodine contrast agents can increase the risk of lactic acidosis.

Adverse reactions to iodine contrast agents are rare, especially with the newer non-ionic dimers of low osmolality. Adverse reactions range from mild effects, such as nausea and vomiting, to moderate effects, such as erythema and chest pain. Severe adverse reactions, for example bronchospasm and anaphylactic shock, always need emergency treatment.

Nephrotoxicity is a risk with frequent administration of contrast agents. Patients with conditions predisposing to nephrotoxicity and therefore at high risk should be examined with another method of imaging if possible.

Other adverse reactions can be due to the contrast agent binding to body proteins, particularly enzymes.

Radiographers who inject contrast agents must be trained in emergency treatment of severe adverse reactions and basic life support. The most severe adverse reaction is anaphylactic shock. This is treated as an emergency with adrenaline, atropine and oxygen. Antihistamines, aminophylline, salbutamol and corticosteroids may also be necessary.

Adjuncts to radiography include peripherally acting analgesics and possibly codeine for pain after injection or other procedures and topical anaesthetics prior to injection.

For abdominal examinations using barium contrast agents, laxatives and bowel evacuants are necessary to empty the bowel. The most commonly used preparation for this purpose is Picolax, which is a combination of a stimulant laxative and an osmotic laxative.

Buscopan or glucagon is used to inhibit gastrointestinal activity while imaging is taking place. Metoclopramide is used to speed up transit time of barium in follow-through examination

Following therapeutic radiography, antiemetics and loperamide may be necessary to counteract nausea and vomiting or diarrhoea respectively. Wound care may also be needed with topical antibacterials or with systemic antibiotics if severe with cellulitis (see Chapter 8).

Useful websites

www.mhra.gov.uk Medicines and Healthcare Products Regulatory Agency. Use of contrast agents.

http://www.medcyclopaedia.com/library/topics GE Healthcare Medical Diagnostics. Encyclopaedia of medical imaging.

www.radiology.co.uk Scottish Radiological Society. General information on contrast agents.

www.radiologyinfo.org Radiological Society of North America and American College of Radiology. Patient information leaflets.

www.rcrad.org.uk Royal College of Radiographers. Authentication required for entry to web site.

www.sor.org.uk Society of Radiographers. Professional and educational web site.

Case study

This is a case study of a patient who has been assessed for angiography.

Mr Butler is a 55-year-old patient with a history of heart disease. He had a heart attack 10 years ago and was successfully treated with by-pass surgery. Up until quite recently, the patient was relatively well. The present diagnosis is angina (mild on exertion) and high blood pressure controlled by drug therapy. Mr Butler developed type 2 diabetes two years ago.

Mr Butler also has a history of manic depression of about 15 years. He appears to be well stabilized on lithium.

The patient does not work, but has many interests and he tries to get regular exercise in the form of walking. He admits that he possibly drinks too much.

Mr Butler is in for angiography to investigate the cause of his angina and to assess whether he would benefit from angioplasty or other intervention.

Mr Butler's current drug therapy includes the following drugs. You may need to consult the *BNF*.

Amlodipine 5 mg once a day
CoAprovel irbesartan 150 mg and hydrochlorothiazide 12.5 mg once a day
Isorbide mononitrate 60 mg (modified release) one every morning
Bisoprolol 5 mg one every morning
Simvastatin 40 mg two at night
Aspirin 75 mg once a day
Esomeprazole 40 mg once a day
Lithium carbonate 400 mg (modified release) two at night
Metformin 500 mg three times a day with meals

Discuss the treatment of this patient using the questions below as a guide. They are intended to stimulate discussion, not limit debate.

- Determine which drugs are for which condition.

- Is there anything about the combination of drugs that might concern you?

- What questions should you ask the patient prior to administering contrast agent?

- Are there any special precautions that should be taken before administering contrast agent to this patient?

- Would this patient need a local anaesthetic and why?

Chapter review questions

You should be able to answer the review questions below using the information in this chapter.

1. Explain the importance of osmolality in reducing the risk of adverse effects with the use of intravenous contrast agents.

2. Name six conditions in which intravenous contrast agents should be used with caution, and explain why.

3. Give three mild and three moderate adverse reactions to iodine contrast agents. What should be done about them?

4. Anaphylactic shock is a serious adverse reaction. Describe emergency treatment that is necessary if this happens.

5. Briefly describe the four types of laxatives and explain how they work.

6. Explain the actions and uses of Buscopan, glucagon, metoclopramide and loperamide as adjuncts to radiography.

Part III
Prescribing and the law

14

Medicines, the law and health care professionals

14.1 Chapter overview

This chapter deals with the legislation surrounding the sale, supply and use of medicinal products and the prescribing of medicines by members of the health care professions. The Medicines Act 1968 and the Misuse of Drugs Act 1971 are discussed with emphasis on areas relevant to health care professionals and including specific exemptions for health care professionals.

The Medicines Act divides all medicinal products into three categories: general sale list (GSL) items; pharmacy medicines (P); and prescription-only medicines (POM). The Act sets out legal requirements associated with each category of medicine and the administration of medicines to patients, including the use of patient specific directions (PSDs) and patient group directions (PGDs).

The history and development of non-medical prescribing is explained together with consideration of the different forms this can take, for example supplementary and independent prescribing.

The current situation regarding access, supply and prescription of medicines by physiotherapists, radiographers and podiatrists is discussed.

14.2 Legislation

There are a number of pieces of legislation concerned with the sale, supply and use of drugs. The major ones are the Medicines Act 1968 and the Misuse of Drugs Act 1971 plus associated Schedules, Orders, Regulations and European Directives.

These Acts of Parliament deal with substances that are known as medicinal products within the meaning of the Act(s).

Pharmacology for the Health Care Professions Christine M. Thorp
© 2008 John Wiley & Sons, Ltd

14.2.1 The medicines act 1968

The Medicines Act 1968 begins by prohibiting almost all dealings with medicinal products, and then sets out exemptions that allow, for example the manufacture and sale of medicinal products under licence. The Act exempts various activities of professionals, thus allowing doctors to prescribe and pharmacists to dispense.

The law says who can and cannot prescribe medicines. It also allows local arrangements to be developed so medicines can be administered to certain types of patients in certain circumstances, for example PSDs and PGDs (see below). Podiatrists and some other health care professionals are specifically mentioned in the Act as being exempt from some of its provisions. For example paramedics can administer certain named drugs in emergency situations.

Podiatrists and chiropodists can administer (and supply) certain POM, including local anaesthetics.

The Medicines Act divides drugs into three major categories:

1. General sale list (GSL)

2. Pharmacy medicines (P)

3. Prescription-only medicines (POM)

Each category is subject to different legal requirements.

14.2.1.1 General sale list (GSL)

This is a list of all medicines that have been produced under licence or are composed of materials that are exempt from the licensing arrangements.

GSL items may be supplied from premises other than a registered pharmacy, for example a shop or supermarket and by persons other than a pharmacist and without supervision of a pharmacist. For this reason, they are often known as over-the-counter (OTC) drugs.

There are no restrictions on administration or security of storage, neither are there requirements to keep records. There are however requirements as to the labelling of GSL items.

Examples of drugs that are GSL items, subject to quantity and retention of original packaging, are aspirin and paracetamol, antacids and some topical anti-fungal creams.

14.2.1.2 Pharmacy medicines (P)

Pharmacy medicines may only be supplied to the public from a pharmacy or other registered premises by or under supervision of pharmacist.

Pharmacy medicines do not have to be recorded but must comply with labelling requirements.

A large number of medicines fall into this class, unless placed by legislation into either of the other two categories.

14.2.1.3 Prescription-only medicines (POM)

POM are drugs that are specifically referred to in the Prescription Only Medicines (Human Use) Order 1997. This order is regularly amended.

They may only be sold or supplied from a registered pharmacy, by or under supervision of a pharmacist in accordance with the prescription of a doctor, dentist or veterinary practitioner or other qualified prescriber. Nurse independent prescribers, pharmacist independent prescribers and supplementary prescribers can write prescriptions for POM.

Given below are examples of the types of drug that are POM:

- Any drug controlled by the Misuse of Drugs Act 1971 unless it is:

  o codeine, dihydrocodeine, morphine or pholcodeine

  o subject to only one of these being in a product AND that it does not exceed a certain strength

- Certain specific items, for example radiopharmaceuticals

- All products for administration by injection

- All other listed drugs in the order unless there is specific exemption, for example when given by a certain route or when in low concentration.

14.2.1.4 Administration of prescription-only medicines

POM may only be administered by the patient, a practitioner or someone acting in accordance with the directions of a practitioner.

The Prescription Only Medicines (Human Use) Order 1997 allows anyone to administer non-injection (non-parenteral) drugs to the patient for which they are intended and in the manner prescribed.

Normally, drugs for injection can only be administered by the patient or by a medical practitioner. However, in an emergency, when used for the purposes of saving life, certain drugs may be administered parenterally by anyone (see Table 14.1).

In addition ambulance paramedics may administer a range of named injections in emergency situations.

Exemptions are also made to allow the supply and use of POM for research, business and other unusual circumstances.

14.2.1.5 Labelling of POM

All labels for medicines must be indelible, legible and generally in English. POM must also bear specific information. The following information must be written on the POM label:

- Name of patient

- Name and address of supplier

Table 14.1 Drugs for injection. These drugs can be injected by anyone in an emergency

Drug	Use
Chlorphenamine injection and promethazine injection	Antihistamines used in allergic emergency/anaphylaxis
Hydrocortisone injection	Allergic emergency/anaphylaxis
Adrenaline injection	Allergic emergency/anaphylaxis
Atropine sulfate injection	Bradycardia following myocardial infarction
Glucose intravenous infusion	Emergency rehydration
Glucagon injection	Hypoglycaemia with unconsciousness
Dicobalt edetate injection, sodium nitrite injection and sodium thiosulfate injection	Cyanide poisoning
Snake venom antisera	Venomous snakebite
Sterile pralidoxime	Nerve gas/insecticide poisoning

NB the use of 'parenteral' in the Medicines Act is taken to mean drugs given by injection.

- Date of dispensing

- The name of the product, the directions for use and any special cautions, dose, dose frequency and quantity supplied

- 'keep out of reach of children' or similar

- 'for external use' if appropriate

- Any other information the pharmacist considers necessary.

14.2.1.6   Containers

Containers must be clean, sound and fit for the intended purpose. Where possible pharmacies will fit child resistant closures (CRCs) and these must be used for medicines containing aspirin and paracetamol. Fluted bottles must be used for medicines that are for external use only. Patient information leaflets or summary of product characteristics must be included. Their content is set out by law.

14.2.1.7   Patient specific direction (PSDs)

A patient specific direction is the traditional prescription written by a doctor, dentist or other qualified prescriber for medicines to be supplied or administered to a named patient. The majority of medicines are supplied or administered in this way.

A PSD can be a written statement defining the care of a named patient agreed between the doctor and other health care professionals. An example could be the variation of dose range of analgesic in a rheumatology patient by a physiotherapist.

14.2.1.8   Patient group directions (PGDs)

A PGD is a relatively new method (an amendment to the Prescription Only Medicines [Human Use] Order 2000, replacing patient group protocols) by which POM may be supplied or administered without a normal prescription to specified groups of patients. These are written directions made in favour of health care professionals and require the signature of a doctor (or dentist) and a pharmacist. Since 2003, qualified registered podiatrists, dieticians, occupational therapists, orthoptists, paramedics, physiotherapists, prosthetists and orthotists, radiographers and speech and language therapists comprise the list of professionals able to supply medicinal products under PGDs. A PGD can include a variable dose range, so the health care professional can decide on a suitable dose for individual patients. PGDs should only be used where there is clear benefit for patients and by health care professionals with the necessary expertise and competence. PGDs are not suitable where patients need a range of different medicines at the same time.

The use of PGDs must be authorized by the National Health Service (NHS) Trust, Health Authority or Primary Care Trust.

An example of PGD use is in accident and emergency analgesia, where a nurse or physiotherapist would be allowed to administer an analgesic before a patient has been seen by a doctor.

Another example would be the administration of local anaesthetics and anti-inflammatory corticosteroids by intra- and extra-articular injections by physiotherapists involved in the ongoing management of rheumatology patients.

Radiographers have been trained since 1998 to administer select drugs for treatment and management of side effects of radiotherapy under PGDs.

14.2.1.9   Specific exemptions

Some health care professionals, for example podiatrists and chiropodists, midwives and paramedics, are specifically mentioned in the Medicines Act as being exempt from some of its provisions.

The restrictions on the sale and supply of certain medicinal products imposed by the Medicines Act do not apply to registered podiatrists who have successfully completed training, hold a certificate of competence and have the entitlement annotated on the Health Professions Council (HPC) register, providing the following conditions are met:

1. the podiatrist must be registered with the HPC;

2. the sale or supply must be in the course of professional practice as a podiatrist;

3. the product has been manufactured and packed on different premises to where it is sold or supplied.

The medicinal products to which this exemption applies are shown below:

- Any medicinal product for external use that is on the current general sale list

- Any of the following medicinal products all of which are legally classified as Pharmacy Medicines, and are for external use: potassium permanganate crystals or solution, ointment of heparinoid and hyaluronidase

- Any product which contains any of the following substances, providing the stated strength is not exceeded:

  o 9% borotannic complex

  o 10% buclosamide

  o 3% chlorquinaldol

  o 5% diamthazole hydrochloride

  o 1% clotrimazole

  o 10% crotamiton

  o 1% econazole nitrate

  o 1% fenticlor

  o 10% glutaraldehyde

  o 0.4% hydrargaphen

  o 2% miconazole nitrate

  o 10% polynoxylin

  o 2% mepyramine maleate

  o 2% phenoxyporpan

  o 70% pyrogallol

  o 20% podophyllum

  o 70% salicylic acid

  o 0.1% thiomersal

- and recently added (November 2006): 1.0% griseofulvin and 1.0% terbinafine

- Any of the following medicinal products, a most of which are POM:

  o co-dydramol 10/500 tablets where the quantity sold or supplied at any one time does not exceed three days treatment (maximum of 24 tablets)

  o amorolfine hydrochloride cream where the maximum concentration does not exceed 0.25% w/w

  o amorolfine hydrochloride lacquer where the maximum concentration does not exceed 5% w/v

- topical hydrocortisone cream where the maximum concentration does not exceed 1% w/w

- ibuprofen, other than POM preparations, sufficient for three days treatment at a maximum single dose of 400 mg, maximum daily dose of 1200 mg and maximum pack size of 3600 mg

- and recently added (November 2006):

  - silver sulfadiazine

  - amoxicillin

  - erythromycin

  - flucloxacillin

  - tioconazole 28%.

Registered podiatrists who hold a certificate of competence in the use of analgesics approved by the HPC are allowed to administer parenteral analgesics (and some other injectable drugs) in the course of their professional practice.

These substances are POM and would otherwise only be available for use by appropriate medical practitioners. The list of analgesics is given below:

- Bupivacaine hydrochloride

- Bupivacaine hydrochloride plus adrenaline (maximum strength of adrenaline 1 mg in 200 ml)

- Lidocaine hydrochloride

- Lidocaine hydrochloride plus adrenaline (maximum strength of adrenaline 1 mg in 200 ml)

- Mepivacaine hydrochloride

- Prilocaine hydrochloride

- and recently added (November 2006):

  - Levobupivacaine

  - Ropivacaine

- in addition, since November 2006 the following two drugs can also be administered by injection

  - Adrenaline

  - Methyl prednisolone.

There is an online register, published by the HPC, which can be consulted to check if a podiatrist or chiropodist can administer local anaesthetic or supply POM.

14.2.2   Misuse of drugs act 1971

This act controls the manufacture, supply, possession and use of drugs that are dangerous or otherwise harmful. These drugs were known as 'dangerous drugs' but are now referred to as 'controlled drugs'. The use of controlled drugs for medicinal purposes is permitted by the Misuse of Drugs Regulations (2001) and subsequent amendments.

The regulations define classes of people who are authorized to supply and possess controlled drugs during the course of their professional activities.

14.2.2.1   Controlled drugs

Controlled drugs are divided into five Schedules by the Regulations.

See Table 14.2. This is based on therapeutic use and abuse potential.

14.2.2.2   Safe custody

Exhaustive regulations govern the safe custody of controlled drugs (Schedules 1, 2 and 3). Generally controlled drugs are kept in a locked metal, cabinet (with a small number of available keys – in practice only one), which is inside another locked cabinet (which usually contains other drugs).

Table 14.2   Controlled drug schedules

Schedule	Drugs included in schedule
Schedule 1	Drugs of little or no therapeutic use, for example cannabis, lysergic acid diethylamide (LSD)
Controlled Drugs Licence	Possession and supply illegal except by Home Office Licence
Schedule 2	Drugs of high abuse potential with medicinal use, opiates and major stimulants, for example amfetamines and cocaine
Controlled Drugs Register	Subject to full controlled drug requirements under the law
Schedule 3	Drugs of lesser abuse potential with medicinal use for example minor stimulants and barbiturates
Controlled drugs no register	Subject to special prescription requirements, but not other requirements under the law
Schedule 4	Anabolic steroids and related hormones
Part 1 Controlled drugs	Most benzodiazepines and zolpidem
Anabolic steroids	Subject to minimal control requirements
Part 2 Controlled drugs	Sale and supply and possession without a prescription for
Benzodiazepines	personal use an offence
Schedule 5	Drugs of low abuse potential because dispensed or formulated in small amounts, for example low doses of codeine, pholcodeine and morphine (P or POM)
Controlled Drugs Invoice	Only requirement is retention of invoices for two years

Table 14.3 Controlled drug classes

Class	Controlled drugs
Class A	All opiates, hallucinogens, cocaine, injectable amfetamines, cannabinol and coca leaf
Class B	Amfetamines, codeine, pholcodeine and barbiturates
Class C	Milder stimulants and tranquilizers, benzfetamine, benzodiazepines, cannabis and anabolic steroids and related hormones

NB amphetamine is now spelt amfetamine, although the two are synonymous in *BNF*.

14.2.2.3   Destruction

Restrictions on the destruction of controlled drugs apply to those who must keep controlled drugs (Schedule 2) records, for example pharmacists. Individual authorities will have their own codes of conduct or practice in the event of surplus controlled drugs.

These are the minimum legal requirements and they are intended for the safety of all concerned. In addition the Regulations and Orders pertaining to controlled drugs are subject to constant updating and changes. The persons most likely to be best informed of these changes are pharmacy staff.

14.2.2.4   Controlled drug classification

Controlled drugs are further classified according to the degree of danger that misuse of them presents and for determining penalties for offences under the Act (see Table 14.3).

14.3   Non-medical prescribing

Non-medical prescribing is the term applied to prescribing by members of the health care professions who are not 'medically' qualified. Prior to 1994, the only professions allowed to prescribe medicinal products in the United Kingdom were doctors, dentists and veterinary practitioners.

In 1994, district nurses, midwives and health visitors were allowed to prescribe from a limited formulary of dressings, appliances and some medicines until extended formulary nurse prescribing was introduced in 2002. This allowed registered nurses to prescribe from the nurse prescribers' extended formulary, which included treatment for minor ailments, minor injuries, health promotion and palliative care. This formulary was gradually expanded over the next few years and listed in the *British National Formulary* (*BNF*).

Meanwhile, in 1999 a Review of Prescribing, Supply and Administration of Medicines for the Department of Health (1999) recommended two types of prescriber:

- the independent prescriber who would be responsible for assessment of patients with undiagnosed conditions and for decisions about clinical management required, including prescription;

- the dependent prescriber (subsequently called the supplementary prescriber) who would be responsible for the continuing care of patients already assessed by the independent prescriber, which may include prescribing.

Over the next few years, following the Health and Social Care Act 2001, supplementary prescribing by nurses and pharmacists was introduced. After much consultation with the medical, pharmacy and nursing professions and the Department of Health and the Medicines Control Agency as well as meetings of the Committee on Safety of Medicines and the Medicines Commission, amendments were made to the legislation to allow nurses and pharmacists to become supplementary prescribers as from April 2003.

A similar process occurred with the podiatry, physiotherapy and radiography professions and led to changes to the NHS Regulations in April 2005 extending supplementary prescribing to these professions.

Extended formulary nurse prescribing was discontinued in 2006 and replaced by qualified nurse independent prescribing. Nurses can now prescribe any licensed medicine, including some controlled drugs, if they are qualified to do so. At the same time, pharmacists became eligible to train as pharmacist independent prescribers, being able to prescribe all licensed medicines but, as yet, no controlled drugs.

14.3.1　Supplementary prescribing

Supplementary prescribing is a form of non-medical prescribing available to qualified, registered, experienced and suitably trained podiatrists, physiotherapists and radiographers, pharmacists, nurses and midwives.

Prescribing can be defined in the following three ways:

1. to order in writing the supply of a POM for a named patient;

2. to authorize by means of an NHS prescription the supply of any medicine (POM, P or GSL) at public expense;

3. to advise a patient on suitable care or medication including over-the-counter drugs and therefore with no written order.

Supplementary prescribing is currently defined as 'a voluntary partnership between an independent prescriber (a doctor or dentist) and a supplementary prescriber to implement an agreed patient-specific clinical management plan (CMP) (see below) with the patient's agreement'.

Supplementary prescribing is intended to improve patient access to care, make it easier for patients to get the medicines they need, make the best use of clinical skills of health care professionals and enhance professional relationships. Time spent on developing a CMP should save time when the patient returns for review with the supplementary prescriber rather than the doctor.

Supplementary prescribers prescribe in partnership with an independent prescriber who must be a doctor or dentist. They are able to prescribe for the full range of medical conditions, providing they do so under the terms of a patient-specific CMP. Discussion between the independent prescriber and the supplementary prescriber will determine which patients would benefit from supplementary prescribing.

Following diagnosis of the patient by an independent prescriber and following consultation and agreement between the independent prescriber and the supplementary prescriber an individual CMP can be drawn up. This must be agreed with the patient before supplementary prescribing begins.

14.3.1.1 The clinical management plan (CMP)

Templates are available for CMPs, which can be adapted or individuals can develop their own. CMPs should be kept as simple as possible. The list below shows what a CMP must include.

- Name of the patient

- Illness/conditions which may be treated by supplementary prescribing

- Date on which the plan takes effect

- When the plan is to be reviewed by the doctor/dentist

- Reference to the class/description of medicines or types of appliances which may be prescribed or administered under the plan

- Any restrictions or limitations of strength/dose of any medicine which may be prescribed or administered under the plan

- Any period of administration or use of any medicine/appliance which may be prescribed or administered under the plan

- Relevant warnings about known sensitivities of the patient or known difficulties of the patient with particular medicines/appliances

- Arrangements made for notification of suspected/known reactions of clinical significance to any medicine prescribed or administered under the plan, or suspected or known clinically significant adverse reactions to any other medicine taken at the same time as any prescribed or administered under the plan

- Circumstances in which the supplementary prescriber should refer to, or seek the advice of the doctor/dentist.

In addition to the CMP the independent prescriber and supplementary prescriber must share access to a common patient record. Ideally, this would be electronic, but paper records or patient held records can also be used.

The independent prescriber will determine the level of responsibility the supplementary prescriber has under the CMP. This will take into account the experience and expertise of the supplementary prescriber.

The CMP would come to an end at any time at the discretion of either the independent prescriber or the supplementary prescriber or at the request of the patient. It must also be renewed at the time specified for review of the patient or if the independent prescriber changes for whatever reason. Supplementary prescribing must not continue until a new agreement has been made with the patient and a new independent prescriber.

14.3.1.2   Medicines that can be prescribed under a CMP

Any GSL, P or POM medicine (including controlled drugs since July 2006, but only where there is a patient need and the doctor has agreed in a patient's CMP) can be prescribed under a CMP.

This includes the prescribing of:

1. anti-microbials;

2. 'black triangle' drugs and those in the *BNF* described as 'less suitable' for prescribing;

3. drugs used outside their United Kingdom licensed indications;

4. unlicensed drugs.

(Black triangle drugs are those recently licensed and still being monitored for adverse reactions; unlicensed medicines are those not licensed in the United Kingdom.)

Before agreeing to the prescribing of high-risk drugs (that is those with known dangerous side effects) in a CMP, the independent prescriber must be confident that the supplementary prescriber has the necessary skills, knowledge and competence.

A supplementary prescriber should not agree to prescribe any medicine unless they are confident that it falls within their knowledge and competence.

14.3.1.3   Relationship between independent prescriber and supplementary prescriber

A professional relationship between the independent prescriber and the supplementary prescriber is paramount to safe and effective supplementary prescribing.

The two should be able to communicate easily, share access to the same common patient record and any guidelines used in the CMP, agree common understanding of and access to the written CMP and ideally review the patient's progress together at agreed intervals.

The responsibilities of the independent prescriber are given below:

• Initial clinical assessment and diagnosis of the patient

• Agreement with the supplementary prescriber about the limits of their responsibility

• Provision of advice and support to the supplementary prescriber

• Review of patient's progress at appropriate intervals with the supplementary prescriber if possible

- Sharing patient's records with supplementary prescriber

- Reporting of adverse drug reactions.

The responsibilities of the supplementary prescriber are:

- Contributing to development of the CMP

- Prescribing for the patient according to the CMP

- Altering medicines and/or dosages within limits agreed in the CMP if appropriate

- Monitoring and assessing patient's progress

- Working at all times within clinical competence and professional code of conduct

- Recognizing when not competent to act

- Consulting the independent prescriber when necessary

- Accepting professional accountability and clinical responsibility for prescribing

- Passing prescribing responsibility back to independent prescriber as appropriate

- Reporting adverse reactions and inform independent prescriber of them

- Informing independent prescriber of any clinically significant events

- Recording prescribing and monitoring activity in the CMP

14.3.1.4   Training to be a supplementary prescriber

In addition to the responsibilities shown above, the supplementary prescriber must be a registered professional with the HPC and have a minimum of three years professional experience. It is up to individuals to negotiate with their employer that supplementary prescribing should form part of their job. Before becoming eligible to prescribe, they must successfully complete an approved supplementary prescribing training course including all assessments and the period of learning in practice.

The supplementary prescriber must ensure that the supplementary prescribing qualification is recorded on the HPC professional register and have access to continuing professional development thereafter. In the course of prescribing, the supplementary prescriber must enter into a prescribing partnership and agree the CMP with the independent prescriber. They must also arrange for access to prescribing pads and a budget to meet the costs of prescriptions and any other costs.

Individuals selected and trained to be supplementary prescribers must have the opportunity to prescribe in partnership with an independent prescriber on completion of training.

Approved training programmes are at degree level and consist of at least 26 taught days (normally over three to six months and no longer than 12 months) provided by an Institution of Higher Education and at least 12 days learning in practice. The period of learning in practice must be under supervision of a designated medical practitioner who will provide the student with support and opportunities to develop competence in prescribing practice.

An outline curriculum has been developed and agreed with the HPC who also accredits programmes provided by Higher Education Institutions. The pharmacology content of the training courses will include general principles of pharmacology: absorption, distribution, metabolism and excretion (Chapter 2); individual variation to drug therapy, adverse drug reactions, drug–drug interactions, interactions with other diseases and drug interaction with receptors (Chapter 3) and the legal basis for prescribing, supply and administration of medicines, including regulatory aspects of controlled drugs (this chapter).

Employers are expected to recognize the need for private study time and to generally provide support.

Training for supplementary prescribing is now incorporated into nurse and pharmacist independent prescribing as multidisciplinary training.

It is quite likely that legislation will change in the future to allow other health care professionals to train to become independent prescribers.

In summary, supplementary prescribing is likely to be most suitable with patients who have chronic conditions and can be managed by a supplementary prescriber between reviews by the doctor. That is, providing the supplementary prescriber is competent to manage the patient's condition and there is a close working relationship between the independent prescriber and the supplementary prescriber who can share the same patient records.

Supplementary prescribing would not be appropriate in emergency, urgent or acute conditions because a CMP has to be agreed before prescribing can begin.

14.4 Summary

The sale, supply and use of medicinal products are governed by the Medicines Act 1968 and the Misuse of Drugs Act 1971, plus associated Schedules, Orders, Regulations and European Directives. The Medicines Act divides all medicinal products into three categories: GSL items; P; and POM.

The law says who can and cannot prescribe medicines and allows health care professionals to administer and supply medicines to certain patients under PGDs.

Podiatrists are specifically mentioned in the Medicines Act as being exempt from some of its provisions, which means that they can administer and supply certain POM from a specified list providing they are registered and qualified to do so.

Non-medical prescribing is a relatively new means by which members of the health care professions who are not 'medically' qualified can prescribe drugs to patients under certain conditions.

Nurses, midwives and pharmacists can now train to prescribe any licensed medicine as independent prescribers to patients.

Supplementary prescribing is a form of non-medical prescribing open to qualified, registered, experienced and suitably trained podiatrists, radiographers, physiotherapists, nurses and midwives. Supplementary prescribing is currently defined as 'a voluntary partnership between an independent prescriber (a doctor or dentist) and a supplementary prescriber to implement an agreed patient-specific CMP with the patient's agreement'. Supplementary prescribing is intended to improve patient access to the care and medicines that they need. The CMP must be drawn up and agreed with the independent prescriber, the supplementary prescriber and the patient before supplementary prescribing can begin.

Training to be independent prescribers and/or supplementary prescribers is provided by Higher Education Institutions in the United Kingdom according to a curriculum developed and agreed with the HPC.

Supplementary prescribing is likely to be most suitable with patients who have chronic conditions and can be managed by a supplementary prescriber between reviews by the doctor. Legislation may change again in the future to allow other health care professionals, in addition to nurses and pharmacists, to train to become independent prescribers.

Reference

Department of Health (1999) Review of Prescribing, Supply and Administration of Medicines; Final report (Crown II), Department of Health, London.

Useful websites

http://www.bnf.org/bnf/ British National Formulary.

www.chre.org.uk Council for Healthcare Regulatory Excellence. Established April 2003.

www.cks.library.nhs.uk National Library for Health. Clinical knowledge summaries.

http://www.dh.gov.uk/en/PolicyAndGuidance/MedicinesPharmacyAndIndustry/Prescriptions/index.htm. Guidance on Department of Health Prescribing Policy.

http://www.dh.gov.uk/en/PolicyAndGuidance/MedicinesPharmacyAndIndustry/Prescriptions/TheNonMedicalPrescribingProgramme/index.htm. Department of Health Non-medical Prescribing.

http://www.dh.gov.uk/en/Publicationsandstatistics/Publications/PublicationsPolicyAndGuidance/DH064325 Department of Health. Medicines Matters July 2006.

http://www.dh.gov.uk/en/Publicationsandstatistics/Publications/PublicationsPolicyAndGuidance/DH4110032 Supplementary prescribing by nurses, pharmacists, chiropodists/podiatrist, physiotherapists and radiographers within the NHS in England. A Guide for Implementation 2005.

www.hpc-uk.org/ Health Professions Council. Professional registers.

http://www.hpcheck.org/lisa/onlineregister/MicrositesSearchInitial.jsp Online Health Professions Council registers for member status.

www.library.nhs.uk National Library for Health.

www.mhra.gov.uk Medicines and Healthcare Products Regulatory Agency.

www.nmc-uk.org Nursing and Midwifery Council.

www.npc.co.uk National Prescribing Centre.

www.portal.nelm.nhs.uk National Electronic Library for Medicine.

http://www.ppa.org.uk/ppa/edt_intro.htm Prescription Pricing Authority. NHS Drug Tariff for England and Wales.

www.prescribing.info General information about non-medical prescribing (London Metropolitan University).

www.prodigy.nhs.uk Patient information leaflets and guidance on common conditions.

www.rpsgb.org.uk Royal Pharmaceutical Society of Great Britain.

Case studies

The following case studies show situations that could be suitable for drug administration under PGDs or by supplementary prescribing.

The first two case studies are patients who might be seen by a physiotherapist, although other health care professionals might see similar patients for different reasons. The third situation is where a radiographer might administer or prescribe drugs to counteract the effects of therapeutic radiography.

A possible case for supplementary prescribing by a podiatrist is given at the end of Chapter 9, page 175.

Case study 1

This is a case study of a patient who might be seen by a physiotherapist working in a Falls Prevention Service. This is quite a complicated case and you may need to refer to previous chapters and the *BNF*.

A 72-year-old female patient attends the Falls Clinic for investigation into the cause of falls she has been having recently. She has had three falls over the past 18 months and has sustained soft tissue injuries each time and a fractured wrist 12 months ago.

This lady has a 40-year history of depression and is under the care of psychiatrists. She has hypertension controlled by medication.

She has osteoarthritis in her left knee, is under the care of an orthopaedic consultant, and is being considered for total knee replacement.

The patient drinks approximately one to two units of alcohol each evening and is an ex-smoker, having given up four years ago. Until then she smoked 20 cigarettes a day for most of her life.

The patient tells you that she has recently begun to get shortness of breath on walking up hill. Since a shopping expedition at the weekend, the patient has noticed pain and swelling of the left knee.

In addition, the patient has slight impairment of vision and high levels of anxiety, which causes her to rush.

The patient is on the following medication:

- Amitriptyline 50 mg, three tablets at night;
- Diazepam 5 mg as required;
- Bendroflumethiazide 2.5 mg in the morning.

Consider the case and using the following questions as a guide, discuss how the patient could be helped with advice, referral or change in medication.

1. What actions could be taken immediately to help this patient reduce her risk of more falls?
2. What could be done in the longer term?
3. Is there anything about the patient's current medication that might contribute to her risk of falls?
4. What should you do to investigate the shortness of breath?
5. Is there anything you can advise to deal with the shortness of breath?
6. What could be the problem with the left knee and what could be done about it?
7. This lady is subsequently found to be mildly osteoporotic. What could be done to treat this condition?
8. Are there likely to be interactions between any of the patient's medications?
9. Are there any cautions about this patient's medication because of her age?
10. Do you think this is a suitable patient for treatment under PGDs or supplementary prescribing?

Case study 2

The following case study shows a situation where it would be appropriate for a physiotherapist supplementary prescriber to prescribe medication for a patient.

A 50-year-old male patient has been referred to a chronic pain clinic with a six month history of low back pain and leg pain for pain management. The patient was triaged to the physiotherapy practitioner for assessment for pain management.

Surgery has been excluded for this patient by an orthopaedic surgeon.

The patient was diagnosed by the physiotherapy practitioner as having nociceptive low back pain and neuropathic leg pain, which was confirmed by very mild nerve root compression as seen on magnetic resonance imaging (MRI) scan. He scored 4/10 for low back pain and 6/10 for his leg pain, which also scored positive on the painDETECT questionnaire for neuropathic pain. The patient was otherwise fit and well, but has not worked for six months. The patient was complaining of constant pain and gastric irritation.

The patient's current medication is:

- Co-codamol 8/500, up to a maximum of eight tablets per day;

- Diclofenac, maximum 150 mg per day.

The patient's pain management plan included listing for epidural therapy; rationalization of his medication; and referral to a concurrent pain management programme.

As part of the pain management plan was to rationalize medication while the patient waited for his epidural therapy, a consultant pain physician was called to confirm diagnosis.

It was agreed between the physiotherapist, the patient and consultant that the physiotherapist would manage his medication according to a predefined CMP.

- Discuss what would go into the CMP for this patient.

- Discuss how the physiotherapist could advise the patient and how the physiotherapist might alter the patient's medication within the CMP.

- Suggest a prescription that the physiotherapist could write for this patient.

- Apart from the consultant pain physician, who else should be informed of the CMP and prescription?

- When and by whom should the patient be reviewed?

Case study 3

A patient has been receiving abdominal radiotherapy for a period of time for the treatment of colon cancer. She has suffered quite serious nausea as a result of the treatment. You have supplied metoclopramide (in 28 packs of 10 mg tablets) to the patient on a regular basis under PGDs. The patient is now complaining that this drug does not seem to work any more. She has also been experiencing facial muscle spasms and wonders if this could be a side effect of her radiotherapy or the drug. In addition, the patient has developed a painful skin rash over the treatment area.

The patient has Parkinson's disease for which she takes amantadine.

Discuss this patient's case and consider whether you can suggest an alternative to metoclopramide and a suitable treatment for the skin rash. You should consider why the patient is experiencing facial muscle spasms and whether this would have any bearing on what can be prescribed for her. Do you think this patient could be suitable for supplementary prescribing with a patient management plan agreed between the patient, the oncologist and the radiographer?

Chapter review questions

1. The Medicines Act 1968 divides all medicinal products into three classes. Briefly describe each one.

2. Explain what is meant by PSDs.

3. Explain what is meant by PGDs.

4. List the nine health care professionals who are allowed to supply medicinal products under PGDs.

5. What are the conditions that must be satisfied in order for a podiatrist to legally access and supply certain medicinal products from a list specified in the Medicines Act 1968?

6. What is the Misuse of Drugs Act 1971 intended to prevent?

7. Define the term 'non-medical prescribing'.

8. What is the essential difference between independent prescribing and supplementary prescribing?

9. Which health care professionals are currently allowed to train as supplementary prescribers?

10. What conditions must be met in order for them to train to be supplementary prescribers?

15

Prescribing in practice

15.1 Chapter overview

Legislation surrounding the sale, supply and use of medicines is discussed in Chapter 14. Health care professionals are increasingly being involved in the administration and prescription of medicines to patients. This chapter is intended to illustrate 'prescribing in practice' by podiatrists, radiographers and physiotherapists and has been written largely by members of those professions. Given that prescribing can be considered to include advising a patient on suitable care or medication including over-the-counter drugs as well as the more familiar written orders or prescriptions, there is considerable scope for health care professionals to be involved in patient medication. There are in fact five ways in which suitably qualified and registered health care professionals can supply medicines to patients: patient group directions (PGDs), patient specific directions (PSDs), supplementary prescribing, independent prescribing and specific exemptions to the Medicines Act. The first three of these require a working partnership with medically qualified professionals; the last two do not.

Two sections presented here have been written by practitioners in podiatry and radiography. In addition, practitioners in podiatry and physiotherapy have made verbal contributions to this chapter. Collectively they have described the use of various forms of access, supply, administration and prescription of medicines in their professions today and considered future developments in the light of the recent legislation allowing pharmacists and nurses to train as independent prescribers. Hopefully, this will give the reader a realistic view of what is currently happening and what might happen in non-medical prescribing.

15.2 Podiatry

The following section has been written by Anthony Waddington, FCPodS BSc (Hons) HND, Podiatric Surgeon with Herefordshire Primary Care Trusts (PCTs) and Lecturer Practitioner at the University of Salford.

Pharmacology for the Health Care Professions Christine M. Thorp
© 2008 John Wiley & Sons, Ltd

15.3 Extension of access to prescription-only medicines in podiatry and podiatic surgery

15.3.1 Glossary of terms

15.3.1.1 Podiatrist

A podiatrist has studied for three years to obtain a degree in podiatric medicine and Registration. Podiatrists are independent clinicians, qualified to diagnose and treat foot problems. They may specialize in particular areas of work, for example diabetology, rheumatology or sports medicine. With the exception of nail surgery, podiatrists undertake the treatment of foot problems by non-invasive methods (until recently podiatrists were known as chiropodists).

15.3.1.2 Podiatric surgeon

A podiatric surgeon is a non-medically qualified specialist in the treatment of all foot problems by both surgical and non-surgical methods. Podiatric surgeons qualify as podiatrists initially and then train for a further five years to specialize in the surgical management of foot problems. Podiatric surgeons are not doctors (that is Registered Medical Practitioners). Podiatric surgeons have specialized throughout their training, purely in the treatment of one area of the body, the foot and ankle. The podiatric surgeon has a special insight into the management of foot problems because of a unique training and background. This is combined with an especially close cooperative approach to the management of foot pathology, which includes general practitioners (GPs) and other medical specialists, other podiatrists, physiotherapists and crucially the patients themselves in working towards their cure or improvement.

The Health Professions Council (HPC) is the statutory body responsible for the registration of podiatrists in the United Kingdom. Podiatric surgeons are a specialist group of podiatrists registered by the HPC. In order to practice podiatric surgery in the United Kingdom, a podiatric surgeon must complete a Bachelor of Science Degree in Podiatry/Podiatric Medicine at one of the approved universities and be registered with the HPC. In addition, a podiatric surgeon must have obtained a Masters Degree in Podiatric Surgery (or equivalent, including a pharmacology module at masters level) and been successful in the Part D Fellowship in Podiatric Surgery examination.

15.3.1.3 Podiatric surgery

Podiatric surgery is a specialized form of foot surgery performed by podiatric surgeons.

15.3.1.4 Patient group directions (PGD)

PGDs are written instructions for the supply or administration of medicines to groups of patients who may not be individually identified before presentation for treatment.

15.3.2   The purpose of prescribing

15.3.2.1   Formulary development

The formulary of drugs appropriate in the management of podiatric conditions is limited. The range of drug classes requested by podiatrists in the United Kingdom and prescribed by podiatrists in the United States of America in podiatric general practice are anti-microbials, local anaesthetics, non-steroidal anti-inflammatory drugs (NSAIDs), analgesics, corticosteroids and anxiolytics/sedatives.

It should be noted that podiatrists in the United States of America have had prescribing rights for a full range of prescription-only medicines (POM) since the mid-1990s. Even with access to a large range of drugs, the podiatric drug requirement for patient management in the United States of America appears to be limited.

15.3.2.2   Enhanced patient outcomes

The role of the podiatric surgeon is to provide specialist management of foot pathology by surgery (where appropriate) and alternative measures, either directly or by referral to podiatrists or other healthcare professionals. In the United Kingdom, podiatrists have had access to a range of POMs by means of an exemption order (since 1980 for certain parenterally administered local anaesthetics) for administration, sale or supply (and recently, November 2006, extended). These medicines include injectable local anaesthetic drugs, a number of antibiotics and some NSAIDs.

15.3.2.3   Professional indemnity

Indemnity provision is provided via the Society of Chiropodists and Podiatrists and has been modified in view of the extended list of POMs. Discussions with the Society's indemnity insurance provider indicated that no issues with indemnity were apparent.

15.3.2.4   National and international trends

Originally, the Medicines Act 1968 and associated secondary legislation allowed only doctors and dentists to write prescriptions for POM. The Prescription Only Medicines (Human Use) Order 1997, known as the POM Order, contains some specific exemptions that allow for the sale or supply and administration of certain POM directly to patients, without the directions of a doctor or dentist. These exemptions, which continue to apply, relate to midwives, ambulance paramedics, optometrists, and podiatrists and chiropodists.

The desire to increase access to pharmacotherapeutic treatment available to podiatrists and podiatric surgeons reflects national and international experience that a wider group of health professionals with direct access to POM results in benefits to individual patients and the community. In fact, podiatry (reflecting the work of podiatric surgeons) was specifically mentioned in the Crown Report (Department of Health, 1999) as one of the first groups thought to be suitable for extension of supplementary prescribing.

15.3.3   Patient group directions (PGDs)

Following a Department of Health report in 1998 on the Supply and Administration of Medicines under Group Protocols, group protocols became known as PGDs. PGDs are written instructions for the supply or administration of medicines to groups of patients, who may not be individually identified before presentation for treatment. The report recommended that the legal position needed clarification, and in August 2000 the relevant medicines legislation was amended. The majority of clinical care should still be provided on an individual, patient-specific basis. The supply and administration of medicines under PGDs should be reserved for those limited situations where this offers an advantage for patient care without compromising patient safety, and where it is consistent with appropriate professional relationships and accountability.

It may be necessary or convenient for a patient to receive a medicine (that is, have it supplied and/or administered) directly from a health care professional other than a doctor. Unless already covered by exemptions to the Medicines Act (see Chapter 14), there are two ways of achieving this: by PGD or by patient specific directions.

In essence, a PGD allows a range of specified health care professionals, including podiatrists, to supply and/or administer a medicine directly to a patient with an identified clinical condition without them necessarily seeing a medical prescriber. So, patients may present directly to health care professionals using PGDs in their services without seeing a doctor. Alternatively, a doctor may have referred the patient. However the patient presents, the health care professional working within the PGD is responsible for assessing that the patient fits the criteria set out in the PGD (see Appendix C for examples). In general, a PGD is not meant to be a long-term means of managing a patient's clinical condition. This is best achieved by a health care professional prescribing for an individual patient on a one-to-one basis (that is by PSD).

An example of a service using PGDs, which was cited in the Crown Report (Department of Health, 1999), is Podiatric Surgery. For example following day-case foot surgery, a podiatric surgeon can use a PGD to give a patient a supply of NSAIDs for post-operative pain.

Podiatrists have gained considerable experience with prescription-only drugs such as anti-microbials, NSAIDs, injectable corticosteroids, anti-emetics and other medicines by working collaboratively with a patient's general medical practitioner to obtain the suitable prescription drug or by means of PGDs.

15.3.3.1   Limitations of PGDs

Not all podiatrists and podiatric surgeons work exclusively in the National Health Service (NHS) and initiation of appropriate drug therapy by means of PGDs can be delayed while arrangements are made for GP consultation. At present, even in the NHS, with good cooperation of all concerned there can be delays in providing necessary POM, due to problems in liaison or communication. For some patients this delay is a mere inconvenience, but for patients who are surgical cases, delays in the provision of treatment for infection present a significant concern. These delays can have serious consequences, such as the development of septicaemia and amputation. Patients with significant pain also deserve rapid relief to minimize their morbidity. The best practice is for the clinician with the direct responsibility for the patient, if suitably trained, to be able to initiate immediately the required therapeutic regime. For podiatrists and podiatric surgeons to have access to a relevant range of POM in common use, as part of their overall surgical and non-surgical patient management, will minimize delays in the commencement of treatment and reduce delays in providing diagnostic or therapeutic interventions.

15.3.3.2 *The law and patient group directions*

The relevant modifications to the Medicines Act 1968 are contained in the following pieces of legislation: the Prescription Only Medicines (Human Use) Amendment Order 2000; the Medicine (Pharmacy and General Sale – Exemption) Amendment Order 2000; and the Medicines (Sale and Supply) (Miscellaneous Provisions) Amendment (No2) Regulations 2000.

These modifications apply to the NHS, including private and voluntary sector activity funded by the NHS. Therefore, treatment provided by NHS Trusts, PCTs, Health Authorities (including Strategic Health Authorities), GP or dentist practices, walk-in centres and NHS funded family planning clinics are all included. In April 2003, further legislation was passed to cover non-NHS provision. Independent hospitals, agencies and clinics registered under the Care Standards Act 2000; prison health care services; police services and defence medical services can now use PGDs.

The PGD must be signed by a senior doctor (or a dentist) and a senior pharmacist, both of whom must be involved in developing the direction. In addition, the relevant body, as set out in the legislation, must authorize the PGD. In practice, Clinical Governance Leads are likely to be appropriate, but they can only do so as named individuals. The information that must be contained in all PGDs is listed below:

- Name of the business to which the direction applies;
- Date the direction comes into force and the date it expires;
- Description of the medicine(s) to which the direction applies;
- Class of health professional who may supply or administer the medicine;
- Signature of a doctor (or dentist) and a pharmacist;
- Signature on behalf of an appropriate organization;
- Clinical condition or situation to which the direction applies;
- Description of those patients excluded from treatment under the direction;
- Description of the circumstances in which further advice should be sought from a doctor (or dentist), and arrangements for referral;
- Details of appropriate dosage and maximum total dosage, quantity, pharmaceutical form and strength, route and frequency of administration, and minimum or maximum period over which the medicine should be administered;
- Relevant warnings, including potential adverse reactions;

- Details of any necessary follow-up action; and the circumstances;

- Statement of the records to be kept for audit purposes.

Examples of PGDs relevant to podiatric surgery are shown in Appendix C.

15.3.4   Exemptions to the medicines act

Podiatrists are already exempt from certain requirements of the Medicines Act. These exemptions allow them to administer or supply certain specified medicines listed in Chapter 14 without the directions of a doctor. This arrangement continues and is not affected by provisions for PGDs.

The list of medicines available to podiatrists was extended in November 2006. Since extended prescribing and supplementary prescribing (see below) are relatively new and because PGD usage is now integral to practice in many organizations, confusion has arisen about the difference between them and which is the most appropriate in which circumstances. While the definition of a PGD is straightforward, local interpretation has allowed them to be used creatively in a wide range of circumstances. It is important that all podiatrists and their employers, understand the scope and limitations of PGDs as well as the wider context into which they fit when designing safe, effective services for their patients.

In summary, PGDs are most suitable when the use of medicines follows a predictable pattern and is not tailored for individuals with complicated conditions. They are generally most appropriate to manage a single treatment episode or series of episodes of the same condition. Nevertheless, an appropriate person should be identified within any healthcare organization to ensure that only fully competent, qualified and trained health care professionals are allowed to use PGDs.

15.3.5   Supplementary prescribing

The working definition of supplementary prescribing is 'a voluntary partnership between an independent prescriber (a doctor or dentist) and a supplementary prescriber to implement an agreed patient-specific Clinical Management Plan (CMP) with the patient's agreement'.

Supplementary prescribing has its basis in the recommendations of the final report of the Review of Prescribing, Supply and Administration of Medicines (Department of Health, 1999), which recommended that there be two types of prescriber:

1. the independent prescriber who would be responsible for the assessment of patients with undiagnosed conditions and for decisions about the clinical management required, including prescribing

2. the dependent prescriber who would be responsible for the continuing care of patients who have been clinically assessed by an independent prescriber.

This continuing care might include prescribing, which would usually be informed by clinical guidelines and be consistent with individual treatment plans, or continuing established treatments by issuing repeat prescriptions, with the authority to adjust the dose or dosage form according to the patients' needs. The Review recommended that there should be provision for regular clinical review by the assessing clinician (note: dependent prescriber is now referred to as a supplementary prescriber, see Chapter 14).

15.3.5.1 *Legal basis of supplementary prescribing*

Following the Health and Social Care Act 2001 prescribing responsibilities were extended to a wider group of health care professions. New types of prescriber were recognized, including the concept of a supplementary prescriber. Originally, in April 2003 amendments to the Prescription Only Medicines Order and NHS regulations allowed supplementary prescribing by suitably trained nurses and pharmacists. Further amendments to the Prescription Only Medicines Order and NHS regulations in April 2005 extended the opportunity to train as supplementary prescribers to podiatrists, physiotherapists and radiographers.

15.3.5.2 *The aim of supplementary prescribing*

Supplementary prescribing is intended to improve patient care by providing quicker and more efficient access to the medicines they need and to make the best use of the clinical skills of experienced health care professionals.

Supplementary prescribing is designed for the medium to longer-term management of individuals by health care professionals other than doctors. According to the Department of Health definition, a supplementary prescriber 'forms a voluntary partnership with an independent prescriber (a doctor or dentist)'. A CMP is agreed with individual patient and the supplementary prescriber manages the patient's clinical condition, including prescribing, according to the CMP.

Supplementary prescribers have few restrictions on the medicines that they can prescribe. As long as the medicines can be prescribed by a doctor (or dentist) at NHS expense and are referred to in the patient's CMP, they can be prescribed.

Supplementary prescribers prescribe in partnership with a doctor (or dentist). Health care professionals (podiatrists, physiotherapist and radiographers) are able to prescribe all medicines, including unlicensed medicines and controlled drugs since July 2006. All supplementary prescribers may prescribe for the full range of medical conditions, provided that they do so under the terms of a patient-specific CMP and within the limits of their competence.

Supplementary prescribing may only commence following assessment and diagnosis by the independent prescriber and the development of a written CMP agreed between the independent prescriber, the supplementary prescriber and the patient. The independent prescriber is responsible for the initial diagnosis of a patient and the contents and conditions of their CMP, although they need not write it personally. Within the limits of a CMP, the supplementary prescriber may be able to alter the choice of medicine, the dosage, the frequency of administration and other variables. As an alternative to listing medicines individually, the CMP may refer to recognized and accepted local or national clinical guidance in written or electronic form. Any guidelines referred to should be easily available to all concerned in the patient's care (see Chapter 14 for information that a CMP must include).

Supplementary prescribing must include arrangements for regular clinical review of the patient's progress by the independent prescriber, at appropriate predetermined intervals, depending on the patient's condition and the medicines prescribed for them. The intervals should be no longer than one year apart and more frequent if antibiotics are included in the CMP. The independent prescriber may, at any time, review the patient and/or resume full responsibility for their care. For the safety of patients, it is essential that the independent prescriber and the supplementary prescriber use and share access to the same common patient record and jointly keep it up to date. The key to safe and effective supplementary prescribing is the relationship between the individual independent prescriber and the individual supplementary prescriber (see Chapter 14).

There are likely to be many nurses, pharmacists and other health care professionals who meet the criteria for training to become supplementary prescribers. The four key principles that should be used to prioritize potential applicants are:

1. patient safety;

2. maximum benefit to patients and the NHS by provision of quicker and more efficient access to medicines for them;

3. improved quality of care;

4. better use of professionals' skills.

All health care professionals who are involved with supplementary prescribing will require continuing professional development (CPD). Practitioners have a responsibility to keep themselves up to date with clinical and professional developments. Since 2005, podiatrists (and others) have also had to meet the requirements of the Standards for Continuing Professional Development of the Health Professions Council. This is a self-declaration, which has to be kept current with practice within an individual's practice. It is subject to periodic audit, requiring the submission of evidence of CPD to the HPC. Employers should allow practitioners access to relevant education and training. The Department of Health has commissioned CPD support for supplementary prescribers through the National Prescribing Centre (see www.npc.co.uk).

In light of the recent extension to the exemptions list with regards to local anaesthetics, it has also been highlighted that podiatrists and podiatric surgeons also require keeping up to date with their resuscitation skills, with immediate life support being the minimum level.

For further details of how supplementary prescribing will work and information about which health care professionals can undertake supplementary prescribing, which medicines can be prescribed under supplementary prescribing arrangements and training, see Chapter 14.

Acknowledgements

I would like to take this opportunity to thank Mr Thomas Galloway, Consultant Podiatric Surgeon, Hereford PCT for his input in helping me produce this section.

References

Department of Health (1999) Review of Prescribing, Supply and Administration of Medicines (Crown Report 2), Department of Health London.

http://www.dh.gov.uk/en/PolicyAndGuidance/MedicinesPharmacyAndIndustry/ Prescriptions/TheNon-MedicalPrescribingProgramme/index.htm Department of Health Non-medical Prescribing.

http://www.mhra.gov.uk/home Statutory Instrument 2005 No. 766 The Medicines (Pharmacy and General Sale – Exemption) Amendment Order 2005.

www.npc.nhs.uk National Prescribing Centre.

www.scpod.org The Society of Chiropodists and Podiatrists.

15.4 Podiatry in the community

The following account illustrates the work of a health care professional using supplementary prescribing in the community.

The healthcare professional is a consultant podiatrist in diabetology, who is employed by a PCT in the north-west of England and she is one of the first podiatrists in the country who trained to be a supplementary prescriber. This consultant podiatrist is using supplementary prescribing in primary care, mostly with older patients many of whom are living in nursing homes. She prescribes from a personal formulary of drugs within her own area of competence. In practice, this means that the majority of her prescribing is for antibiotics and dressings for the treatment of infected diabetic foot ulcers. The consultant podiatrist is a member of a rapid access referral team for the treatment of patients with serious skin ulcers. As a team, they use Scottish Intercollegiate Guidelines Network (SIGN) guidelines for antibiotic therapy. Members of the team are also responsible for drug monitoring for adverse effects and drug–drug interactions and, if necessary, referral of patients for liver and renal blood tests. The clear advantages of such a scheme are that patients benefit from treatment at the point of access, can have the treatment immediately and can be given a prescription without the need to refer back to their GP. Initial referral from GPs of these high-risk patients includes the drawing up of pre-arranged CMPs, with the patients' consent, before consultation with the podiatrist. This arrangement saves time and for the patient can mean the difference between recovery and amputation through development of gangrene, which can be rapid. The consultant podiatrist is intending to extend her area of competence in the future to include prescription for smoking cessation and the treatment of peripheral vascular disease with vasodilators. These are important treatment options for diabetic patients, which could offer patients a better quality of life as well as reducing the risk of amputation.

Supplementary prescribing is still new, and practitioners are still finding ways of working out procedures to fit into existing structures, like the development of pre-arranged CMPs described above. There is a non-medical prescribing network in the north-west, members of which are working collaboratively to address these issues.

Since the law changed in 2005 allowing more health care professionals to train to become supplementary prescribers, low numbers of podiatrists, physiotherapists and radiographers have so far taken up the opportunity in the north-west. This may change in the future.

It seems that, in podiatry at least, supplementary prescribing is most useful for long-term management of disease in the community, rather than in hospital. To a certain extent the usefulness of non-medical prescribing depends on the way in which individuals in community podiatry and podiatry in general work. Exemptions (to the Medicines Act, see Chapter 14) may be most appropriate for some practitioners; PGDs or supplementary prescribing and possibly independent prescribing in the future may be better for others. There is a feeling in the podiatry profession that ultimately independent prescribing would allow them to provide the best care in many situations.

The clear advantages of supplementary prescribing for a consultant podiatrist working in the community now are that a referral system has been set up with the GPs of vulnerable patients in a local area; the patients receive timely and appropriate access to urgent treatment; there is improved compliance for many patients; management of serious foot ulcers follows published protocols; drug monitoring for adverse effects and drug–drug interactions is the responsibility of the prescribing podiatrist; and the patients are likely to need fewer episodes of care.

15.5 Radiography

Peter Hogg (Nuclear Medicine), Geraldine Francis (Radiography), Christina Freeman (Society and College of Radiographers), Dianne Hogg (East Lancashire PCT), Vera Mountain (Superintendent Radiographer), Andy Pitt (Advanced Practitioner), Sam Sherrington (NHS North-west)

The following paper was first published in *Synergy Imaging and Therapy Practice*, December 2007 (pages 26–31) and is reproduced here with kind permission from The Society and College of Radiographers.

15.6 Medicines in radiography: prescription, supply and administration

What is the current position for prescribing, supplying and administering medicines in radiography? What direction will this take in the future? These questions are answered below.

Please note that the comments in this section of the chapter only relate to practice within the NHS; private healthcare is not considered here.

15.6.1 Introduction

The development of non-medical prescribing, from recognition of need to widespread availability, took several years. In Baroness Julia Cumberlege's report, Neighbourhood Nursing (Department of Health and Social Security, 1986), she noted that patient care was often complicated by the nurse's inability, following a full assessment, to prescribe the evidence-based treatment for the GP to follow-up.

A working party headed by Dr June Crown produced a report (Department of Health, 1989) outlining who should prescribe, what they should prescribe and how it should be funded. In the first instance, those nurses with a district nurse or health visitor qualification, working in primary care, were identified to prescribe from a limited formulary designed around common areas of practice. This was to be funded from existing prescribing budgets and, as such, should be substitute prescribing not additional, as still applies today.

Legislation was passed in 1992 (Medicinal Products, 1992), amending the 1968 Medicines Act, permitting nurses to prescribe. Subsequently, pilot studies were implemented across England and Wales, addressing local and regional need, and the position today is that some PCTs are approaching the point of having more non-medical prescribers (for example nurses) than medical prescribers (doctors).

Dr June Crown's working party's second report (Department of Health, 1999) considered who else should be able to take on prescribing responsibilities. The expectation was that extending prescribing would optimize the use of resources, enhance professional relationships and improve patient access to care. The report defined mechanisms for what were to become supplementary prescribing and independent prescribing, and recommended that other groups of professionals should be able to legally prescribe. A process was outlined by which the professional bodies might make a case for their members and recommendations were also made about what was to become PGDs.

15.6.2 Definitions

Arising from the two Crown reports, there are currently four ways in which non-doctors/ non-dentists can prescribe/supply/administer medicines (there are exemptions not covered here).

15.6.2.1 Patient group direction (PGD)

The majority of prescription, supply and administration of medicines should be provided on a named patient basis. Under a PGD (National Health Service, 2000), the supply and administration of medicine should apply only when this route offers advantages to the patient without compromising safety. Only specific professional groups can work within a PGD, including radiographers, paramedics, nurses, podiatrists, physiotherapists, speech therapists and pharmacists. If a non-permitted professional group uses a PGD then they will break the law. The mechanism for creating and approving a PGD is rigorous, requiring formal input from the Trust as a whole (through Trust medicines management processes). The Department of Health provides templates that can be helpful (www.portal.nelm.nhs.uk/PGD/default.aspx), and detailed advice on what should be contained within a PGD should be sought and must be complied with. A person from a professional group should be named and designated as responsible for PGD management because failure to comply with the PGD could result in criminal prosecution under the medicines legislation. It should be noted that restrictions apply to the applications of PGDs – one will be highlighted later in this text (radiopharmaceuticals).

15.6.2.2 Patient specific direction

This is a mechanism by which an independent prescriber indicates that a patient should be given a particular medicine (indicating amount, route of administration, etc). It can be achieved by writing an instruction within the patient case notes or, for example by a doctor indicating on an X-ray request form that particular medication should be given as part of the examination.

15.6.2.3 Independent prescribing

This is traditionally undertaken by doctors and dentists. In 2002, nurses were given independent prescribing rights and, more recently, legislation has been passed to allow for pharmacists to be independent prescribers (www.dh.gov.uk/en/PolicyAndGuidance/). Independent prescribers take responsibility for: the clinical assessment of the patient; establishing a diagnosis; indicating the clinical management required and prescribing where necessary. They prescribe medicines from the British National Formulary (BNF) (or part thereof) in accordance with their professional competency.

15.6.2.4 Supplementary prescribing

Supplementary prescribing is defined as 'a voluntary prescribing partnership between an independent prescriber and a supplementary prescriber to implement an agreed patient-specific CMP with the patient's agreement' (Department of Health, 2006).

There are a number of criteria that must be met for supplementary prescribing to occur (www.dh.gov.uk/en/Publicationandstatistics/). For instance, the independent prescriber must be a doctor or dentist; the supplementary prescriber must be from a professional group that is legally allowed to be a supplementary prescriber (such as radiographer, nurse, pharmacist, chiropodist, physiotherapist, optometrist); and there must a written CMP relating to a named patient and their specific condition(s).

Supplementary prescribing is competency led, so there are no restrictions – other than the skills and knowledge of the practitioner – on the clinical conditions which supplementary prescribers may treat. Consequently, there is no specific formulary or list of medicines for supplementary prescribing. Provided the medicines are prescribable by a doctor or dentist at NHS expense, and are referred to in the patient's CMP, supplementary prescribers are able to prescribe from the complete *BNF*.

Supplementary prescribing requires a prescribing partnership and a CMP for the patient before it can begin, and hence it is likely to be most useful in dealing with long-term medical conditions such as asthma, diabetes, coronary heart disease and thrombosis management. The processes of setting it up are time intensive, and would need to be off-set against the benefits gained. However, it will be for the independent and supplementary prescribers to decide when it is appropriate.

15.6.3 Historical context for radiography

15.6.3.1 Diagnostic

Nuclear medicine was likely to be the first field in which there was widespread uptake of medicines administration by radiographers. Only doctors or dentists were (indeed still are) legally permitted to practice this domain of medicine and legislation (Medicines Regulations, 1978) indicates that any person acting under their direction may do so, within the bounds of radioactive product certification. 'Acting under the direction of the dentist or doctor clinically responsible for nuclear medicine' allowed for non-doctors to administer radiopharmaceuticals, but, strictly speaking, this did not apply to the administration of adjunct medicines (such as diuretics).

Figure 15.1 demonstrates the growth of radiographers performing intravenous (IV) injections.

It is highly probable that the level of activity shown on the graph until the late 1980s is due to IV administrations associated with nuclear medicine procedures. Anecdotally, it

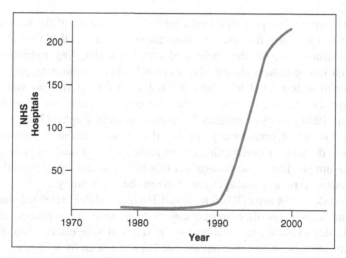

Figure 15.1 Intravenous administrations by radiographers.
Adapted from Price, R.C., Miller, L.R. and Mellor, F. (2002), Longitudinal changes in extended roles in radiography. Radiography, 8 223–234, with permission from Elsevier

is known that, by 1990, radiographers were starting to administer other medicines, IV or otherwise, but it is likely that such practices were not within the then current legislation.

Changes to legislation permitted the adoption of new roles by radiographers and, during the 1990s, the supply and administration of medicines increased rapidly – the most notable being IV administrations of contrast agents for urinary tract investigations. By the early-mid 1990s, peer-reviewed literature (Loughran, 1993) had started to document the medicines administration role of radiographers. This was supported by professional body literature, which continues to be produced today (Society of Radiographers, 2001 and Society of Radiographers, 2003).

Alongside the clinical demand for radiographers to administer medicines came demand for university and hospital based education and training programmes to meet the training need. In this respect, the most popular courses were those approved by The College of Radiographers (Society and College of Radiographers, 2005a) as having met the requirements laid out in Course of Study for the Certification of Competence in Administering Intravenous Injections.

Administering medicines, especially by IV routes, was therefore considered a post-registration activity. The turn of the millennium brought with it a revision of undergraduate curricula and the inclusion of formal post-registration clinical competencies. Many undergraduate radiographer curricula now cover aspects of medicines management, and some address in detail the theory and practice of IV injections (www.qaa.ac.uk/).

15.6.3.2   Therapy

The practice and attitudes of therapeutic radiographers have always placed the patient experience at the centre of care and responsibility. In 1995, this philosophy gained strategic importance with the publication of the Calman Hine Report (Department of Health, 1995).

This importance was, in part, due to the report's recognition of the need for seamless care, because it was noted that therapy radiographers were well placed to enhance the delivery of seamless care in the support of patients undergoing radiotherapy. It also recognized that radiographers should play a central role in monitoring patients' physical and psychosocial welfare and in identifying the need for referral, as well as ensuring that routine reviews were undertaken during the course of external beam radiotherapy. Throughout the 1980s, many experienced radiotherapy radiographers became increasingly aware of the amount of unnecessary waiting that patients endured because of lack of timely access to doctors for prescription and/or review of their medicines for radiotherapy toxicity management. They also recognized that they had untapped potential and could contribute more to care and management if given the opportunity.

Published work in this area (Westbrook and Hodgetts, 1997) revealed that the role of the doctor (oncologist) was often unnecessary in the patient review process and, out of 79 recorded activities to obtain medication for the relief of side effects from radiotherapy, there was only one case where the patient needed to be seen by a doctor.

For therapy radiographers, supply and administration of medicines is not limited to the adverse consequences of radiotherapy. These radiographers have a significant involvement in the production of images that inform radiotherapy planning and a growing number of these imaging examinations require the use of X-ray contrast agents. As such, since the 1980s, therapy radiographers have become more involved with the administration of such medicines as part of the pre-treatment imaging processes.

15.6.4   Medicines management

The use of medicines is not always optimized, so there can be waste and patients can suffer harmful effects. Consequently, Trusts have medicines management policies in place to address issues such as:

- managing risk;

- minimizing potential of harm;

- maximizing the use of advanced competencies;

- improving the patient experience;

- maximizing cost-effective use of resources.

Medicines management encompasses all aspects of the use of medicines in NHS Trusts, and includes the selection, procurement, delivery, prescription, dispensing and administration of medicines. Figure 15.2 illustrates examples of Trust committees that play a part in the management of medicines.

The structure, format and terms of reference of medicines management structures and committees vary between Trusts because they are locally agreed, but the existence of medicines management procedures is mandatory.

The Drug and Therapeutics Committee plays a critical role in the evaluation of the evidence surrounding a new medicine that is being considered for use within the Trust. This committee has significant support from senior pharmacy and medical staff and it often has the power to review local processes, documentation and systems.

As far as radiography is concerned, this committee can have a direct impact on practice. For instance, the committee may choose to adopt the use of a new medicine because

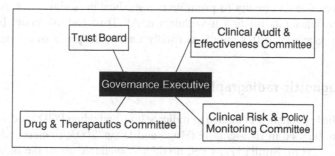

Figure 15.2 Medicines management structure. An example showing committees concerned solely with medicines management

of aspects such as efficacy and/or cost implications, and could therefore replace one medicine with another. If radiographers are using the 'old medicine' as part of a PGD then that PGD would need urgent updating.

An important aspect of medicines management is audit, because the Trust will monitor compliance with its policies and address deficiencies as required. In this respect, it is important that radiographers pay due attention to policies and comply as required with Trust governance. Radiographers, and other staff, should be adequately trained and educated in support of reducing risk to patients and also improving care and management (Society and College of Radiographers, 2002 and www.hpc-uk.org). Trusts are expected to offer in-house training for those who wish to use PGDs and such courses are likely to include:

- stated outcomes (examples are indicated below);

- assessment that verifies the outcomes have been attained;

- remedial procedures for those who do not pass.

The stated outcomes of PGDs are:

- Understand the concept of patient group directions

- Understand the historical and legal background to patient group directions

- Understand the meanings of pharmacodynamics and pharmacokinetics in relation to the drugs that the radiographer is able to supply and administer

- Understand the pathophysiological processes in the conditions under which the patient group directions can be used

- Understand the protocols and guidelines that exist to support the use of patient group directions.

Supplementary prescribing courses are offered by universities and these are open to all professions that are legally permitted to hold these responsibilities. Presently, ring-fenced monies from the Department of Health are managed by Strategic Health Authorities to fund these courses in an attempt to promote non-medical prescribing. There will come a point, however, when this funding is withdrawn. All Trust and university-based training offered to radiographers in this area is normally on a multi-professional basis.

15.6.5 Diagnostic radiography and medicines

There is very little published evidence to indicate how and when PGDs and supplementary prescribing can be used in radiography (Hogg and Hogg, 2003; Francis and Hogg, 2006; Hogg and Hogg 2006). Similarly, as yet, no data is available about the impact that these practices have had on radiography patient care and management. Consequently, only anecdotal evidence can be used here.

15.6.5.1 PGDs

Many diagnostic imaging departments have implemented PGDs for patients who have the same investigations. For many aspects of diagnostic imaging, the patient pathway through the examination follows a fairly set format, and only in infrequent cases are deviations required. When the procedure requires the intervention of medicine, usually as an adjunct to the imaging process, the fashion in which the medicine is given also follows a specific and predetermined format. Below are three examples of medicinal products that illustrate this.

1. Prior to examination: barium enema patients require bowel cleansing, achieved by oral laxatives.

2. During imaging: certain examinations require the introduction of contrast enhancers, including barium sulfate and iodine based liquids.

3. During imaging: there may be a requirement to administer a medicine to improve diagnostic efficacy, including diuretics, cardiac stimulation agents and smooth muscle relaxants.

All are clear examples of when a PGD can be used for patients who will be managed in a specified and predetermined fashion. As such, the indicated medicinal products are commonly incorporated into PGDs within diagnostic imaging departments. The Society and College of Radiographers has set up a discussion forum and download area on its website to allow for the sharing of radiography-specific PGDs (www.sor.org).

15.6.5.2 Supplementary prescribing

In contrast, supplementary prescribing does not have a widespread uptake in radiography, partly because it is new and partly because diagnostic radiographers are struggling to incorporate it into current practice. A radiographer could become a supplementary prescriber, and it is highly likely that radiologists would act as the independent prescribers. In theory, a CMP could be drawn up and funded training is available for radiographers.

The problem surrounding lack of uptake into diagnostic radiography appears to be related to the very nature of supplementary prescribing – it is most useful in dealing with long-term medical conditions and diagnostic imaging, in the main, involves one-off examinations. These tend to be quick, typically less than six hours, and the vast majority do not involve serial imaging at various future time points, so the radiographer often meets each patient only once for that examination.

Unless the radiographers redefine their role to be consistent with the patient's care-pathway, and that care-pathway is for a chronic disease that requires medicinal intervention, then it is unlikely that diagnostic radiography patients would benefit from radiographer supplementary prescribing. There are, however, limited areas of diagnostic practice in which serial imaging of chronic disease does take place, for example cancer imaging (nuclear medicine) and bone densitometry. In these cases, the radiographer could serially image a patient over several years but it may be as little as once a year and since the CMP should be reviewed at least yearly, it becomes clear that there is limited value in a diagnostic radiographer becoming a supplementary prescriber.

15.6.5.3 *Independent prescribing*

There is a potential role for independent prescribers in diagnostic radiography, and this is particularly evident in areas of acute medical practice. There are examples of radiographers working in accident centres whose role involves diagnosis and on the basis of this they propose treatment and discharge patients (Snaith, 2007). This treatment can involve the supply and administration of medicines, particularly for pain relief.

The current arrangements are that these radiographers work within PGDs. However, on examining the definition of an independent prescriber, it is clear that they meet the requirements. The interesting aspect of this practice is that a major growth area of radiographer role advancement currently is the reporting of accident centre images by radiographers (Price and Le Masurier, 2006), which is a precursor to radiographer-led discharge and the associated supply and administration of medicines.

15.6.6 Diagnostic – where is it going?

Anticipating that radiographers will gain independent prescribing responsibilities at some stage, then the following is suggested.

PGDs will continue to be used where large groups of patients undergo imaging procedures that routinely require medicinal administration in a set format, as indicated earlier. This method of working minimizes the need to train large groups of staff to a high level while still maintaining sensitivity to individual patient needs. The majority of radiographer administrations will employ this route, with practitioners, advanced practitioners and consultant practitioners all using this form of administration.

Supplementary prescribing will have a limited range of applications because of the reasons indicated above.

Independent prescribing will have niche roles which will be limited to radiographers who hold consultant or advanced practitioner roles. Clearly, the use of independent prescribing will be limited to where the clinical demand exists and this case needs to be fully explored.

15.6.7 Radiotherapy and oncology and medicines

15.6.7.1 *PGDs*

Since the late 1990s, therapy radiographers have been trained to administer medicines for the management of radiotherapy side effects, such as pain; rectal symptoms, constipation, diarrhoea; skin reactions, wound care; oral care; nausea and vomiting. They are also beginning to use PGDs for X-ray contrast agents.

Practical implementations of the above often involve the therapy radiographers conducting a formal consultation with the patient; completing a patient care plan; and advising and providing medicine. They must recognize when and when not to provide medication and, for the former, knowing which PGD is to be used (and medicine therein). It is essential that the therapy radiographer recognizes when to refer the patient on to the doctor (oncologist), so they need to differentiate between tumour- and treatment-related signs and symptoms.

It is not known how widespread the use of PGDs is within the therapy community, but an unpublished national survey of radiotherapy service managers indicated that 45 radiographers (in 12 out of 29 departments) were involved in the administration of medicines using PGDs. In accordance with national advice, the survey revealed that more than half of them had been trained in-house while the remainder had attended a university-based post-registration programme. The care of patients in pre-treatment (for example contrast-based X-ray examinations) also offers scope for the use of PGDs similar to those for diagnostic radiographers, but again it is not known how widespread this is.

15.6.7.2 *Supplementary prescribing*

The philosophy of supplementary prescribing is well suited to the management of therapy patients. Cancer is a chronic condition and it is highly feasible to set up a CMP between the independent prescriber (oncologist) and supplementary prescriber (therapy radiographer). In many respects, supplementary prescribing is better than PGDs and the values of this mechanism with respect to toxicity management are indicated below:

- PGDs are inflexible, being restricted to specific medicines at specific doses. Doses cannot be altered and medication cannot be changed to an alternative medicine (within a specific PGD);

- PGDs require regular updating in accordance with Trust policies, and the associated processes can be labour intensive;

- supplementary prescribing allows therapy radiographers to have broad professional latitude within their competence in the medicines they prescribe and the amount they administer. This allows for more responsive and individualized care and management.

Notwithstanding the above, therapy radiographers have yet to fully realize the importance of supplementary prescribing in their practice and until this is incorporated and evaluated it is likely that widespread uptake will be limited. Nonetheless, medicines that might be used to good effect using supplementary prescribing include:

- prophylactic anti-emetics for patients receiving high- and moderate-risk emetogenic radiotherapy;

- codeine phosphate for pain or diarrhoea;

- Entonox and micro enemas for brachytherapy patients.

In addition, all medicines for toxicity management, as indicated earlier under PGDs, specifically:

- pain (paracetamol, ibuprofen, and co-proxamol tablets);

- constipation (bisacodyl tablets/suppositories, ispaghula husk sachets);

- diarrhoea (loperamide capsules);

- rectal symptoms (Anusol cream, lidocaine 2% gel);

- skin reactions (hydrocortisone 1% cream, chlorpheniramine tablets);

- oral care (chlorhexidine mouthwash, nystatin mixture, aciclovir cream);

- nausea and vomiting (metoclopramide tablets/IV injection, domperidone tablets/ suppositories, granisetron tablets/IV injection, dexamethasone tablets/IV injection).

(See Chapter 13.)

15.6.7.3   Independent prescribing

As with diagnostic radiography, this is likely to have specialist areas of application. It might be that therapy radiographers, having fully understood the significance of supplementary prescribing, may realize that independent prescribing could have particular values in the management of certain side effects and for certain patient groups. This needs to be explored further.

15.6.8   Therapy – where is it going?

Although the context of practice may change and improve, it is likely that most radiotherapy services will continue to be organized in such a way that the patients' needs will be partly met through the use of PGDs and partly through supplementary prescribing. Clinical circumstances will dictate when each would be used and a rule of thumb could be the following.

Minor professional latitude is required for one-off events (for example the use of x-ray contrast agents for pre-treatment imaging associated with planning). PGDs would be appropriate in these situations.

Broader professional latitude is required so as to be more sensitive to patients on a serial/chronic basis (for example management of treatment side effects). Supplementary prescribing would be appropriate in these situations.

Independent prescribing would grow out of supplementary prescribing for specific conditions and particular patient groups.

15.6.9   Nuclear medicine – the special case

Nuclear medicine is worthy of special mention for three reasons:

1. the clinical service is delivered by technicians and radiographers;

2. additional legislation is in force that legally permits an alternative to PGDs (The Medicines for Human Use Miscellaneous Amendments Order, 2006);

3. PGDs cannot be used for radiopharmaceuticals.

While working within nuclear medicine, technicians and radiographers often have similar responsibilities. This is particularly true for activities involving the scanning of patients where there is frequently a requirement to administer adjunct medicines, such as diuretics. For this activity, radiographers have been working within PGDs or in limited cases using PSDs. Technicians are not legally permitted to use PGDs and as such cannot work within them. Neither can the use of a PGD be delegated by a radiographer. As a consequence of this, until recently, each adjunct medicine would have been approved by an independent prescriber (doctor) on a patient-specific basis or the doctor themselves would have administered it.

Not surprisingly, this was a time-consuming process and as a result, on the advice of Administration of Radioactive Substances Advisory Committee (www.arsoc.org.uk), new legislation has been passed. The British Nuclear Medicine Society (BNMS) has interpreted this legislation and offered guidance (www.bnms.org.uk) by providing clarification about its possible implementation. BNMS explains that it allows locally written protocols to be used for the administration of adjunct medicines and in essence, these might look similar to PGDs. The only difference is that technicians (and others) can work within these new protocols, thereby allowing for a more efficient method of working. It is worth remembering that radiographers still have PGDs available and ideally (in line with all other non-medical prescribing groups) these should remain in place. During January 2007, comments were informally solicited from a wide number of nuclear medicine departments about the use of this new legislation. Some had already implemented protocols as indicated in BNMS guidance but many had not complied with local medicines management arrangements for risk assessment and Trust approval. An informal discussion with the Department of Health confirmed that Trusts retain the right to approve or veto the adoption of legislation into working practice, amplifying the requirement to seek formal Trust approval prior to implementation. It also adds weight to the need for radiographers to continue using PGDs as these are firmly embedded within Trust medicines management procedures and indemnity arrangements.

The administration of radiopharmaceuticals remains the same as that outlined earlier and as illustrated in the 1978 Medicine Act (radioactive substances) amendment.

15.6.10 Nuclear medicine – where is it going?

It is anticipated that technicians will soon become HPC registered and shortly after that they are likely to lobby for a range of professional responsibilities, including the use of PGDs. When that time arrives, it would be advisable for technicians to come in line with all other non-medical Trust supply and administration arrangements and switch from the newly-approved protocol system to the more widely recognized PGD arrangements.

Supplementary prescribing is unlikely to have any value in nuclear medicine because of the arguments outlined earlier. Independent prescribing will only have value if the 1978 Medicines (Administration of Radioactive Substances) Regulation is brought into a competency-based rather than profession-specific ethos. This move is supported by the Society and College of Radiographers (2005b).

15.6.11 Suboptimal practice

There is no rigorous published information about the clinical use of PGDs or supplementary prescribing by radiographers, so only anecdotal information can be drawn on to inform debate. Based on anecdotal information, the following anomalies have been noted.

A small-scale analysis (Burke, 2007) of gastrointestinal practitioners within the northwest of England revealed marked differences between them, the most notable being in the detail included in PGDs and in practitioner understanding about PGDs. Some PGDs were superficial, others were quite in-depth. Discussions with those holding responsibility for creating the PGDs revealed that there was no mechanism for sharing paperwork between Trusts. It is not surprising, therefore, that PGDs for the same application can look remarkably different in different Trusts. A further consequence is that considerable time can be wasted by people repeating the same development work time and time again.

While there is fairly widespread uptake of PGDs in diagnostic radiography, there is also a wide range of good and suboptimal practice. At a non-medical prescribing conference for radiographers (Hogg, 2006, unpublished), it was noted that some PGDs intended for use by radiographers were being used by administrative staff. Two examples were offered – laxatives and topical anaesthetic cream. In both cases, the administrator was personally responsible for posting out these medicines to the patient, in accordance with a PGD.

There is a second problem associated with this arrangement, regarding the use of the postal service as part of a PGD arrangement. Where does liability rest? With the administrator or with the postman (who unknowingly pushes the medicine through the letter box)? Whichever professional is considered, it should be noted that neither of these groups can legally work within a PGD. Also, any regulations and related guidance must be complied with strictly.

Another example noted in some PGDs is that working practices may not be in line with Summary of Product Characteristics for that medicinal product (www.medicines.org.uk/). For example some PGDs have indicated that the amount of Omnipaque (an X-ray contrast agent) to be administered can exceed that outlined on the product information sheet. The only evidence used to support this is that some peers do this in other Trusts. Not only does this practice fall short of minimizing risk but it might also fail one component of the Bolam test (*Bolam v Friern Hospital Management Committee*, 1957 and *Bolitho v City and Hackney Health Authority*, 1997).

In some departments, it is clear that the implementation and development of prescription, supply and administration of medicines by radiographers has occurred in isolation/semi-isolation to the Trust as a whole. Consequently, good practice that is regarded as quite basic to other health professions has been overlooked and not implemented. These problems can be minimized/eliminated by utilizing multi-disciplinary approaches to training and seeking support through Trust medicines management processes.

15.6.12 Recommendations

1. A multi-professional mechanism for sharing practice (for example PGDs) should be used.

2. Radiographers should be encouraged to share their practices in public forums (for example through the Society and College of Radiographers website).

3. Suboptimal PGD practice should be identified and eliminated.

4. Compliance with local governance arrangements is essential.

5. Supplementary prescribing should be taken up within therapy.

6. The case for independent prescribing for radiographers (therapy and diagnostic) should be made.

7. Nuclear medicine technicians should be supported to gain supply, administration and prescription responsibilities (after HPC registration confirmed).

References

Bolam v Friern Hospital Management Committee (1957) *All England Law Reports 2*, 118.

Bolitho v City and Hackney Health Authority (1997) *All England Law Reports 4*, 771.

Burke, L. (2007) PGDs – do you know what they are saying? *Synergy*, April, 12–14.

Department of Health (1989) *Report of the Advisory Group on Nurse Prescribing (Crown I)*, HMSO, London.

Department of Health (1995) *Calman Hine Report Policy Framework for Commissioning of Cancer Services*, HMSO, London.

Department of Health (1999) *Review of Prescribing, Supply and Administration of Medicines; Final report (Crown II)*, Department of Health, London.

Department of Health (2006) *Medicines Matters: A Guide to Mechanisms for the Prescribing, Supply and Administration of Medicines*, National Practitioner Programme Department of Health Core Prescribing Group, London.

Department of Health and Social Security (1986) *Neighbourhood Nursing; a Focus for Care (Cumberlege Report)*, HMSO, London.

Francis, G. and Hogg, D. (2006) Radiographer prescribing: enhancing seamless care in oncology. *Radiography*, **12**(1), 3–5.

Hogg, P. (2006) *Radiography Prescribing Conference Manchester*, (unpublished).

Hogg, D. and Hogg, P. (2003) Radiographer prescribing: lessons to be learnt from the community nursing experience Editorial. *Radiography*, **9**(4), 263–65.

Hogg, P. and Hogg, D. (2006) Prescription, supply and administration of drugs in radiography. *Synergy*, March, 4–8.

Loughran, C.F. (1993) Intravenous Urography – injection of contrast agents by radiographers. *Clinical Radiography*, **47**, 368.

Medicines Regulations (1978) Administration of Radioactive Substances.

The Medicines for Human Use Miscellaneous Amendments Order (2006) *No 2807*.

Office of Sector Information (1992) Medicinal Products Prescription by Nurses etc. Act, HMSO, London.

Price, R. and Le Masurier, S.B. (2006) Longitudinal changes in extended roles in radiography: a new perspective. *Radiography*, **13**(1), 18–29.

Snaith, B. (2007) Radiographer-led discharge in accident and emergency – the results of a pilot project. *Radiography*, **13**(1), 13–17.

Society and College of Radiographers (2002) Statement of Professional Conduct.

Society and College of Radiographers (2005a) Course of Study for the Certificate in Competence in Administering Intravenous Injections.

Society and College of Radiographers (2005b) Nuclear Medicine Practice.

Society of Radiographers (2001) *Prescribing for Radiographers: A Vision Paper*.

Society of Radiographers (2003) *Role Development Revisited: The Research Evidence*.

Westbrook, K. and Hodgetts, A. (1997) *Development of a Limited Formulary Prescribing Scheme for Therapy Radiographers and Oncology Nurses NHS Executive (S&W) R&D Project Grant: Ref D/12/11.95/Westbrook Abstract Reporting Back the Results of R&D Conference*.

Further reading

Health Service Circular (2000) HSC 2000/026 NHS Executive.

Department of Health (2006) Medicines Matters: A Guide to Mechanisms for the Prescribing, Supply and Administration of Medicines, National Practitioner Programme Department of Health Core Prescribing Group, London.

Price, R.C., Miller, L.R. and Mellor, F. (2002) Longitudinal changes in extended roles in radiography. *Radiography*, **8**, 223–34.

Web sites

www.arsac.org.uk/ Administration of Radioactive Substances Advisory Committee

www.bnms.org.uk/ British Nuclear Medicine Society Administration of Pharmaceuticals in Nuclear Medicine

http://www.dh.gov.uk/en/PolicyAndGuidance/MedicinesPharmacyAndIndustry/ Prescriptions/TheNonMedicalPrescribingProgramme/index.htm http://www.dh.gov.uk/ en/Publicationsandstatistics/Publications/PublicationsPolicyAndGuidance/DH_ 4110032 Department of Health, Supplementary Prescribing by Nurses, Pharmacists, Chiropodists/Podiatrists, Physiotherapists and Radiographers within the NHS in England: A Guide for Implementation 2005

www.hpc-uk.org Health Professions Council Standards of Proficiency

www.medicines.org.uk/ Summary of Product Characteristics

www.portal.nelm.nhs.uk/PGD/default.aspx www.qaa.ac.uk/academicinfrastructure/bench mark/health/Radio-final.asp Subject Benchmark Statements: Health Care Programmes: Radiography

www.sor.org/members/effective-practice/pres_rad_news/news4.htm Society of Radiographers

15.7 Physiotherapy

A physiotherapy perspective of non-medical prescribing is discussed here.

Like podiatrists and radiographers, physiotherapists can, under PSDs, currently supply and/or administer medicines to a named patient and under PGDs they can currently supply and/or administer medicines to groups of patients who may not be individually identified prior to treatment. Many physiotherapists are already using PSDs and PGDs as named individuals within their current practice. There may be many more that are not, but perhaps should be doing so, because often physiotherapy intervention can be improved by timely administration of medicines by a therapist and outcomes of long-term care can also be enhanced. The use of PSDs in physiotherapy practice is illustrated by situations such as the post-operative administration of oxygen by a physiotherapist on the direction of an anaesthetist or the administration of or advice on the use of, analgesia prescribed by a doctor to a patient with rheumatoid arthritis.

In 2003, qualified registered physiotherapists were included in the list of professionals who are able to supply medicinal products under PGDs. A typical example of PGD use is in accident and emergency analgesia, when a physiotherapist would be allowed to supply or administer an analgesic, following a written direction, to groups of patients with an identified clinical condition. Another instance is the intra- and extra-articular administration of local anaesthetics and anti-inflammatory corticosteroids by a physiotherapist involved in the ongoing management of rheumatology patients.

The information that must be included in a PGD for a physiotherapist follows the same general rules that are listed in the podiatric example given earlier in this chapter. Specific examples of drugs that might be included in a physiotherapy PGD are triamcinolone acetonide 10 mg/ml and methylprednisolone acetate 40 mg/lidocaine hydrochloride 10 mg/ml, which are administered by intra- or extra-articular injection. Both of these drugs are used for the treatment of inflammatory conditions such as soft tissue injuries and isolated joint inflammation (see Chapter 7).

Examples of PGDs are accessible on the internet and they follow a similar format to those given for podiatric surgery in Appendix C.

15.7.1 Supplementary prescribing for physiotherapists

Following amendments to the legislation in 2003, which allowed nurses and pharmacists to become supplementary prescribers, further changes to the NHS Regulations in April 2005 extended supplementary prescribing to physiotherapists, podiatrists and radiographers (see Chapter 14). There are no legal restrictions on clinical conditions that can be treated by supplementary prescribers under an agreed CMP. Supplementary prescribing is best suited to the management of chronic conditions and includes the prescribing of general sales list medicines, pharmacy medicines and prescription-only medicines. Thus, physiotherapists are ideally placed in many areas of practice to use supplementary prescribing effectively because they treat patients with long-term conditions.

Supplementary prescribing can be suitable in some areas of community physiotherapy. The advantages of physiotherapy supplementary prescribing in the community are that both patients and physiotherapists benefit in many ways. For example instead of having to be referred back to a doctor and having to make another journey, a patient could be managed by the physiotherapist. This could provide timely and appropriate access to treatment for chronic disease management. A community physiotherapist treating a patient in a remote location who is not responding well to medicines prescribed by a consultant could, under a CMP, alter dosages for that patient's benefit. Specialist knowledge allows the physiotherapist to explain to patients in these situations how to take their medication for best effect and could help improve compliance. A GP or a consultant, whichever is relevant, would have to be involved in the initial development of a CMP and review process so that physiotherapist could prescribe medication as appropriate for individual patients.

Extended scope practitioners are clinical physiotherapy specialists with extended roles who see patients referred for assessment, clinical diagnosis and management. Their work often encompasses tasks that may previously have been undertaken by a medical practitioner. Areas of extended scope practice include rheumatology, paediatrics, orthopaedics, neurology, respiratory care and musculoskeletal medicine. As a supplementary prescriber, an extended scope physiotherapist in any of these areas would be able to prescribe, for example anti-inflammatory drugs and other non-opioid analgesics under an agreed CMP.

Typically an extended scope practitioner may work in an orthopaedic screening service for a PCT. This often involves a non-medical prescribing group and development of a competency framework for prescribing drugs for patients with acute or chronic pain, for example spinal pain (see Chapter 12 for a case study of such a situation). Supplementary prescribing allows the physiotherapist to complete the patient's episode of care and provide timely and effective prescribing when the patient most needs it. Other areas of care suitable for supplementary prescribing are the treatment of musculoskeletal pain in chronic conditions such as rheumatic diseases or in neuro-rehabilitation, women's health and mental health.

It could be that doctors do not always prescribe effectively. For instance, a specialist physiotherapist in neurology may know more about specific use of anti-spasticity medication (Chapter 11); or a respiratory physiotherapist may have extensive experience in the use of bronchodilators and they could use this knowledge for more effective prescribing.

In a hospital setting a consultant physiotherapist may specialize in the rehabilitation of older patients, who have had falls or experienced physical decline following infection. They would be able to provide a better service for patients if they were prescribing under a CMP. A CMP could be written with a doctor as soon as the patient is admitted into hospital, providing the patient agrees. The physiotherapist would then have full access to medical records and medication details. A physiotherapist working in rehabilitation of older patients is likely to find occasion to prescribe analgesics, anticoagulants, antibiotics (including for MRSA) antacids and laxatives. Management of long-term neurological conditions could mean wider prescribing.

The benefits of extended prescribing apply to both patients and physiotherapists.

The prescribing physiotherapist can help to reduce patients' anxieties by ensuring that they are better informed about their medication and better able to discuss their concerns about their treatment. In addition, waiting times for the patients to receive medication could be reduced; treatment can be made more cost-effective by reducing the number of visits for patients to the physiotherapist and/or to other medical practitioners. Job satisfaction for the physiotherapist may be improved through increased independent working while enhancing skills and knowledge. Other benefits may be had through increased direct patient contact, and the opportunity to expand services to patients. Extended roles offering variety and increased levels of responsibility are generally more stimulating and attract recognition and respect from other staff.

15.7.2 Future progress

By February 2007, 13 physiotherapists had qualified as supplementary prescribers in the United Kingdom and have established roles as supplementary prescribers and currently more physiotherapists are in training. Postgraduate certificate courses in Non-medical Prescribing for Health Care Professionals are run by many universities. Most courses are now interdisciplinary and allow training in supplementary prescribing and independent prescribing (independent prescribing includes diagnostic reasoning) and enable supervised practice in different clinical settings as part of the course.

For some, according to the Chartered Society of Physiotherapy, the ultimate goal is independent prescribing with physiotherapists taking responsibility for diagnosis and prescribing. This approach is not currently on the government's agenda. For this to become a reality, the profession needs to build a case for moving towards independent prescribing. Pharmacology would have to be incorporated into undergraduate programmes as it is in podiatry programmes.

It is likely that many physiotherapists will have undertaken postgraduate education in order to use injection of anti-inflammatory drugs, for example to manage musculoskeletal injuries and they may have experiential knowledge of a range of medicines related to their areas of expertise. However, an understanding of the mechanism of action of various different drugs, drug–drug interactions and adverse drug reactions is necessary to appreciate the full importance of prescribing. An overall understanding of major body systems and common medications used in those systems is recommended.

There is a case for the physiotherapy profession to have an exemption order to the Medicines Act similar to that enjoyed by podiatrists. Exemptions to the Prescription Only Medicines (Human Use) Order 1997 allow podiatrists (and some other health care professionals) to administer medicines from a specified list to patients on their own initiative. Development of this for physiotherapy would involve drawing up of a list of medications most used by physiotherapists for their patients, which would then have to be put on an exemption schedule through a change in legislation. A suggestion for the exemption list has been made for the use of corticosteroids in acute musculoskeletal treatment, which would allow non-prescribing physiotherapists to inject joints and tendons without always having to refer to a doctor or a PGD. This would improve patient access to treatment. In women's health, exemptions could be made for the supply and/or administration of anti-cholinergic drugs for overactive bladder symptoms in an NHS setting.

The physiotherapy profession could aim to follow the path taken by the nursing profession, from first being able to prescribe a few drugs, to widening access to where they are today with independent prescribing. In 2006, nurses and pharmacists became able to train as independent prescribers.

References

The Chartered Society of Physiotherapy (2004) *Prescribing Rights for Physiotherapists – an update*, Chartered Society of Physiotherapy, London.

Health Service Circular (2000) *Patient Group Directions (England Only)*, Department of Health, London.

Limb, M. (2006) *Supplementary Question Frontline*, Vol. 12(6), pp. 21–23.

National Health Service (2000) www.npc.nhs.uk.

National Prescribing Centre (2004) Maintaining Competency in Prescribing.

Rotherham NHS PCT (2006) *PGD for the Administration of Methylprednisolone Acetate 40mg/lidocaine Hydrochloride 10mg/ml by Physiotherapists*, http://www.rotherhampct. nhs.uk/healthprofessionals/prescribing/PatientGroupDirections.

Royal United Hospital Bath NHS Trust (2000) Policy and Procedures for PGDs, www.ruh. nhs.uk/.

Shropshire County and Telford and Wrekin NHS PCT (2007) PGDs for Triamcinolone Acetonide 10mg/ml by Physiotherapists, www.telfordpct.nhs.uk/.

15.8 Summary

Non-medical prescribing is becoming increasingly common amongst health care professionals in practice.

Podiatrists have had specific exemptions to the Medicines Act 1968 since the 1980s. These exemptions allow them to access and supply certain medicines (some of which are POM) and administer local anaesthetics, in the course of their professional practice, on their own initiative and without referring to a doctor.

In addition, since 2003, podiatrists, physiotherapists and radiographers have been amongst those health care professionals who are allowed to supply and administer medicines under PGDs. This has proved particularly useful in certain situations. For example PGDs are used for the provision of medicines to counter the side effects of contrast agents used in diagnostic radiography.

Further developments in prescribing for health care professionals came with the advent of supplementary prescribing in 2005. Through an amendment to the Prescription Only Medicines Order and NHS Regulations, podiatrists, physiotherapists and radiographers can now train to prescribe in partnership with and independent prescriber (a doctor or dentist) according to an agreed CMP.

The initial take up of supplementary prescribing has been slow in the three health care professions. However, it seems that supplementary prescribing is appropriate for long-term management of disease in the community and for the management of chronic conditions and rehabilitation in community and hospital settings.

Many health care professionals would eventually like to move towards independent prescribing, although it is acknowledged that non-medical prescribing in its various forms will continue to have different applications in different situations.

Appendices

Appendix A
Drug Names

A quick reference to drugs used as examples in the main body of the text, together with their main therapeutic applications.

This is not intended to be a comprehensive list of drugs available; the student/practitioner is advised to consult the latest edition of the *BNF* for most accurate and up-to-date information.

Abacavir – antiviral
Acarbose – oral hypoglycaemic
Acipimox – lipid lowering drug
Acitretin – oral retinoid; psoriasis
Acrivastine – antihistamine
Acyclovir – antiviral
Adalimumab – disease-modifying anti-rheumatic drug
Adrenaline – emergency treatment of anaphylaxis and circulatory collapse
Albendazole – anthelmintic
Alemtuzumab – monoclonal antibody; leukaemia treatment
Allopurinol – prevention of uric acid production, gout treatment
Alteplase – fibrinolytic; thrombosis
Amantadine – antiviral; Parkinson's disease
Amfetamine (amphetamine) – CNS stimulant (little therapeutic use)
Aminophylline – bronchodilator; asthma, chronic bronchitis
Amiodarone – anti-arrhythmic
Amisulpiride – antipsychotic
Amitriptyline – antidepressant
Amlodipine – calcium channel blocker; hypertension, angina
Amorolfine – antifungal
Amoxicillin – antibiotic
Amphotericin – antifungal
Amprenavir – antiviral; HIV
Anakinra – cytokine inhibitor; rheumatoid arthritis
Antidiuretic hormone – replacement therapy in diabetes insipidus
Antihaemophilic globulin – replacement therapy in haemophilia A
Apomorphine – dopamine receptor agonist; Parkinson's disease
Aspirin – non-steroidal anti-inflammatory drug

Atenolol – beta blocker; hypertension, angina, arrhythmia
Atomoxetine – CNS stimulant; treatment of ADHD
Atorvastatin – statin; lipid lowering drug
Atovaquone – antimalarial
Atracurium – muscle relaxant (non-depolarizing neuromuscular blocker); adjunct to anaesthesia
Atropine – emergency treatment of hypotension with bradycardia
Azathioprine – immunosuppressant; cancer chemotherapy, rheumatoid arthritis
Baclofen – skeletal muscle relaxant; spasticity
Barbiturate – hypnotic (little therapeutic use)
Beclometasone – anti-inflammatory corticosteroid, asthma, chronic bronchitis
Bendroflumethiazide – thiazide diuretic; hypertension, cardiac failure
Benzodiazepine – anxiolytic, hypnotic
Benzoic acid – antifungal
Benztropine – anticholinergic; Parkinson's disease
Benzylpenicillin – antibiotic
Bevacizumab – monoclonal antibody; treatment of colorectal cancer
Bezafibrate – lipid lowering drug
Bicalutamide – chemotherapy of prostate cancer
Bisacodyl – stimulant laxative
Bisoprolol – beta blocker; hypertension, angina, arrhythmia
Bleomycin – chemotherapy of cancers, wart treatment
Borotannic complex – antifungal
Bromocriptine – dopamine receptor agonist; Parkinson's disease
Budesonide – anti-inflammatory corticosteroid, asthma, chronic bronchitis
Bupivacaine – local anaesthetic
Buspirone – anxiolytic
Busulphan – chemotherapy of leukaemia
Calcipotriol – vitamin D analogue; psoriasis
Calcitonin – osteoporosis, Paget's disease
Calcitriol – vitamin D analogue; psoriasis
Capsaicin – counter-irritant for neuropathic pain
Captopril – ACE inhibitor; cardiac failure, hypertension
Carbamazepine – antiepileptic
Carbidopa – dopa decarboxylase inhibitor; Parkinson's disease
Carbimazole – antithyroid drug; hyperthyroidism
Carbocisteine – mucolytic; COPD
Carmustine – chemotherapy of brain tumours
Cefadroxil – cephalosporin antibiotic
Cefuroxime – cephalosporin antibiotic; infection prophylaxis prior to bone surgery
Celecoxib – COX_2 inhibitor; anti-inflammatory
Celiprolol – beta blocker; hypertension, angina, arrhythmia
Chloramphenicol – antibiotic
Chlorphenamine – antihistamine
Chloroquine – antimalarial, disease-modifying anti-rheumatic drug

Chlorpromazine – antipsychotic
Cholestyramine – lipid lowering drug
Christmas factor complex – replacement therapy for haemophilia B
Ciclosporin – immunosuppressant; cancer chemotherapy, psoriasis, rheumatic disease
Cimetidine – histamine antagonist; stomach ulcers
Ciprofloxacin – antibiotic
Cisplatin – chemotherapy of cancers
Clomethiazole – hypnotic
Clonidine – centrally-acting antihypertensive
Clopidrogel – anti-platelet anticoagulant
Clotrimazole – antifungal
Clozapine – atypical antipsychotic
Codeine – opioid analgesic
Colchicine – treatment of gout
Colestipol – lipid lowering drug
Corticosteroid – anti-inflammatory drug; chronic inflammatory disease
Cortisol – anti-inflammatory corticosteroid
Co-trimoxazole – combination of trimethoprim and sulphamethoxazide; pneumocystis pneumonia
Crisantaspase – chemotherapy of cancers
Cyclophosphamide – alkylating agent, cancer chemotherapy
Cytarabine – chemotherapy of leukaemia
Dactinomycin – chemotherapy of cancers
Dantrolene – skeletal muscle relaxant; spasticity
Desflurane – general anaesthetic
Desmopressin – antidiuretic hormone analogue, diabetes insipidus
Dexamfetamine – CNS stimulant; to treat ADHD
Dextran – plasma substitute; fluid replacement
Diamorphine – opioid analgesic
Diazepam – anxiolytic and hypnotic benzodiazepine; premedication sedative
Diclofenac – non-steroidal anti-inflammatory drug
Diethylcarbamazine – anthelmintic
Digoxin, digitoxin – cardiac glycosides; cardiac failure, atrial fibrillation
Dihydrocodeine – opioid analgesic
Diltiazem – calcium channel blocker; angina, hypertension
Diphenhydramine – antihistamine; allergic reactions
Dipyridamole – antiplatelet anticoagulant
Dithranol – psoriasis
Docetaxel – cancer chemotherapy
Docusate – faecal softener laxative
Domperidone – antiemetic
Donepezil – central acetylcholinesterase inhibitor; Alzheimer's disease
Doxazosin – α-adrenoreceptor blocker; hypertension
Doxorubicin – chemotherapy of cancers
Econazole – antifungal

Efalizumab – immunosuppressant; psoriasis
Efavirenz – antiviral; HIV
EMLA – local anaesthetic combination of lidocaine and prilocaine
Entacapone – catechol-o-methyl transferase inhibitor; Parkinson's disease
Enoxaparin – low molecular weight heparin; thrombosis prophylaxis prior to surgery
Epoetin – erythropoietin replacement; anaemia
Erythromycin – antibiotic
Esomeprazole – proton pump inhibitor; hyperacidity and gastric ulcer
Etanercept – cytokine inhibitor; rheumatoid arthritis
Ethambutol – antibiotic; tuberculosis
Ethinylestradiol – chemotherapy of prostate cancer
Ethosuximide – antiepileptic
Etomidate – intravenous general anaesthetic
Ezitimibe – lipid lowering drug
Fenbrufen – non-steroidal anti-inflammatory drug
Fentanyl – opioid analgesic
Ferrous sulphate – iron supplement; anaemia
Flecainide – membrane stabilizer; antiarrhythmic
Flucloxacillin – antibiotic
Fluconazole – antifungal
Flucytosine – antifungal
Fludrocortisone – corticosteroid replacement in Addison's disease
Fluorouracil – chemotherapy of cancers
Fluoxetine – antidepressant
Flupentixol – antipsychotic
Formaldehyde – keratolytic; wart treatment
Furosemide – diuretic; hypertension
Gabapentin – antiepileptic
Ganciclovir – antiviral; cytomegalovirus
Gentamicin – antibiotic
Glibenclamide – oral hypoglycaemic
Gliclazide – oral hypoglycaemic
Glipizide – oral hypoglycaemic
Glucagon – smooth muscle relaxation prior to gastrointestinal imaging
Glutaraldehyde – keratolytics; warts
Glyceryl trinitrate GTN – vasodilator; angina
Goserelin – chemotherapy of breast and prostate cancers
Griseofulvin – antifungal
Guanethidine – adrenergic neurone blocker; antihypertensive
Halofantrine – antimalarial
Haloperidol – antidepressant
Halothane – general anaesthetic
Heparin – anticoagulant
Hirudin – anticoagulant
Hydrocortisone – anti-inflammatory corticosteroid

Hydroxychloroquine – antimalarial, disease-modifying anti-rheumatic drug
Hydroxycobolamin – Vitamin B_{12} replacement
Hydroxyurea – chemotherapy of cancers
Hyoscine – premedication antimuscarinic; to inhibit gastric secretion and intestinal motility prior to gastrointestinal imaging
Ibuprofen – non-steroidal anti-inflammatory drug
Imipramine – antidepressant
Indinavir – antiviral; HIV
Indometacin – non-steroidal anti-inflammatory drug
Infliximab – cytokine inhibitor; rheumatoid arthritis
Insulin – insulin dependent diabetes mellitus
Interferon α – biological response modifier in cancer treatment
Interferon β – immunomodulator; multiple sclerosis
Interleukin-2 – enhances proliferation of leukocytes in cancer treatment
Ipratropium – anti-muscarinic bronchodilator; asthma, chronic bronchitis
Isoflurane – general anaesthetic
Isoniazid – antibiotic; tuberculosis
Isosorbide dinitrate – vasodilator; angina
Isosorbide mononitrate – vasodilator; angina
Itraconazole – antifungal
Ivermectin – anthelmintic
Ketamine – intravenous general anaesthetic
Ketoconazole – antifungal
Lactulose – osmotic laxative
Lamivudine – antiviral; HIV
Lamotrigine – antiepileptic
Lepirudin – anticoagulant
Levamisole – anthelmintic
Levobupivacaine – local anaesthetic
Levodopa – dopamine replacement; Parkinson's disease
Lidocaine – local anaesthetic; antiarrhythmic
Lisinopril – ACE inhibitor; hypertension, cardiac failure
Lithium – mania, manic depression
Loperamide – anti-diarrhoeal
Loprazolam – hypnotic benzodiazepine
Lormetazepam – hypnotic benzodiazepine
Magnesium citrate – osmotic laxative
Magnesium sulphate – osmotic laxative
Maprotiline – antidepressant
Mebendazole – anthelmintic
Mefenamic acid – non-steroidal anti-inflammatory drug
Memantine – glutamate NMDA receptor blocker; Alzheimer's disease
Mepivacaine – local anaesthetic
Mercaptopurine – chemotherapy of leukaemia
Mesna – prevents haemorrhagic cystitis with cyclophosphamide metabolite acrolein

Metformin – oral hypoglycaemic
Methodone – opioid analgesic
Methotrexate – immunosuppressant; disease-modifying anti-rheumatic drug, cancer chemotherapy, psoriasis
Methyl cellulose – bulk-forming laxative
Methyldopa – antihypertensive
Methylphenidate – CNS stimulant; ADHD
Methylprednisolone – anti-inflammatory corticosteroid
Meticillin – antibiotic (unavailable in United Kingdom)
Metoclopramide – stimulates gastric emptying and small intestine transit in barium imaging; antiemetic
Miconazole – antifungal
Midazolam – benzodiazepine; premedication sedative
Mirtazepine – antidepressant
Moclobemide – antidepressant
Montelukast – anti-inflammatory drug; asthma
Morphine – opioid analgesic
Naproxen – non-steroidal anti-inflammatory drug
Nateglinide – oral hypoglycaemic
Nedocromil sodium – mast cell stabilizer; asthma prophylaxis
Nevirapine – antiviral; HIV
Niclosamide – anthelmintic
Nicorandil – potassium channel activator; angina
Nifedipine – calcium channel blocker; angina, hypertension, Raynaud's disease
Nitrazepam – benzodiazepine anxiolytic
Nitrous oxide – general anaesthetic
Nystatin – antifungal
Ondansetron – antiemetic
Olanzapine – atypical antipsychotic
Oseltamivir – antiviral; influenza
Oxazepam – anxiolytic benzodiazepine
Oxprenolol – beta blocker; hypertension, arrhythmia, angina
Oxytetracycline – antibiotic
Paclitaxel – cancer chemotherapy
Pancuronium – muscle relaxant (non-depolarizing neuromuscular blocker); adjunct to general anaesthesia
Paracetamol – peripherally-acting analgesic
Penicillin – antibiotic
Penicillamine – disease-modifying anti-rheumatic drug
Pericyazine – antipsychotic
Pethidine – opioid analgesic
Phenindione – anticoagulant
Phenobarbital – antiepileptic
Phenytoin – antiepileptic
Pioglitazone – oral hypoglycaemic

Piperazine – anthelmintic
Piroxicam – non-steroidal anti-inflammatory drug
Podophyllum – antiviral; wart treatment
Podophyllotoxin – chemotherapy of cancers
Praziquantel – anthelmintic
Prednisolone – anti-inflammatory corticosteroid
Prilocaine – local anaesthetic
Primaquine – antimalarial
Probenecid – uricosuric; gout prophylaxis
Procainamide – antiarrhythmic
Procarbazine – chemotherapy of cancers
Prochlorperazine – antiemetic
Proguanil – antimalarial
Promethazine – antihistamine, antiemetic
Propofol – intravenous general anaesthetic
Propranolol – beta blocker; hypertension, arrhythmia, angina
Pyrazinamide – antibiotic; tuberculosis
Pyridostigmine – anticholinesterase; myasthenia gravis
Pyrimethamine – antimalarial
Quetiapine – atypical antipsychotic
Quinidine – antiarrhythmic
Quinine – antimalarial
Raloxifene – oestrogen receptor modulator; osteoporosis
Reboxetine – antidepressant
Repaglinide – oral hypoglycaemic
Rifampicin – antibiotic; tuberculosis
Riluzole – glutamate inhibitor; motor neuron disease
Risperidone – atypical antipsychotic
Rituximab – monoclonal antibody; rheumatoid arthritis, leukaemia treatment
Rofecoxib – COX_2 inhibitor (withdrawn)
Ropivacaine – local anaesthetic
Rosiglitazone – oral hypoglycaemic
Salbutamol – β_2 receptor stimulant, bronchodilator; asthma, chronic bronchitis
Salmeterol – β_2 receptor stimulant, bronchodilator; asthma, chronic bronchitis
Selegiline – monoamine oxidase-B inhibitor; Parkinson's disease
Sertraline – antidepressant
Sevoflurane – general anaesthetic
Silver sulfadiazine – topical antibacterial
Simvastatin – statin; lipid lowering drug
Sodium aurothiomalate – disease-modifying anti-rheumatic drug
Sodium cromoglicate – mast cell stabilizer; asthma
Sodium etidronate – bisphosphonate; Paget's disease, osteoporosis
Sodium picosulfate – stimulant laxative
Spironolactone – aldosterone analogue; diuretic
Streptokinase – fibrinolytic; thrombosis

Streptomycin – antibiotic; tuberculosis
Sulfadiazine – antibiotic
Sulfasalazine – antibiotic
Sulfinpyrazone – uricosuric; gout prophylaxis
Suxamethonium – neuromuscular blocker (depolarizing); adjunct to general anaesthesia
Tacrolimus – immunosuppressant; severe eczema
Talcacitrol – vitamin D analogue; psoriasis
Tamoxifen – chemotherapy of breast cancer
Temazepam – hypnotic benzodiazepine
Tenoxicam – non-steroidal anti-inflammatory drug
Terbinafine – antifungal
Teriparatide – recombinant fragment of parathyroid hormone; osteoporosis
Tetracycline – antibiotic
Theophylline – bronchodilator; asthma, chronic bronchitis
Thipental – intravenous general anaesthetic
Tiabendazole – anthelmintic
Tioconazole – antifungal
Tizanidine – central α_2 agonist; multiple sclerosis
Tolnaftate – antifungal
Topiramate – antiepileptic
Trastuzumab – monoclonal antibody; treatment of breast cancer
Trazodone – antidepressant
Treosulphan – chemotherapy of ovarian cancer
Tretinoin – induces differentiation in leukaemic cells
Triamcinolone – anti-inflammatory corticosteroid
Tricyclic antidepressants – group of antidepressants
Trimethoprim – antibiotic
Undecenoate – antifungal
Valproate – antiepileptic
Valsartan – angiotensin II inhibitor; antihypertensive
Vancomycin – antibiotic
Venlafaxine – antidepressant
Verapamil – calcium channel blocker; angina, arrhythmia, hypertension
Vigabatrin – antiepileptic
Vinblastine – chemotherapy of cancer
Vincristine – chemotherapy of cancer
Vindesine – chemotherapy of cancer
Voriconazole – antifungal
Warfarin – anticoagulant
Xylometazoline – sympathomimetic; decongestant
Zafirlukast – anti-inflammatory drug; asthma
Zanamivir – antiviral; influenza
Zidovudine – antiviral; HIV
Zolpidem – hypnotic
Zopiclone – hypnotic

Appendix B
Glossary

Acetylation – conjugation with acetyl group

ADP adenosine diphosphate; produced by removal of phosphate group from ATP

Adrenoreceptors – receptors for noradrenaline in the sympathetic nervous system

Adverse reaction – any reaction to a drug that is harmful for the patient

Agonist – drug that activates receptors to produce a response

Anaphylactic shock – sudden wide-spread oedema, bronchospasm, hypotension and circulatory collapse in response to an allergen

Anaemia – insufficiency of haemoglobin

Antagonist – drug that blocks receptors to prevent a response

Antithrombin III – blood protein that inactivates thrombin

Apoptosis – genetically programmed cell death

Arteriosclerosis – hardening of the arteries

Atheroma – lipid deposition and inflammation of inner artery walls

Atopic – susceptible to allergy

ATP – adenosine triphosphate; transfers energy for many cellular reactions, including active membrane transport

Bactericidal – kills bacteria

Bacteriostatic – inhibits bacterial growth

Blood-brain barrier – combination of tight junctions between endothelial cells of brain capillaries and close association of glial cells that restricts entry of substances into the brain

BNF – British National Formulary

Carcinogenesis – causation of cancer

Cardiac output – volume of blood pumped out of the left ventricle in one minute

CHM – Commission on Human Medicines

Chylomicrons – combination of lipids surrounded by protein envelope; transport of lipid from intestine into circulation

Clinical management plan – written agreement between an independent prescriber, a supplementary prescriber and a patient for their care

Controlled drugs – drugs that are particularly dangerous and have abuse potential

COPD – chronic obstructive pulmonary disease, combined effects of chronic bronchitis and emphysema

CSM – Committee on Safety of Medicines (replaced by CHM)

Cytokines – chemicals produce by cells

Cytostatic – drug that halts cell division

Depolarization – change in membrane potential when stimulated

Diabetes insipidus – production of large volumes of dilute urine due to deficiency of antidiuretic hormone

Diastole – relaxation phase of heart beat

DMARDs – disease-modifying antirheumatic drugs

DNA polymerase – enzyme necessary for DNA synthesis

Dose-response curve – plot of drug dose against response to it

DSM-IV – Diagnostic and Statistical Manual of Mental Disorders 4th Edition Revised

ECG – electrocardiogram – record of electrical activity in the heart as it contracts

Embolus – air or part of a blood clot blocking a small blood vessel

Endogenous – from within the body

Enteral – to do with the gastro-intestinal tract

Enterohepatic shunting – cycling between intestine to liver and back to intestine via bile

Enzyme induction – increased activity of enzymes

Enzyme inhibition – decreased activity of enzymes

Exogenous – from without the body

Extra-pyramidal side effects – abnormalities of movement due antipsychotic side effect

First pass metabolism – metabolism on first passage through the liver

Fungicidal – kills fungi

Fungistatic – inhibits fungal growth

GABA (gamma amino butyric acid) – inhibitory neurotransmitter

Gastrectomy – removal of the stomach

Gene polymorphism – different versions of the same gene

Generic name – official name of a drug

Glial cells – cells in the brain, other than neurons

Gluconeogenesis – formation of glucose from non-carbohydrate sources

Glutamate – excitatory neurotransmitter

Glycogenolysis – break down of glycogen to form glucose

Glycosuria – appearance of glucose in the urine

G protein – a membrane protein that acts as a 'go-between' between receptors and cellular effectors

Gram stain – used to distinguish different types of bacteria

Granuloma – abnormal collection of macrophages and lymphocytes, usually in response to injury

GSL – general sales list medicines

GTP – guanine triphosphate – transfers phosphate groups in cellular reactions

GTPase – enzyme that converts GTP into GDP by removal of phosphate group

HDL high-density lipoprotein – small lipoprotein circulating in blood; mops up cholesterol

5-hydroxtryptamine, 5HT, serotonin – neurotransmitter

Hepatic – of the liver

HOCA – high osmolar contrast agent

HMG-CoA reductase – 3-hydroxy-3-methylglutaryl-CoA reductase, enzyme involved in cholesterol synthesis in the liver

HRT hormone replacement therapy – usually with oestrogen and progesterone

Hydrolysis – splitting of a molecule by addition of water

Hydroxylation – addition of hydroxyl group to a molecule

Hypercalcaemia – high blood calcium levels

Hyperglycaemia – high blood glucose levels

Hyperlipidaemia – high blood lipid levels

Hyperpolarization – increase in resting membrane potential

Hyperproliferation – increase in rate of cell division

Hypersensitivity – adverse immune reaction (allergy)

Hypertension – abnormally high blood pressure

Hypertrophy – increase in size

Hyperuricaemia – high blood uric acid levels

Hypoglycaemia – low blood glucose levels

Hypokalaemia – low blood potassium levels

ICD-10 – International Classification of Diseases 10th Edition

IDDN – insulin dependent diabetes mellitus

Idiosyncrasy – individual peculiarity in response to a drug

Independent prescriber – practitioner who diagnoses and prescribes medicines

Intrinsic factor – factor produced in the stomach essential for vitamin B_{12} absorption

Ischaemia – reduced blood flow to a region

Keratinocyte – epidermal cell that produces keratin

Kinetic energy – intrinsic energy of motion

LDL low-density lipoprotein – medium lipoprotein in blood circulation; deposits in artery walls

Lipodystrophy syndrome – insulin resistance, hyperglycaemia, fat redistribution and raised lipid levels as adverse effect of antiretroviral therapy

Lipolysis – break down of lipid

LOCA – low osmolar contrast agent

Mediators – chemicals released by cells that make something happen

Medicines Act 1968 – legislation that deals with all medicines

MHRA – Medical and Healthcare products Regulatory Agency

Microangiopathy – disease of small blood vessels

MIMS – Monthly Index of Medical Specialities

Misuse of Drugs Act 1971 – legislation that deals with controlled drugs

Molecular weight – the mass of one molecule

Muscarinic receptor – receptor in the parasympathetic nervous system, characterized by response to muscarine

Myalgia – muscle pain

Mycoses – fungal infections

Myopathy – disease of muscle

Myositis – muscle inflammation

Na^+/K^+ ATPase pump – active transport mechanism in cell membranes for transport of sodium and potassium ions

Neuroleptic malignant syndrome – adverse reaction to antipsychotic drugs, hyperthermia, fluctuating loss of consciousness, muscular rigidity and autonomic dysfunction

Neuropathy – disease of nerve

Neurotransmitter – chemical that transmits nerve impulse from one neuron to another
NICE – National Institute for Health and Clinical Excellence
NIDDM – non-insulin dependent diabetes mellitus
Non-medical prescribing – prescribing by health care professionals who are not medically qualified
NSAID – non-steroidal anti-inflammatory drug
Nucleoside – nucleic acid base plus sugar molecule
Nucleotide – nucleic acid base plus phosphorylated sugar molecule
Oedema – abnormal accumulation of tissue fluid
Oncogene – formed by mutation of proto-oncogene
Onychomycosis – fungal nail infection
Osteoblast – cell that produces bone
Osteoclast – cell that destroys bone
Oxidation – addition of oxygen to a molecule
P – pharmacy medicine
Paraesthesia – abnormal sensation, pins and needles
Parenteral – drug administration other than via gastrointestinal tract
Patient group directions – written directions for POM to be supplied without normal prescription to specific groups of patients
Patient specific direction – traditional prescription, or other written direction, for a named patient
pH – measure of the hydrogen ion concentration of a solution
pKa – pH at which there is 50% ionization
Pharmacodynamics – study of effects of drugs on the body
Pharmacokinetics – study of effects of the body on drugs
Plasma protein binding – attachment of a drug to a plasma protein
Plasmin – blood protein that digests blood clots as healing takes place
Plasminogen – inactive form of plasmin
Polydipsia – excessive drinking behaviour
Polypharmacy – the use of many drugs at the same time
Polyuria – production of excessive amounts of urine
POM – prescription-only medicine
Postsynaptic membrane – neuronal membrane immediately after a synapse
Presynaptic membrane – neuronal membrane immediately before a synapse
Prodrug – drug that must be converted to active form in the body
Proprietary name – name given to a drug by the manufacturer
Prothrombinase – enzyme that converts prothrombin to thrombin
Prostacyclin – produced by undamaged blood vessel endothelium, prevents unnecessary platelet aggregation
Proto-oncogene – codes for growth factors, can mutate into oncogene
Purine – components of DNA and RNA (adenine and guanine)
Purple skin striae – skin stretch marks
Pyrimidine – components of DNA (cytosine and thymine) and RNA (cytosine and uracil)
Reduction – removal of oxygen or addition of hydrogen to a molecule

Refractory period – the time interval when a second contraction of heart muscle or a second nerve impulse cannot occur.

Retinopathy – vascular disease of the retina

RNA polymerase – enzyme necessary for synthesis of mRNA

Second messenger – component of a cell that mediates the effects of a neurotransmitter or hormone (the first messenger)

Sequestration – distribution of a drug into a particular tissue

Sickle cell disease – inherited disorder of abnormal haemoglobin

Sinoatrial node – heart pacemaker, group of cells in the right atrium that spontaneously contract

Stevens-Johnson syndrome – vasculitis, arthritis, nephritis, central nervous system abnormalities and myocarditis

Stroke volume – volume of blood ejected by the left ventricle in one contraction

Supplementary prescriber – prescribing in partnership with an independent prescriber

Suppressor oncogene – alternative to tumour suppressor gene

Systole – contraction phase of heart beat

Tardive dyskinesia – abnormalities of movement due to long-term antipsychotic use

TC – total plasma cholesterol

Teratogenesis – causation of damage to developing fetus

Thalassaemia – inherited disorder of abnormal haemoglobin

Tachycardia – abnormally high heart rate

Therapeutic ratio – ratio of maximum non-toxic dose and minimum effective dose of a drug

Thrombin – enzyme that converts fibrinogen to fibrin

Thromboxane A_2 – vasoconstrictor and platelet aggregation enhancer

Thrombus – inappropriate blood clot

Thymectomy – removal of thymus gland

Thyrotoxicosis – effects of hyperthyroidism

Tinea capitis – ringworm infection of the scalp

Tinea corporis – ringworm infection of the body

Tinea pedis – ringworm infection of the foot

Tinea unguium – ringworm infection of the nail

Tubulin – tubular protein that forms the cytoskeleton of neurons and spindle during cell division

Tumour lysis syndrome – caused by death of many cells in a tumour, includes high levels of uric acid

Tumour suppressor genes – stop cell division if cell is damaged

Tyrosine kinase – membrane enzyme, part of Type 3 receptor

Uricosuric – drug that increases uric acid excretion

Urticaria – itchy rash

VLDL – very low-density lipoprotein – large lipoprotein, transports lipids from liver to cells

Volume of distribution – describes the body compartments into which a drug could be distributed

Appendix C
Examples of Patient Group Directions

Example A

Following trauma or surgery, the control of swelling and pain associated with the episode is important to aid recovery and tissue healing. There is a range of non-steroidal anti-inflammatory drugs to enable this. The drug listed is commonly used due to its well-documented tolerability and efficacy.

Diclofenac Sodium 50 mg

Table A.1 Clinical condition or situation to which the direction applies

Indication	Treatment of moderate postoperative pain following foot surgery
Criteria for inclusion	Anticipated or actual post operative pain following foot surgery
Criteria for exclusion	History of peptic ulceration
	History of aspirin or NSAID hypersensitivity/intolerance
	Asthma
	Pregnancy
	Warfarin
	Lithium
Action if excluded	Discuss with clinical lead, or the patient's GP.
	Supply co-dydramol 10/500 as alternative.
Action if patient declines treatment	Supply co-dydramol 10/500 as alternative
	Inform or refer to GP as appropriate.
Reference to national/local policies or guidelines	Current edition of *BNF* www.bnf.org/bnf
	Summary Product Characteristics www.emc.medicines.org.uk

Table A.2   Description of treatment

Name, strength and formulation of drug	Diclofenac sodium 50 mg tablets
Legal status	POM – Prescription-only medicine
Dose/Dose range	50 mg
Method/Route	Oral
Frequency of administration	Eight hourly
Maximum total dose	Seven d
Patient advice/Follow-up treatment/Referral to doctor	Patient telephoned first day following surgery (unless an in-patient) Patient information leaflet from pack is given to all patients In out-patient clinic within eight d (unless an in-patient) Avoid alcohol Report any apparent adverse effects
Identification and management of adverse reactions	The staff covered by this PGD are trained in recognizing the above symptoms and maintain ILS certification
Reporting procedure of adverse reactions	Any adverse events that may be attributable to diclofenac sodium should be reported to the CSM using the yellow card system All suspected adverse drug reactions occurring as a result of treatment under this patient group direction must be reported to a senior medical practitioner responsible for the area in which the PGD is used and the patient's general practitioner
Additional facilities	
Special considerations/ Additional information	Patient information leaflet
Records	Entry made in medical and/or podiatry notes Audit done 6/12 mo

Table A.3   Characteristics of staff

Qualifications required	Podiatric surgeon or podiatrist with POM certificate
Additional requirements	Employed by PCT or a General Practice within PCT area Staff must be authorised and have signed a record of agreement/ declaration of competence Act within the Primary Care Trust Drugs Policy Professional registration must be maintained
Continued training requirements	Maintenance of own level of updating with evidence of continued professional development Staff who agree to administer drugs under Patient Group Directions must record their agreement by signing the attached Record of Agreement, and complete their personal profile

Example B

Patients who are to undergo bone surgery, whether it be Podiatric or from another specialty, may require antibiotic prophylaxis depending on their past medical history. Significant histories include rheumatic fever, heart valve replacement and previous joint replacement surgery. Reasons to provide prophylaxis for primary surgery include planned joint replacement. There is a range of antimicrobials available for preoperative prophylaxis besides the example given including clindamycin and erythromycin.

Cefuroxime 1.5 g

Table B.1   Clinical condition or situation to which the direction applies

Indication	Antibiotic Prophylaxis in adults: When the patient is at an increased risk of infection due to a systemic pathology When the nature of the surgery lends itself to increasing the risk of infection, such as a total joint implant
Criteria for inclusion	Patients undergoing podiatric surgery with a clinical indication for antibiotic prophylaxis
Criteria for exclusion	Hypersensitivity to cephalosporins Hypersensitivity to penicillins except when there is clear evidence that previous use of cephalosporins resulted in NO hypersensitivity reaction Pregnancy or breastfeeding Porphyria
Action if excluded	Discuss treatment alternatives with microbiologist Refer to patient's GP
Action if patient declines treatment	Discuss with clinical lead or refer to GP
Reference to national/local policies or guidelines	Current edition of *BNF* www.bnf.org/bnf Summary product characteristics www.emc.medicines.org.uk

Table B.2   Description of treatment

Name, strength and formulation of drug	Cefuroxime
Legal status	POM – Prescription-only medicine
Dose/Dose range	1.5 g
Method/Route	Single dose at commencement of surgery
Frequency of administration	Single dose
Maximum total dose	Single dose
Patient advice/Follow-up treatment/Referral to doctor	Patient information leaflet from pack is given to all patients
Identification and management of adverse reactions	The staff covered by this PGD are trained in recognizing the above symptoms and maintain ILS certification

(*continued overleaf*)

Table B.2 Continued

Reporting procedure of adverse reactions	Any adverse events that may be attributable to cefuroxime should be reported to the CSM using the yellow card system All suspected adverse drug reactions occurring as a result of treatment under this patient group direction must be reported to a senior medical practitioner responsible for the area in which the PGD is used and the patient's general practitioner
Additional facilities	–
Special considerations/Additional Information	Patient information leaflet
Records	Entry made in medical and/or podiatry notes Audit done 6/12 mo

Table B.3 Characteristics of staff

Qualifications required	Podiatric surgeon or podiatrist with POM certificate
Additional requirements	Employed by PCT or a General Practice within PCT area Staff must be authorised and have signed a record of agreement/declaration of competence Act within the Primary Care Trust Drugs Policy Professional registration must be maintained
Continued training requirements	Maintenance of own level of updating with evidence of continued professional development Staff who agree to administer drugs under Patient Group Directions must record their agreement by signing the attached Record of Agreement, and complete their personal profile

Example C

Patients who are to undergo invasive surgery whether it be Podiatric or from another specialty who have a history of thrombo-embolic disease or other high-risk factors such as malignancy; prolonged post operative immobilisation; or direct family history of thrombo-embolic disease require prophylaxis against potentially life-threatening clinical sequellae such as deep vein thrombosis or pulmonary embolism. The drug listed is only one of a selection of low molecular weight heparins available.

Enoxaparin 2000 units

Table C.1   Clinical condition or situation to which the direction applies

Indication	Prophylactic anticoagulation for patients undergoing podiatric surgery deemed to be at increased risks of post-operative thrombosis/thrombo embolism
Criteria for inclusion	In accordance with thromboprophylaxis guidelines and GP/consultant haematologist
Criteria for exclusion	In accordance with thromboprophylaxis guidelines and GP/consultant haematologist
	History of allergy to enoxaparin
	Not to be used in pregnancy
Action if excluded	Discuss with clinical lead, or the patient's GP
	Refer to consultant haematologist
	Defer surgery until assessment and management agreed with consultant haematologist
Action if patient declines treatment	Refer to consultant haematologist
	Defer surgery until assessment and management agreed with consultant haematologist Inform or refer to GP as appropriate
Reference to national/local policies or guidelines	Current edition of *BNF* www.bnf.org/bnf
	Summary Product Characteristics www.emc.medicines.org.uk

Table C.2   Description of treatment

Name, strength and formulation of drug	Enoxaparin 2000 IU subcutaneous injection
Legal status	POM – Prescription-only medicine
Dose/Dose range	2000 IU (0.2 ml syringe) once daily
Method/Route	Subcutaneous injection
Frequency of administration	Once daily
Maximum total dose	As per *BNF* Guidelines – 2000 units once daily for up to 10 d post op
Patient advice/Follow-up treatment/Referral to doctor	After discharge, continued administration by district nurse (unless an in-patient)
	Platelet counts recommended for patients receiving Heparin for longer than 5 d STOP IMMEDIATELY if 50% reduction of platelet count or development of thrombocytopenia
	Patient telephoned day after surgery, seen in clinic 5–8 d post operatively (unless an in-patient)
	Review in accordance with shared care protocol(s) attached
	Avoid alcohol Report any apparent adverse effects
	Use alternative contraception if on oral contraceptive pill

(*continued overleaf*)

Table C.2   Continued

Identification and management of adverse reactions	The staff covered by this PGD are trained in recognizing the above symptoms and maintain ILS certification
Reporting procedure of adverse reactions	Any adverse events that may be attributable to enoxaparin should be reported to the CSM using the yellow card system
	All suspected adverse drug reactions occurring as a result of treatment under this patient group direction must be reported to a senior medical practitioner responsible for the area in which the PGD is used and the patient's general practitioner
Additional facilities	–
Special considerations/Additional information	Patient information leaflet
Records	Entry made in medical and/or podiatry notes

Table C.3   Characteristics of staff

Qualifications required	Podiatric surgeon or podiatrist with POM certificate
Additional requirements	Employed by PCT or a General Practice within PCT area
	Staff must be authorised and have signed a record of agreement/ declaration of competence
	Act within the Primary Care Trust Drugs Policy
	Professional registration must be maintained
Continued training requirements	Maintenance of own level of updating with evidence of continued professional development
	Staff who agree to administer drugs under Patient Group Directions must record their agreement by signing the attached Record of Agreement, and complete their personal profile

Bibliography

Appelbe, G.E. and Wingfield, J. (2005) *Dale and Applebe's Pharmacy Law and Ethics*, 8th edn, Pharmaceutical Press, London.

American Psychiatric Association (2000) *Diagnostic and Statistical Manual of Mental Disorders*, 4th edn Revised, Washington, DC.

Begg, E.J. (2003) *Instant Clinical Pharmacology*, Blackwell, Oxford.

British National Formulary (2007), No 52, BMJ Publishing Group Ltd and RPS Publishing, London.

Chapman, S. and Nakielny, R. (2001) *A Guide to Radiological Procedures*, 4th edn, Saunders, China.

Cutler, S. (2006) Contrast media, in *Medical Imaging Techniques Reflection and Evaluation*, (eds E. Carver and B. Carver), Churchill Livingstone, London.

Dale, M.M. and Haylett, D.G. (2004) *Pharmacology Condensed*, Churchill Livingstone, Edinburgh.

Department of Health (1999) *Review of Prescribing, Supply and Administration of Medicines*, (Crown Report 2), Department of Health, London.

Department of Health (2006) *Medicines Matters: A Guide to the Mechanisms for the Prescribing, Supply and Administration of Medicines*, Department of Health, London.

Ehrlich, R.A. (2004) *Patient Care in Radiography*, 6th edn, Mosby, USA.

Hardman, J.G., Gilman, A.G. and Limbird, L.E. (eds) (1995) *The Pharmacological Basis of Therapeutics*, 9th edn, McGraw-Hill, New York.

Jenkins, G.W., Kemnitz, C.P. and Tortora, G.J. (2007) *Anatomy and Physiology from Science to Life*, Wiley, Chichester.

Kumar, P. and Clark, M. (2005) *Clinical Medicine*, Saunders, Edinburgh.

Merrills, J. and Fisher, J. (2001) *Pharmacy Law and Practice*, Blackwell Science, Oxford.

Monthly Index of Medical Specialities (2006, January), Haymarket Publishing Services Ltd, London.

Morgan, R. and Johnson, M. (2000) *Pharmacology for Podiatrists*, Blackwell Science, Oxford.

Neal, M.J. (2005) *Medical Pharmacology at a Glance*, 5th edn, Blackwell Science, Oxford.

Ng, R. (2004) *Drugs from Discovery to Approval*, John Wiley & Sons, Hoboken, New Jersey.

Porth, C.M. (2007) *Essentials of Pathophysiology: Concepts of Altered Health States*, 2nd edn, Lippincott Williams and Wilkins, Philadelphia.

Pharmacology for the Health Care Professions Christine M. Thorp
© 2008 John Wiley & Sons, Ltd

Rang, H.P., Dale, M.M., Ritter, J.M. and Moore, P. (2003) *Pharmacology*, 5th edn, Churchill Livingstone, Edinburgh.

Roach, S. (2005) *Pharmacology for Health Professionals*, Lippincott Williams and Wilkins, Baltimore.

Royal Pharmaceutical Society of Great Britain (2007) *Medicines, Ethics and Practice No 31*, Pharmaceutical Press,London.

Royal Pharmaceutical Society of Great Britain *Pharmaceutical Journal (relevant editions)*, RPS Publishing,London.

Sherwood, L. (1989) *Human Physiology from Cells to Systems*, St Paul MN, West St Paul.

Simonsen, T., Aarbakke, J., Kay, I. *et al.* (2006) *Illustrated Pharmacology for Nurses*, Hodder Arnold, London.

Williams, B., Poulter, N.R., Brown, M.J. *et al.* (2004) The BHS working party British Hypertension Society guidelines for hypertension management, 2004-BHS IV: Summary. *British Medical Journal*, **328**, 634–40.

Useful websites

At the time of writing, the following websites were accessible.

www.arc.org.uk/ Arthritis Research Campaign, Information for Medical Professionals.

www.askaboutmedicines.org General information about medicines and useful links to other sources of information.

www.bhsoc.org British Hypertension Society, Hypertension guidelines and 10-year risk of cardiovascular disease estimation.

http://www.biomedcentral.com/bmcendocrdisord/ Biomedical Central, Endocrine Disorders e-journal.

http://www.bnf.org/bnf/ British National Formulary.

www.brit-thoracic.org.uk British Thoracic Society, Management guidelines.

www.cancerhelp.org.uk Cancer Research UK, Credible sources of information on cancer chemotherapy.

www.cancerresearch.org Cancer Research Institute USA, Useful links to information sources.

www.cftrust.org.uk Cystic Fibrosis Trust, General information about cystic fibrosis.

www.chre.org.uk Council for Healthcare Regulatory Excellence, Established April 2003 to oversee regulation of healthcare professionals, including the Health Professions Council.

www.cks.library.nhs.uk National Library for Health, Clinical Knowledge Summaries.

www.csp.org.uk The Chartered Society of Physiotherapy.

www.dermatology.co.uk/ Leeds General Infirmary Dermatology Department, Information about eczema, psoriasis, warts, skin infections, and their treatment.

http://www.dh.gov.uk/en/PolicyAndGuidance/MedicinesPharmacyAndIndustry/Prescriptions/index.htm Guidance on Department of Health Prescribing Policy.

http://www.dh.gov.uk/en/PolicyAndGuidance/MedicinesPharmacyAndIndustry/Prescriptions/TheNon-MedicalPrescribingProgramme/index.htm Department of Health, Non-medical Prescribing.

http://www.dh.gov.uk/en/Publicationsandstatistics/Publications/PublicationsPolicyAndGuidance/DH_064325 Department of Health, Medicines Matters July 2006.

http://www.dh.gov.uk/en/Publicationsandstatistics/Publications/PublicationsPolicyAndGuidance/DH_4110032 Department of Health, Supplementary Prescribing by Nurses, Pharmacists, Chiropodists/Podiatrists, Physiotherapists and Radiographers within the NHS in England A Guide for Implementation 2005.

Pharmacology for the Health Care Professions   Christine M. Thorp
© 2008 John Wiley & Sons, Ltd

http://www.diabetes.org.uk/Professionals Diabetes UK, Healthcare Professionals and Scientists page.

www.emc.medicines.org.uk Electronic Medicines Compendium, Patient Information Leaflets and Summaries of Product Characteristics.

www.emea.europa.eu/ European Medicines Agency, European Agency for the Evaluation of Medicines.

www.feetforlife.org Society of Chiropodists and Podiatrists.

www.frca.co.uk/ Anaesthesia UK, Local Anaesthetic Pharmacology.

http://www.frca.co.uk/SectionContents.aspx?sectionid=66 Anaesthesia UK, Section on Pharmacology.

http://guidance.nice.org.uk National Institute for Health and Clinical Excellence Guidelines. Useful for guidance on cancer chemotherapy and other topics.

http://www.hpa.org.uk/infections/topics_az/malaria Health Protection Agency, Malaria prevention and treatment guidelines.

www.hpc-uk.org/ Health Professions Council, Professional registers.

http://www.hpcheck.org/lisa/onlineregister/MicrositesSearchInitial.jsp Online Health Professions Council register for membership status.

www.library.nhs.uk National Library for Health, Page for health care professionals.

http://www.medcyclopaedia.com/library/topics GE Healthcare Medical Diagnostics, Encyclopaedia of medical imaging.

www.medicines.org.uk Datapharm Communications Ltd, Up-to-date information on UK medicines, electronic medicines compendium.

http://www.merck.com/mmhe/sec02.html The Merck Manuals Online Medical Library, Section on Drugs.

http://www.merck.com/mmhe/sec17.html The Merck Manuals Online Medical Library, Section on Infections.

http://www.merck.com/mmhe/sec18.html The Merck Manuals Online Medical Library, Section on Skin Disorders.

www.mgauk.org/ Myasthenia Gravis Association, Medical Guides for Patients.

www.mhra.gov.uk Medicines and Healthcare Products Regulatory Agency, Use of contrast agents and all medicines.

http://www.mhra.gov.uk/home Statutory Instrument 2005 No. 766 The Medicines (Pharmacy and General Sale - Exemption) Amendment Order 2005.

www.mndassociation.org/ Motor Neurone Disease Association, Section for Professionals.

www.mssociety.org.uk/ Multiple Sclerosis Society, Professional Resources.

www.nice.org.uk/ National Institute for Health and Clinical Excellence, Guidelines on ADHD expected in 2008.

http://www.nice.org.uk/CG020 National Institute for Health and Clinical Excellence, Guidelines on Epilepsy 2004.

http://www.nice.org.uk/CG022 National Institute for Health and Clinical Excellence, Guidelines on Anxiety 2007.

http://www.nice.org.uk/CG023 National Institute for Health and Clinical Excellence, Guidelines on Depression 2007.

http://www.nice.org.uk/CG0235 National Institute for Health and Clinical Excellence, Guidelines on Parkinson's disease 2006.

http://www.nice.org.uk/CG034 National Institute for Health and Clinical Excellence Hypertension Guidelines 2006.

http://www.nice.org.uk/CG36 National Institute for Health and Clinical Excellence, Atrial Fibrillation Guidelines 2006.

http://www.nice.org.uk/TA58 National Institute for Health and Clinical Excellence, Anti-viral Therapy Guidelines 2003.

http://www.nice.org.uk/TA43 National Institute for Health and Clinical Excellence, Guidelines on Schizophrenia 2002.

http://www.nice.org.uk/TA111 National Institute for Health and Clinical Excellence, Guidelines on Alzheimer's disease 2007.

http://www.nice.org.uk/TA125 National Institute for Health and Clinical Excellence, Psoriatic Arthritis (moderate to severe) – adalimumab Guidance 2007.

http://www.nice.org.uk/TA126 National Institute for Health and Clinical Excellence, Rheumatoid Arthritis (refractory) – rituximab Guidance 2007.

http://www.nos.org.uk/professionals.htm National Osteoporosis Society, Web page for Healthcare Professionals.

www.npc.co.uk National Prescribing Centre.

www.npsa.nhs.uk National Patient Safety Agency.

http://www.paget.org.uk/guidelines.pdf National Association for the Relief of Paget's Disease, Guidelines for the Management of Paget's Disease of Bone.

www.painrelieffoundation.org.uk The Pain Relief Foundation Research into Chronic Pain.

www.parkinsons.org.uk Parkinson's Disease Society, Section for Healthcare Professionals.

www.portal.nelm.nhs.uk National Electronic Library for Medicine.

http://www.ppa.org.uk/ppa/edt_intro.htm Prescription Pricing Authority, NHS Drug Tariff for England and Wales.

www.prescribing.info general information about non-medical prescribing, (London Metropolitan University).

www.prodigy.nhs.uk Patient Information Leaflets and guidance on common conditions.

www.radiology.co.uk Scottish Radiological Society, General information on contrast agents.

www.radiologyinfo.org Radiological Society of North America and American College of Radiology, Patient Information Leaflets.

www.rcpsych.ac.uk/ Royal College of Psychiatrists, Mental Health Information.

www.rcrad.org.uk Royal College of Radiographers, Authentication required for entry to website.

www.rheumatology.org.uk British Society for Rheumatology, Clinical Guidelines.

www.rpsgb.org.uk Royal Pharmaceutical Society of Great Britain.

www.scpod.org The Society of Chiropodists and Podiatrists:.

http://www.sign.ac.uk/guidelines/ Scottish intercollegiate Guidelines network, Section 13 Oral anticoagulants.

www.sor.org.uk Society of Radiographers, Professional and educational website.

www.thepainweb.com The Pain Web, Website for Health Professionals dealing with Pain.

http://www.who.int/topics/malaria/en/, Guidelines for the treatment of malaria.

www.yellowcard.gov.uk Yellow card scheme for reporting adverse drug reactions, Healthcare Professions and Patients.

Index

α-glucosidase inhibitors, 110
$α_1$-antagonists, 67
5-HT *see* serotonin
abacavir, 163
absorption of drugs, 13–17, 32, 36
acarbose, 110
ACE *see* angiotensin-converting enzyme
acetylcholine, 41, 194
acetylcholinesterase inhibitors, 221
aciclovir, 162
acitretin, 144
acquired immunodeficiency syndrome *see*
     HIV/AIDS
ACTH *see* adrenocorticotrophic hormone
action potentials, 53–4
active transport, 15, 24
acute coronary thrombosis, 57
adalimumab, 122
Addison's disease, 107
adenosine triphosphate (ATP), 15
ADH *see* antidiuretic hormone
ADHD *see* attention deficit hyperactivity
     disorder
adhesion molecules, 179
adjuncts
     anaesthesia, 231, 234–6
     radiography, 253, 260–4
ADME studies *see* drug disposition;
     pharmacokinetics
administration of drugs, 9–12
adrenal glands, 105–7, 112
adrenaline, 194, 277
adrenergic neurone blockers, 68
adrenocorticotrophic hormone (ACTH),
     101–3, 105, 198

affective disorders, 191,
     194–201
afterload, 52
age effects
     analgesia, 246
     drug disposition, 23
     pharmacodynamics, 31, 33,
          36–9
AIDS *see* HIV/AIDS
alcohol, 22, 32, 39
aldosterone, 105, 112
alemtuzumab, 187
alkylating agents, 182–3
allergic rhinitis, 87, 92–3
allergies, 34–5, 239
allopurinol, 125
alteplase, 73
alveoli, 86–7
Alzheimer's disease, 191, 221–3
amantadine, 94, 163–4, 215
ambulance paramedics, 273
amfetamine, 25
aminoglycosides, 159–60
aminophylline, 90, 92, 260
amiodarone, 67
amisulpiride, 204
amitriptyline, 198
amlodipine, 66
ammonium chloride, 25
amorolfine, 11, 149, 167, 276
amoxicillin, 93, 159, 277
amphotericin, 94, 166
amprenavir, 163
anabolic steroids, 278
anaemias, 73–5

Pharmacology for the Health Care Professions   Christine M. Thorp
© 2008 John Wiley & Sons, Ltd

anaesthesia, 229–43, 248–9
  adjuncts, 231, 234–6
  administration, 12
  case studies, 249–50
  complications and risks, 238–41
  general anaesthesia, 229–36
  historical development, 230
  inhalation anaesthetics, 231–2
  intravenous anaesthetics, 232–4, 237
  local anaesthesia, 236–43
  mechanism of action, 230–1, 237–8
  pharmacodynamics, 35, 44
  pharmacokinetics, 25
  premedication, 231, 234–6
  review questions, 250–1
  techniques, 236–7
anakinra, 122–3
analgesia, 229, 243–9
  adjuncts to radiography, 261
  case studies, 249–50
  centrally acting analgesics, 247
  neuropathic pain, 247–8
  pathophysiology of pain, 243–4
  peripherally acting analgesics, 244–6
  premedication and adjuncts in
    anaesthesia, 234, 235–6
  review questions, 250–1
anaphylaxis, 34–5, 260, 274
androgens, 106
angina pectoris, 57–8, 64
angiogenesis, 179
angiotensin II receptor antagonists, 63
angiotensin-converting enzyme (ACE)
    inhibitors, 56, 63, 79, 140
ankylosing spondylitis, 116
anogenital warts, 145
anterior pituitary gland, 100–2
anthelmintics, 170–3
antibiotics, 155, 156–61
  bacterial resistance, 156–7, 159–60
  cancer chemotherapy, 184
  cell wall synthesis inhibitors, 158–9
  cytotoxic, 184
  nucleic acid synthesis inhibitors,
    157–8

protein synthesis inhibitors, 159–60
respiratory disorders, 93–4
skin disorders, 143
tuberculosis, 160–1
wound care, 264
anticholinergic drugs, 215
anticholinesterase drugs, 129–30
anticoagulants, 72
antidepressants, 197–200
antidiuretic hormone (ADH), 11, 99–101
antiemetics, 263–4
antifungals, 166–7
antihelmintics, 155
antihistamines, 92–3, 95, 139, 197, 210,
    260
anti-inflammatory drugs, 88
antimalarials, 168, 169–70
antimetabolites, 183–4
antimuscarinics, 90, 234–5
antipsychotics, 201, 203–5, 209
antisense oligonucleotides (ON), 187
antivirals, 162–5
anxiety, 191, 205–6, 207–9
anxiolytics, 207–9
apomorphine, 214
apoptosis, 178
aqueous solubility, 18
arrhythmias, 58, 64–7
arteriosclerosis, 60
arthrography, 255
aspartate, 194
aspirin
  analgesia, 245–6
  blood disorders, 73
  musculoskeletal disorders, 118, 124
  natural sources, 1
  pharmacodynamics, 32, 45
  pharmacokinetics, 18–19, 21–2, 25
asthma, 87, 88, 91–2
astringents, 139
asystole, 58
atenolol, 65
athlete's foot, 147
atomoxetine, 211
atopic eczema, 139

ATP *see* adenosine triphosphate
atracurium, 235
atrial fibrillation, 58
atrial flutter, 58
atrioventricular (AV) node, 53, 58, 61
atropine, 260
attention deficit hyperactivity disorder
    (ADHD), 191, 210–11
atypical antipsychotics, 205
augmented adverse reactions, 30–2
AV *see* atrioventricular
azathioprine, 130

β-adrenoreceptors, 41–3
β-blockers
    cardiovascular disorders, 64–5
    central nervous system disorders, 206,
        209
    skin disorders, 140
β2-adrenoceptor stimulants, 89
baclofen, 131
bacterial infections
    chemotherapy, 155–61
    respiratory disorders, 93
    skin disorders, 150–1
bacterial pneumonia, 93
bacterial resistance, 156–7, 159–60
BAN *see* British Approved Name
barium sulphate, 256
baroreceptors, 56
beclometasone, 91, 92
bendroflumethiazide, 62
benzamide derivatives, 204
benzatropine, 215
benzimidazoles, 171–2
benzodiazepines
    central nervous system disorders, 201,
        206–8, 210, 217, 219
    legislation, 278
    skin disorders, 143
benzoic acid, 148
benzylpenicillin, 150
bevacixumab, 188
bezafibrate, 78
Bier's block, 237

biguanides, 109–10
bile acid sequestrants, 78
biological response modifiers, 187
bipolar depression, 196–8, 200–1
bisacodyl, 262
bisoprolol, 65
bisphosphonates, 126–7
biventricular failure, 57
bizarre adverse reactions, 32–5
black triangle drugs, 282
bleomycin, 146, 184
blood disorders, 68–79
    anaemias, 73–5
    coagulation, 69–73
    haemostasis, 69–70
    lipid metabolism, 75–9
blood flow, 18
blood pressure, 51, 55–60, 62–8
blood vessel reactions, 69
blood-brain barrier, 18
BNF *see* British National Formulary
BNMS *see* British Nuclear Medicine
    Society
Bolam test, 312
borotannic complex, 147
bowel evacuants, 263
Bowman's capsule, 23–4
BP *see* British Pharmacopoeia
breast feeding, 40
British Approved Name (BAN), 4
British National Formulary (BNF), 3, 4
    pharmacodynamics, 30, 39, 40
    prescribing, 279, 282, 302–3
    respiratory disorders, 89
British Nuclear Medicine Society
    (BNMS), 311
British Pharmacopoeia (BP), 3
British Society for Rheumatology, 117
bromocriptine, 214
bronchioles, 86–7
bronchitis, 87, 88–9, 91
bronchodilators, 88, 89–90, 94–5
buccal administration, 11
budesonide, 91
bulk-forming laxatives, 261

bupivacaine, 241–2, 277
buscopan, 263
buspirone, 208–9
busulphan, 183
butyrophenones, 204

calcipotriol, 141
calcitonin, 103, 104–5, 126, 127
calcium channel blockers, 65–6, 79
Calman Hine Report, 304
cancer chemotherapy, 177–89
    alkylating agents, 182–3
    antimetabolites, 183–4
    biological response modifiers, 187
    biology of cancer, 177–80
    case studies, 189
    cytotoxic antibiotics, 184
    DNA structure, replication and protein
        synthesis, 179–80
    hormones, 185–6
    new approaches, 187–8
    principles, 180–2
    resistance, 181–2
    review questions, 189
    vinca alkaloids, 185
captopril, 63
carbamazepine, 22, 201, 217–18
carbimazole, 35, 103–4
carbocisteine, 91
carcinogenesis, 35
cardiac failure, 57, 62, 65
cardiac glycosides, 61–2
cardiac output, 52
cardiovascular disease (CVD), 77
cardiovascular disorders, 51–68
    arrhythmias, 58, 64–7
    blood pressure, 51, 55–7, 58–60,
        62–6, 67–8
    cardiac failure, 57
    case studies, 80–2
    drug therapies, 60–8
    ischaemic heart disease, 57–8, 63–6,
        79
    physiology, 51–5
    review questions, 82–3

Care Standards Act 2000, 295
carmustine, 183
carrier proteins, 44–5
catechol-o-methyl transferase (COMT)
    inhibitors, 214
cefuroxime, 337–8
celecoxib, 118
celiprolol, 65
cell-mediated reactions, 35
cellulitis, 150–1
central core, 191–2
central nervous system (CNS) disorders,
    191–227
    affective disorders, 191, 194–201
    Alzheimer's disease, 191, 221–3
    anxiety, 191, 205–6, 207–9
    attention deficit hyperactivity disorder,
        191, 210–11
    brain physiology, 191–3
    case studies, 223–7
    dementia, 191
    depression, 191, 195–201
    epilepsy, 191, 216–20
    insomnia, 191, 206–7, 209–10
    mania, 195–8, 200–1
    neurological disorders, 191, 211–21
    neurotransmitters, 193–4
    Parkinson's disease, 191, 192, 211–16
    psychological disorders, 191, 194–211
    psychoses, 191, 201–11
    review questions, 227
    schizophrenia, 191, 192, 201–5
centrally acting analgesics, 247
centrally acting antihypertensives, 67–8
cephalosporins, 158–9
cerebral cortex, 193
Chartered Society of Physiotherapy, 318
chemotherapy
    bacterial infections, 155–61
    cancer, 177–89
    case studies, 174–6
    fungal infections, 165–7
    infectious diseases, 155–76
    parasitic worms, 170–3
    principles, 180–2

protozoan infections, 168–70
review questions, 176
viral infections, 161–6
child resistant containers (CRCs), 274
children *see* paediatrics
chiropody, 272, 277
chloramphenicol, 159–60
chloroquine, 121–2, 140, 169–70
chlorpromazine, 40, 203
CHM *see* Commission on Human
    Medicines
cholesterol, 75–6
cholestyramine, 78
chronic bronchitis, 88–9
chronic obstructive pulmonary disease
    (COPD), 89
ciclosporin, 144
cimetidine, 22, 32, 149
ciprofloxacin, 93, 157–8
cisplatin, 183
Clinical Management Plans (CMPs),
    280–4, 296–300, 303, 307–8,
    316–17
clomethiazole, 209, 210
clonidine, 67–8
clopazine, 205
clopidogrel, 73
CMPs *see* Clinical Management Plans
CNS *see* central nervous system
coagulation, 69–73
coal tar preparations, 142
co-dydramol, 276
colchicine, 124–5
colestipol, 78
Commission on Human Medicines
    (CHM), 3, 30
community podiatry, 299–300
compliance, 4, 38
computed tomography (CT), 198, 255–6
COMT *see* catechol-*o*-methyl transferase
conduction block, 237
Conn's syndrome, 107, 112
contact dermatitis, 139
containers, 274

continuing professional development
    (CPD), 298
contrast agents, 253–60
controlled drugs, 278–9
COPD *see* chronic obstructive pulmonary
    disease
corticosteroids
    adverse reactions, 119–20
    drug disposition, 12
    musculoskeletal disorders, 116,
        118–20, 129–31
    respiratory disorder, 91, 92
    skin disorders, 139, 143
corticotrophin releasing hormone (CRH),
    102, 105
cortisol, 198
co-trimoxazole, 158, 170
COX *see* cyclo-oxygenase
CPD *see* continuing professional
    development
CRCs *see* child resistant containers
CRH *see* corticotrophin releasing
    hormone
crisantaspase, 186
cromoglicate, 90, 92
cryotherapy, 146
CT *see* computed tomography
Cushing's syndrome, 107, 112
CVD *see* cardiovascular disease
cyclo-oxygenase (COX) inhibitors,
    117–18, 245
cyclophosphamide, 182–3
cyclosporins, 140
cystic fibrosis, 87, 93, 95
cytarabine, 184
cytokine inhibitors, 122–3
cytotoxic antibiotics, 184
cytotoxic reactions, 35

dactinomycin, 184
dantrolene, 131
deep brain nuclei, 192
dementia, 191
dependent prescribers *see* supplementary
    prescribing

depolarizing neuromuscular blockers,
235
depression, 191, 195–201
dermis, 137–8
desflurane, 231–2
destruction of controlled drugs, 279
dexamphetamine, 211
diabetes mellitus, 106–11, 112
diagnostic radiography, 303–4, 306, 308,
312
Diagnostic and Statistical Manual of
Mental Disorders (DSM-IV), 195,
201, 206
diaphragm, 85–7
diastolic pressure, 51, 55
diclofenac, 236, 246, 335–6
diffusion, 13–15
digitalis, 1
digoxin, 22, 61–2
diltiazem, 66
dipyridamole, 73
discoid lupus erythematosus, 122
disease states, 31, 33
disease-modifying antirheumatic drugs
(DMARDs), 117, 118–19, 121–3
disposition to drugs see drug disposition
distribution of drugs, 17–20, 37
district nurses, 3
dithranol, 141–2
diuretics, 62, 79
DMARDs see disease-modifying
antirheumatic drugs
DNA structure, 179–80
docetaxel, 185
domperidone, 263–4
donepezil, 221
dopamine
central nervous system disorders, 194,
197, 201–5, 213–15
endocrine disorders, 102
receptor agonists, 214–15
release stimulators, 215
dopaminergic drugs, 213–14
dosages, 4
dose–response curves, 30

doxorubicin, 184
drug disposition, 9–27
absorption, 13–17
administration of drugs, 9–12
age effects, 23
case studies, 26–7
distribution, 17–20
excretion, 19, 23–6
metabolism, 19, 20–3
pharmacokinetics, 9, 10
response variations, 36–7
review questions, 27
drug nomenclature, 4–5
drug–drug interactions, 2
anaesthesia, 240
endocrine disorders, 110
pharmacodynamics, 32, 33, 38
plasma protein binding, 18–19
skin disorders, 148, 150
tubular secretion, 24–5
DSM-IV see Diagnostic and Statistical
Manual of Mental Disorders

ECG see electrocardiograms
ectopic beats, 58–9
eczema, 138–40
EDV see end diastolic volume
efalizumab, 144
elderly patients see geriatrics
electrocardiograms (ECG), 53–5
electrosurgery, 146
emollients, 139, 141
end diastolic volume (EDV), 52
end systolic volume (ESV), 52
endocrine disorders, 99–114
adrenal glands, 105–6, 112
case studies, 112–13
diabetes mellitus, 106–11, 112
pancreas, 106–8, 112
parathyroid glands, 104–5, 111
pituitary gland, 99–102, 111
review questions, 113–14
thyroid gland, 102–4, 111
enhanced patient outcomes, 293
enoxaparin, 338–40
enterohepatic shunting, 17, 25

enzymes
  induction, 22, 32
  inhibition, 22, 32
  targetting, 45
epidermis, 137–8
epidurals, 12, 237
epilepsy, 191, 216–20
ergosterol, 166–7
erythrocytes, 74–5
erythrodermic psoriasis, 140
erythromycin
  bacterial infections, 150, 159–60
  drug disposition, 22
  legislation, 277
  respiratory disorders, 93
  skin disorders, 150
essential hypertension, 59–60
ESV see end systolic volume
etanercept, 122
ethambutol, 161
ethnicity, 63, 65
ethosuximide, 219
etomidate, 233
etoposide, 185
excision of warts, 146
excretion of drugs, 19, 23–6, 32, 37
extra-pyramidal side effects, 203–4
ezitimibe, 77–8

facilitated diffusion, 14–15
faecal softeners, 261
fenbrufen, 118, 246
fentanyl, 236
foetal tissue transplants, 216
fibrates, 78
fibrinolytics, 73
first pass metabolism, 17, 37
flecainide, 67
flucloxacillin
  bacterial infections, 150, 156, 158–9
  legislation, 277
  respiratory disorders, 93
  skin disorders, 150
fluconazole, 148, 167
flucytosine, 167
fluorouracil, 183–4

fluoxetine, 199
folic acid, 74–5, 157, 183, 219
follicle stimulating hormone (FSH),
  101–3
formulary development, 293
FSH see follicle stimulating hormone
fungal infections
  chemotherapy, 165–7
  respiratory disorders, 94
  skin disorders, 146–50
fungal pneumonia, 94
furosemide, 62

G-proteins, 41–3
GABA see gamma-amino butyric acid
gabapentin, 220
gadolinium contrast agents, 259
gamma-amino butyric acid (GABA)
  anaesthesia, 230, 232
  analgesia, 243
  central nervous system disorders, 194,
    197, 206–9, 217
ganciclovir, 162
gastrointestinal tract (GIT), 15–17
general anaesthesia, 229–36
general practitioners (GPs), 292,
  299–300
general sale list (GSL) items,
  271, 272
generalized epilepsy, 216
genetic variations, 31–2, 33
geriatrics
  drug disposition, 23
  pharmacodynamics, 31, 33, 36–9
GH see growth hormone
GIT see gastrointestinal tract
glibenclamide, 109
gliclazide, 109
glipizide, 109
glitazones, 111
glomerular filtration, 23–4
glucagon, 263
glucocorticosteroids, 105–6, 186
glutamate, 194, 221, 232
glyceryl trinitrate (GTN), 64

glycine, 194
glycopeptides, 158–9
GnRH *see* gonadotrophin releasing
    hormone
gold salts, 121
gonadotrophin releasing hormone
    (GnRH), 102
gout, 115, 123–5
GPs *see* general practitioners
granisetron, 264
Grave's disease, 103–4
griseofulvin, 19, 149–50, 166
growth hormone (GH), 101–3
GSL *see* general sale list
GTN *see* glyceryl trinitrate

haemorrhagic anaemias, 73
haemostasis, 69–70
hair follicles, 138
halofantrine, 169
haloperidol, 204
halothane, 231, 235
Hashimoto's thyroiditis, 104
hay fever, 87, 92–3
HDLs *see* high-density lipoproteins
Health Professions Council (HPC), 2, 3
    chemotherapy, 159–60
    podiatry, 147, 292
    prescribing, 275–7, 283–4
Health and Social Care Act 2001, 280,
    297
health visitors, 3
heart attacks, 57
heart block, 58–9
helminths, 170–3
heparin, 35, 72
hepatitis C, 164
herceptin, 187
high osmolar contrast agents (HOCAs),
    255, 258
high-density lipoproteins (HDLs), 75–6
hirudin, 72
HIV/AIDS, 157, 164–5, 173–4
HOCAs *see* high osmolar contrast agents
hormone replacement therapy (HRT), 128

hormones
    cancer chemotherapy, 185–6
    central nervous system disorders,
        198
    endocrine disorders, 99–105, 111
    targets for drug action, 43–4
HPC *see* Health Professions Council
HPV *see* human papilloma virus
HRT *see* hormone replacement therapy
human immunodeficiency virus *see*
    HIV/AIDS
human papilloma virus (HPV),
    144–6
hydrocortisone, 277
hydrolysis, 21
hydroxychloroquine, 121–2
hydroxyurea, 186
hyoscine, 235
hyperglycaemia, 106–7
hyperlipidaemias, 62, 76–9
hypersecretion of thyroid hormone,
    103–4, 111
hypersensitivity, 34–5
hypertension, 51, 57, 58–60, 62–6,
    67–8
hyperuricaemia, 124
hypnotics, 207
hypoglycaemia, 108–9, 110
hypokalaemia, 62
hypomania, 195–6, 200
hyposecretion of thyroid hormone, 104,
    111
hypothalamic hormones, 99–102
hypothalamus, 192
hypothyroidism, 104

ibuprofen, 118, 236, 246, 277
ICD-10 *see* International Classification of
    Diseases
IDDM *see* diabetes mellitus
idiosyncrasy, 33
IL-1 *see* interleukin-1
imidazoles, 148, 167
imipramine, 197, 198
immune complex reactions, 35

immunosuppression
  musculoskeletal disorders, 122,
    129–30
  skin disorders, 140
independent prescribers, 279, 282–3,
  296–8, 302, 308, 310
indinavir, 163
indometacin, 246
infants *see* paediatrics
infections *see* bacterial infections; fungal
  infections; parasitic worms;
  protozoan infections; viral
  infections
infiltration, 236
infliximab, 122
inhalation administration, 12, 231–2
injection administration, 12,
  232–4, 237
insensitivity, 33
insomnia, 191, 206–7, 209–10
insulin, 106–9, 111
  resistance, 62
integrins, 179
interferons, 131, 164, 187
interleukin-1 (IL-1) inhibitors, 122–3
interleukin-2, 187
International Classification of Diseases
  (ICD-10), 195, 201
intolerance to drugs, 33
intra-arterial injections, 12
intra-articular injections, 12
intramuscular injections, 12
intraspinal injections, 12
intravenous injections, 12, 232–4, 237
iodine
  contrast agents, 254–5, 257, 259
  radioactive, 186
ion channels, 44
iopamidol, 255–6
ipratropium, 90, 92
iproniazid, 197
iron deficiencies, 74
ischaemic heart disease, 57–8, 63–6, 79
isoflurane, 231–2
isoniazid, 23, 74, 94, 161

isosorbide dinitrate/mononitrate, 64
itraconazole, 94
ivermectin, 171, 173

juvenile rheumatoid arthritis, 116

keratinocytes, 137–8, 140
keratolytics, 142, 145–6
ketamine, 233–4, 248

labelling, 273–4
lamivudine, 163
lamotrigine, 220
laser treatments, 146
laxatives, 261–3
LDLs *see* low density lipoproteins
left ventricular failure, 57
legislation, 271–9, 286–9
  prescribing, 295–6, 301, 311–12
leukotriene receptor antagonists, 91, 92,
  94–5
levobupivacaine, 241–3, 277
levodopa, 213–14
Lewy bodies, 213, 221
LH *see* luteinising hormone
lidocaine
  anaesthesia, 237, 240, 241–2, 248
  cardiovascular disorders, 66
  legislation, 277
lifestyle changes, 111
lipids
  lipid-lowering drugs, 77–9
  metabolism, 75–9
  solubility, 19
lipodystrophy syndrome, 163
lipoproteins, 110
lisinopril, 63
lithium, 40, 140, 200–1
LOCAs *see* low osmolar contrast agents
loop diuretics, 62
loperamide, 264
low density lipoproteins (LDLs), 75–6,
  78, 110
low osmolar contrast agents (LOCAs),
  255, 258
lubricants, 261

lupus, 116
luteinising hormone (LH), 101–3

macrolides, 159–60
magnesium citrate/sulphate, 262–3
magnetic resonance imaging (MRI),
    255–6
malaria, 168, 169–70
mania, 195–8, 200–1
MAO-B see monoamine oxidase-B
MAOIs see monoamine oxidase
    inhibitors
maprotiline, 199
mast-cell stabilizers, 90
mebendazole, 171–2
Medicines Act, 1968 1, 2, 271–7
    1978 amendment, 311–12
    prescribing, 293, 295, 301, 318, 319
Medicines and Healthcare Products
    Regulatory Agency (MHRA), 3,
    30, 122
mefenamic acid, 246
meglumine salts, 254
membrane stabilizers, 66–7
membrane transport mechanisms, 13–15
mepivacaine, 241–2, 277
mercaptopurine, 183
messenger RNA (mRNA), 180
metabolism of drugs, 19, 20–3, 37
metformin, 109–10
methadone, 248
methotrexate, 32, 122, 143–4, 183
methyl cellulose, 261
methyldopa, 45, 67
methylphenidate, 211
methylprednisolone, 119, 277
meticillin resistant Staphylococcus aureus
    (MRSA), 156, 159
metoclopramide, 263
MHRA see Medicines and Healthcare
    Products Regulatory Agency
miconazole, 148, 167
midwives, 3
MIMS see Monthly Index of Medical
    Specialities

mirtazapine, 199
Misuse of Drugs Act, 1971 271, 278–9
moclobemide, 200
monoamine oxidase inhibitors (MAOIs),
    197, 200
monoamine oxidase-B (MAO-B)
    inhibitors, 215
monoclonal antibodies, 123, 187, 188
montelukast, 91, 92
Monthly Index of Medical Specialities
    (MIMS), 3, 4, 30, 89
morphine, 236, 247
motor neuron disease, 115, 130–1
MRI see magnetic resonance imaging
mRNA see messenger RNA
MRSA see meticillin resistant
    Staphylococcus aureus
mucolytics, 91
multiple sclerosis, 115, 131
muscle relaxants, 234, 235
musculoskeletal disorders, 115–35
    case studies, 133–5
    gout, 115, 123–5
    motor neuron disease, 115, 130–2
    multiple sclerosis, 115, 131–2
    myasthenia gravis, 115, 128–30
    osteoarthritis, 115, 125–6
    osteomalacia, 115, 128
    osteoporosis, 115, 127–8
    Paget's disease, 115, 126–7
    review questions, 135
    rheumatic diseases, 115–23
myasthenia gravis, 115, 128–30
Mycobacterium tuberculosis, 160–1

Na$^+$/K$^+$ ATPase pumps, 61
naloxone, 247
naproxen, 118, 246
nasal administration, 11
National Institute for Health and Clinical
    Excellence (NICE), 3, 122–3, 131,
    163–4
National Prescribing Centre, 298
National Reporting and Learning System
    (NRLS), 30

nephrogenic systemic fibrosis, 259
nerve block, 237, 248
neuraminidase inhibitors, 164
neurological disorders, 191, 211–21
neuromuscular blockers, 235
neuropathic pain, 247–8
neurotransmitters, 193–4
NICE *see* National Institute for Health
    and Clinical Excellence
niclosamide, 171, 173
nicorandil, 64
nicotinic acetylcholine receptors, 41
nicotinic acid derivatives, 78–9
NIDDM *see* diabetes mellitus
nifedipine, 66
nitrates, 63–4
nitrous oxide, 230, 232
non-depolarizing neuromuscular blockers,
    235
non-medical prescribing, 279–84,
    287–9
non-steroidal anti-inflammatory drugs
    (NSAIDs)
    anaesthesia, 236
    analgesia, 245
    musculoskeletal disorders, 116–18,
        124–6
    prescribing, 293
    skin disorders, 140, 143
    see also aspirin
noradrenaline, 194, 197–8
NRLS *see* National Reporting and
    Learning System
NSAIDs *see* non-steroidal
    anti-inflammatory drugs
nuclear medicine, 310–12
nucleoside analogues, 162–3
nurses, 3, 280, 298
nystatin, 166

olanzapine, 205
omega-3 triglycerides, 78
ON *see* antisense oligonucleotides
oncology, 308–10
opioid analgesics, 235–6, 247–8

oral
    administration, 10
    contraceptives, 25, 74
    hypoglycaemics, 109–10
oseltamivir, 94, 164
osmotic laxatives, 261–2
osteoarthritis, 115, 125–6
osteomalacia, 115, 128
osteoporosis, 115, 127–8
overdose, 31, 33, 239, 246, 247
oxidation reactions, 21
oxytocin, 99–100

*p*53, 178
pacemakers, 53, 59
paclitaxel, 185
paediatrics
    analgesia, 246
    drug disposition, 23
    pharmacodynamics, 31, 33, 39
Paget's disease, 115, 126–7
palmoplantar psoriasis (PPP), 140
pancreas, 106–8, 112
    see also diabetes mellitus
pancuronium, 235
paracetamol, 21–2, 246
paramedics, 273
parasitic worms, 170–3
parathyroid glands, 104–5, 111
parathyroid hormone (PTH), 104
Parkinson's disease, 191, 192,
    211–16
paroxysmal supraventricular tachycardia,
    58
partial epilepsy, 216–17
patient compliance, 4, 38
patient group directions (PGDs), 271,
    272, 275
    examples, 335–40
    physiotherapy, 315–16, 318
    podiatry, 291–2, 294–6, 300
    radiography, 301–2, 306–13
patient information leaflets (PILs), 3
patient specific directions (PSDs), 271,
    272, 274, 291, 302, 315

penicillamine, 121
penicillin, 24–5, 35, 158–9
peripherally acting analgesics, 244–6
PET *see* positron emission tomography
pethidine, 37
PGDs *see* patient group directions
pharmacists, prescribing, 3, 298
pharmacodynamics
    adverse reactions, 29–35, 38
    age effects, 31, 33, 36–9
    case studies, 46–7
    infants and children, 39
    response variations, 36–40
    review questions, 47
    targets for drug action, 40–5
pharmacokinetics
    absorption, 13–17
    distribution, 17–20
    drug disposition, 9, 10
    excretion, 19, 23–6
    metabolism, 19, 20–3
    response variations, 36–7
pharmacy medicines (P), 271, 272
phase I/II reactions, 20–2
phencyclidine, 234
phenobarbital, 40, 217, 219
phenothiazines, 203
phenytoin, 22, 74, 217–18
photochemotherapy, 142–3
phototherapy, 142–3
physiotherapy
    anaesthesia and analgesia, 229, 242
    prescribing, 2–3, 315–19
PILs *see* patient information leaflets
pioglitazone, 111
piperazine, 171–2
pituitary gland, 99–102, 111
plasma protein binding, 18–19
Plasmodium falciparum, 168, 169–70
platelet inhibitors, 72–3
platelet reactions, 69
pleural membranes, 85
pneumocystis pneumonia, 168, 170
pneumonia, 87, 93–5
podiatric surgery, 292

podiatry
    anaesthesia and analgesia, 229
    community, 299–300
    glossary of terms, 292
    legislation, 272, 277
    patient group directions, 291–2,
        294–6, 300
    prescribing, 2–3, 291–300
    skin disorders, 145, 147, 149–50
podophyllotoxin, 185
podophyllum, 146
polypharmacy, 2, 38
    see also drug–drug interactions
POM *see* prescription-only medicines
positive inotropic effect, 61
positron emission tomography (PET), 198
posterior pituitary gland, 99–100
potassium channel activators, 64
PPP *see* palmoplantar psoriasis
prandial glucose regulators, 110
praziquantel, 171, 173
prednisolone, 91, 119
pregnancy, 35, 39–40
premedication in anaesthesia, 231, 234–6
prescribing, 291–319
    independent, 279, 282–3, 296–8, 302,
        308, 310
    non-medical, 279–84, 287–9
    nuclear medicine, 310–12
    physiotherapy, 315–19
    podiatry, 291–300
    purpose, 293
    radiography, 300–15
    reference texts, 3
    supplementary, 280–4, 287–9, 296–8,
        302–3, 307–10, 316–17
    training and legislation, 2–3
    see also patient group directions
Prescription Only Medicines (Human
    Use) Order 1997, 273, 293, 318,
    319
prescription-only medicines (POM),
    271–4, 276–7, 280, 293–5
PRH *see* prolactin releasing hormone
prilocaine, 241–2, 277

primaquine, 169
probenecid, 24–5, 32, 39, 125
procarbazine, 186
prodrugs
    metabolism, 20
    pharmacodynamics, 45
professional indemnity, 293
proguanil, 170
prolactin releasing hormone (PRH), 102
propanolol, 23
propofol, 233, 235
prostaglandins, 117–18
protease inhibitors, 163, 165
protein binding, 32
protein synthesis, 179–80
proto-oncogenes, 178–9
protozoan infections, 168–70
PSDs *see* patient specific directions
Pseudomonas aeruginosa, 156
psoralen, 142–3
psoriasis, 140–4
psoriatic arthritis, 116, 122
psychological disorders, 191, 194–211
psychoses, 191, 201–11
psychostimulants, 211
PTH *see* parathyroid hormone
pustular psoriasis, 140
pyrazinamide, 94, 161
pyridostigmine, 130

quetiapine, 205
quinidine, 67
quinine, 1, 169
quinolones, 157–8

radioactive iodine, 186
radiography, 253–67
    adjuncts, 253, 260–4
    anaesthesia and analgesia, 229
    analgesia, 261
    case studies, 266–7
    complications and risks, 257–60
    contrast agents, 253–60
    diagnostic, 303–4, 306, 308, 312
    glossary of terms, 301–3
    historical development, 303–5

laxatives and bowel evacuants, 261–3
legislation, 275
medicines management, 305–6
nuclear medicine, 310–12
patient group directions, 301–2,
    306–13
prescribing, 2–3, 300–15
review questions, 267
suboptimal practice, 312–13
therapeutic, 304–5, 308–10
raloxifene, 127
RAS *see* reticular activating system
ras gene, 179
reboxetine, 199
receptors, 40–4
recommended International
    Non-proprietary Name (rINN), 4
rectal administration, 11
reduction reactions, 21
renin-angiotensin-aldosterone pathway,
    56
replication of DNA, 179–80
respiratory disorders, 85–97
    allergic rhinitis, 87, 92–3
    asthma, 87, 88, 91–2
    bronchitis, 87, 88–9, 91
    case studies, 95–6
    cystic fibrosis, 87, 93, 95
    drug treatments, 89–94
    hay fever, 87, 92–3
    physiology, 85–8
    pneumonia, 87, 93–5
    review questions, 97
    tuberculosis, 87, 94–5
response variations, 36–40
reticular activating system (RAS), 192,
    206, 243
retinoids, 144
retroviruses, 163
reverse transcriptase inhibitors, 163, 165
rheumatic diseases, 115–23
rheumatoid arthritis, 116
ribosomal RNA (rRNA), 180
rickets, 128
rifampicin, 94, 161

right ventricular failure, 57
riluzole, 131
ringworm, 147
rINN *see* recommended International
    Non-proprietary Name
risperidone, 205
rituximab, 123, 187
rofecoxib, 118
ropivacaine, 241–3, 277
rosiglitazone, 111
rRNA *see* ribosomal RNA

SA *see* sinoatrial
safe custody of controlled drugs, 278
salbutamol, 12, 89, 92, 260
salicylic acid, 142, 148
salmeterol, 89
schizophrenia, 191, 192, 201–5
Scottish Intercollegiate Guidelines
    Network (SIGN), 299
secondary hypertension, 59
sedatives, 234
selective serotonin re-uptake inhibitors
    (SSRIs), 199
selective toxicity, 155
selegiline, 215
serotonin, 194, 197–9, 202, 208
    receptor blockers, 199
sertraline, 199
severe acute asthma, 92
sevoflurane, 231–2
short-contact treatment, 142
sickle cell disease, 75
SIGN *see* Scottish Intercollegiate
    Guidelines Network
silver sulfadiazine, 150–1, 157, 264, 277
simple diffusion, 13–14
sinoatrial (SA) node, 53, 58
skin disorders, 137–53
    bacterial infections, 150–1
    case studies, 152–3
    eczema, 138–40
    fungal infections, 146–50
    psoriasis, 140–4
    review questions, 153
    structure of skin, 137–8

viral infections, 144–6
warts, 144–6
sodium bicarbonate, 25
sodium etidronate, 127
sodium ion channel blockers, 217
sodium picosulfate, 262
SPCs *see* summaries of product
    characteristics
spinal anaesthesia, 237
SSRIs *see* selective serotonin re-uptake
    inhibitors
Staphylococcus aureus, 156
statins, 77
status asthmaticus, 92
steroid hormone receptors, 43–4
stimulant laxatives, 262
streptokinase, 73
streptomycin, 161
stroke volume, 52
subcutaneous injections, 12
sublingual administration, 11
substance P, 243–4
sulfasalazine, 121
sulfinpyrazone, 125
sulfonylureas, 109, 110
sulphonamides, 157
summaries of product characteristics
    (SPCs), 3
supplementary prescribing, 280–4,
    287–9
    physiotherapy, 316–17
    podiatry, 296–8
    radiography, 302–3, 307–10
suppressor oncogenes, 178
suxamethonium, 23, 235
systemic antipsoriatic drugs, 143–4
systemic lupus erythematosus, 116
systolic pressure, 51, 55

tacrolimus, 140
talcacitrol, 141
targets for drug action, 40–5
taxanes, 185
TENS *see* transcutaneous electrical nerve
    stimulation

teratogenesis, 35
terbinafine, 148–9, 167
teriparatide, 127–8
tetracyclines, 19, 159–60
thalamus, 192
thalassaemia, 75
theophylline, 22, 40, 90
therapeutic radiography, 304–5, 308–10
therapeutic ratio (TR), 30
thiazide diuretics, 62
thiopental, 232–3
thioxanthines, 204
thromboembolic disease, 71–2
thromboxane A_2, 69
thyroid gland, 102–4, 111
thyroid hormone receptors, 43–4
thyroid stimulating hormone (TSH),
    101–3
thyrotrophin releasing hormone (TRH),
    102
tinea, 147
tioconazole, 277
tissue sequestration, 19
tizanidine, 131
TNF-α see tumour necrosis factor
tonaftate, 147
topical
    administration, 11–12
    anaesthesia, 236
    corticosteroids, 139, 143
topiramate, 220
toxicity, 1
toxoplasmosis, 157, 168, 170
TR see therapeutic ratio
transcription, 180
transcutaneous electrical nerve
    stimulation (TENS), 248
transfer RNA (tRNA), 180
translation, 180
trastuzumab, 187
trazodone, 199
treosulphan, 183
tretinoin, 187
TRH see thyrotrophin releasing hormone
tricyclic antidepressants, 198–9

trihexyphenidyl, 215
trimethoprim, 157–8, 183
tRNA see transfer RNA
TSH see thyroid stimulating hormone
tuberculosis, 87, 94–5, 160–1
tubular reabsorption, 25
tubular secretion, 24–5
tumour lysis syndrome, 181
tumour necrosis factor (TNF-α)
    inhibitors, 122–3
Type A adverse reactions, 30–2, 33
Type B adverse reactions, 32–5
tyrosine kinase receptors, 43

ultraviolet (UV) radiation, 142–3
undecenoates, 147
unipolar depression, 195
uric acid, 123–5
UV see ultraviolet

valproate, 217, 218–19
vancomycin, 158–9
variation in response, 36–40
Vaughan–Williams classification, 58–9
venlafaxine, 199
ventricular fibrillation, 58
verapamil, 65
verrucas, 144
very low-density lipoproteins (VLDLs),
    75–6, 78, 110
vigabatrin, 217, 219–20
vinblastine, 185
vinca alkaloids, 185
vincristine, 185
vindesine, 185
viral infections
    chemotherapy, 161–6
    respiratory disorders, 94
    skin disorders, 144–6
viral pneumonia, 94
vitamin A, 144
vitamin B_6, 214
vitamin B_{12}, 74
vitamin D, 128, 141
vitamin K, 69, 71
    antagonists, 72

VLDLs *see* very low-density lipoproteins
volume of distribution, 19–20
voriconazole, 94

warfarin
    blood disorders, 71
    pharmacodynamics, 32, 39, 40
    pharmacokinetics, 18–19, 22
warts, 144–6
Wilson's disease, 212
wound care, 264

X-rays *see* radiography
xanthine bronchodilators,
    90
xylometazoline,
    92

zafirlukast, 91
zanamivir, 164
zidovudine, 163
zolpidem, 209, 210
zopiclone, 209, 210